U.
S.
R.
APSHERON
PENINSULA
AZERBAIJAN
COLORADO MOUNTAIN COLLEGE
1 03 0001932272
HD
9576
.N36
M65
1973
HD9
Mosl
D0049218
Mosul
Baba Gurgur
Qaiyara
Kirkuk
Teheran
AFGHANISTAN
Baghdad
IRAQ
(MESOPOTAMIA)
Euphrates
N.Z.
DATE DUE
8.25-97
Demco, Inc. 38-293
GULF OF OMAN
Muscat
SAUDI
ARAB
MASIRA
ISLAND
Mecca
Jidda
Salala
YEMEN
Sana
SOUTHERN
YEMEN
DISCARDED
ARABIAN SEA
Spring Valley Library
Colorado Mountain College
3000 County Road 114
Glenwood Springs, CO 81601
Aden
GULF OF ADEN
ETHIOPIA
TERRITORY OF
AFARS & ISSAS
SOMALIA
JEAN PAUL TREMBLAY

Books by LEONARD MOSLEY

Duel for Kilimanjaro: An Account of the East African Campaign, 1914–1918

The Last Days of the British Raj

The Glorious Fault: The Life of Lord Curzon

The Cat and the Mice

Report from Germany: 1945

Gideon Goes to War: The Story of Wingate

Downstream: 1939

Castlerosse

So Far So Good

Haile Selassie: The Conquering Lion

Faces from the Fire: The Biography of Sir Archibald McIndoe

Hirohito: Emperor of Japan

On Borrowed Time: How World War II Began

The Battle of Britain

Backs to the Wall: The Heroic Story of the People of London during World War II

Power Play: Oil in the Middle East

Power Play

The site on which the Abadan Refinery now stands, as it looked in 1901.
PHOTO: BP COMPANY LTD.

Random House New York

Abadan Refinery in 1946.
PHOTO: BP COMPANY LTD.

Power Play

Oil in the Middle East

by LEONARD MOSLEY

Published in the United States by Random House, Inc., New York.

Originally published in Great Britain by Weidenfeld & Nicolson, London.

Library of Congress Cataloging in Publication Data

Mosley, Leonard, 1913–
Power play,

Bibliography: p.
1. Petroleum industry and trade—Near East
I. Title.
HD9576.N36M65 1973 338.2′7′2820956
ISBN 0–394–47050–8 72–13789

Grateful acknowledgment is made for permission to reprint excerpts from the following publications:
Discovery!, by Wallace Stegner, published by Middle East Export Press Ltd., Beirut, Lebanon. Reprinted by permission of Brandt & Brandt.
The Wind of Morning by Colonel Sir Hugh Boustead. Reprinted by permission of Chatto and Windus Ltd.
Arabian Jubilee by H. St. John Philby, published by Robert Hale & Company. Reprinted by permission of David Higham Associates, Ltd.
Portrait in Oil (British title *Pantaraxia*) by Nubar Gulbenkian. Copyright © 1965 by N. S. Gulbenkian. Reprinted by permission of Simon and Schuster and Hutchinson Publishing Group Ltd.
Oil in the Middle East, 3rd Edition, by Stephen Longrigg. Published by Oxford University Press under the auspices of the Royal Institute of International Affairs. Reprinted by permission.
Middle East Oil by George Stocking. Published by Vanderbilt University Press, 1970. Reprinted by permission.

Designed by Andrew Roberts

9 8 7 6 5 4 3 2

First American Edition

"Imperialism is for us a long cow with its head in the Middle East, where it is being fed, and its udders in America, where it is being milked of its oil."

—ADIL HUSAIN, an Egyptian nationalist

"If one is to be anybody in the world oil business, one must have a footing in the Middle East."

—JEAN PAUL GETTY

"They do not tell us what is going on outside the production phase. They say this is complicated and is the business of the parent companies. They treat us like children."

—SHEIK ABDULLAH TARIKI, former Saudi Arabian oil minister

"In one hand bread, in the other a sword."

—ARAB SAYING

"By the end of the century, we could be looking down the muzzle of a gun."

—SIR DAVID BARRAN, Shell Trading Company

Foreword

EVER SINCE MAN CAN REMEMBER, through the cap of earth and water covering the area nowadays known as the Middle East, the bitumen, natural gas and oily scum that betray the presence of petroleum underneath have been leaking onto the surface of its deserts, mountains and seas. There are accounts in ancient documents of fiery outcrops, gaseous eructations or bitumenous seepages all the way from the Caspian Sea in the north to the Persian Gulf in the south, and from the mountains of the Bakhtiaris in the east to the Egyptian-Libyan desert in the west. The Eternal Fires of Iraq have been burning for five thousand years, and have been doused only once in that time. The Egyptians embalmed their dead Pharaohs with the aid of petroleum distillates brought from the Red Sea and the Western Desert. The Hanging Gardens of Babylon were fastened by pitch between the bricks, and it still holds the ruins together today. Wandering Bedouins used bitumen to waterproof their tents from an outcrop lake at Burgan, near Kuwait, whose whereabouts was passed down for generations. And across the Caspian Sea, on the Apsheron Peninsula, in the Caucasus, Persians and Parsees came to worship in hundreds of fire temples which were lit by flames known as the Fires of Zoroaster.

In the Bible, in Pliny, in Herodotus, and in the records of many another ancient writer and traveler in the region, there are references to the presence and use of oil in multifarious forms—as an object of worship, as light, as fuel, as building material, as a massage for weary limbs, as an aphrodisiac for enfeebled lovers, as an ointment for blemished skins, and as a medicine for internal aches and distresses.

Today, in most of the places where Biblical man dug pitch or worshiped fire, there are petroleum wells exploiting the oil beneath them. The big oil companies have, of course, done even better than the ancients, for they have found oil where the tribes never suspected it existed—for instance, beneath the waters of the Persian Gulf, and below the sandy wilderness of the Rub al Khali (the Empty Quarter) of Saudi Arabia, which only a handful of Arabs had ever visited until U.S. oil technicians took them there.

From all these widely scattered fields, both old and newly discovered, no fewer than 24.260 million barrels of crude oil are siphoned out of the earth beneath the Middle East *every day.* * This is enough to satisfy the needs of Western Europe, Japan, Australasia, East and South Africa and two thirds of non-Communist Asia. Altogether 1.213 billion tons of oil are shipped out of the area in a year,† which is more than is produced by the United States and its chief suppliers in the Caribbean (Venezuela, Colombia and Trinidad) put together. The amount of money this makes for the U.S., British, Dutch, French and Japanese oil companies, and for the Arab and Iranian governments who have granted the concessions is so huge that it sounds like a U.S. budget deficit in a particularly bad year. I do not wish to confuse the reader with too many astronomical figures, but anyone with a

* Not counting the small amount used locally.

† These figures are for 1971.

Note: American barrel = 42 American gallons or 35 imperial gallons.
Barrels per day multiplied by 50 = approx. tons per year. Thus 20,000 barrels per day = 1 million tons per year.

pencil will be able to work out that if a barrel of oil is now sold by Middle East producers at $2.373 (which was the price prevailing at the beginning of 1972) then a *daily* sum of $39,866,400—or $14,521,236,000 a year—is being made in sales.

The oil companies paid to the state of Kuwait alone $1.395 billion in taxes and royalties in 1971. Kuwait is only the third largest oil-producing country in the Middle East and has no more than 733,000 citizens, fewer than a medium-sized city.

These are fabulous sums of money in any currency, and they have changed the face of the Middle East, physically and psychologically as well as financially. Some cynics would say that as a result, what were once poor backward lands have simply become rich backward lands, and that all that this fabulous wealth has done is to make Arabs and Iranians more arrogant, more feckless and more venal than ever. Others might reply that no Arab could possibly compete in arrogance, profligacy and venality with the Western oilmen and their governments who have been exploiting them for well over half a century.

This book, however, is not written for cynics but for those who ask questions every time there is a Middle East oil crisis, and who never get adequate answers.

Is the West losing in its fight to hold on to its Middle East oil concessions?

How did the oil companies acquire the concessions in the first place?

Could the Arabs and Iranians run their own oil fields? And if they did, what difference would it make to the world fuel situation?

Can the United States be blackmailed by a Middle East threat to stop oil supplies? How great is America's need for Arab oil?

What happens to all the money the oil companies pay out to the producing countries, and what has it done for the average Arab and Iranian?

These are some of the questions which this book has tried to answer as fully and as honestly as possible.

❖

LIKE GOLD, OIL brings out the cupidity in men and governments, and the sordid double-dealing behind the granting of every oil concession has disfigured the history of the Middle East since the turn of the century. But though this book will shed some new light on the gaudy and garish entrepreneurs who engineered the concessions, and will not neglect the shahs and sheiks with whom they schemed and connived, it would be unbalanced if it failed to recount at the same time the devotion, skill, courage and endurance of the men who first found the oil and brought it out in the face of heat, disease and hostility. There are heroes as well as scoundrels among the Europeans, Americans, Arabs and Iranians in these pages, and their contribution to the story deserves recognition. So this is a book about an area and its people in which, because of its almost limitless petroleum resources, every industrialized country in the world, Communist and non-Communist alike, now has a vested interest. A decision made in Riyadh, Kuwait, Baghdad, Tripoli or Teheran could bring the machines of half the world grinding to a halt. A third of the United States' overseas investments and half of Britain's are tied up in Middle East oil, and it is useful to know something about the men in power in the area who could wrest those holdings from them at a moment's notice.

For the purposes of this narrative, I have included Libya as one of the Middle East oil countries, even though geographically it is in North Africa. Its petroleum policies are closely tied to those of its Arab neighbors, and only too often what its volatile leader decides to do today will be emulated by the sheiks of the Persian Gulf tomorrow. Other countries with which the survey deals are Iran, Iraq, Bahrain, Kuwait, Saudi Arabia, Qatar, Oman and the Federation of Arabian Emirates.* The book also touches on Syria, Lebanon, Israel and Egypt, none of which is a major oil export-

* Comprising Abu Dhabi, Dubai, Sharjah, Al Fujairah, Ajman, Umm al Qaiwain and Ras al Khaimah, a parcel of ex-British protectorates along the Persian Gulf.

ing country but all of which have an influence on the political if not the oil policies of their neighbors.

For technical reasons, I was not able to make a visit to Israel during my researches for this book, and had to confine myself to discussions about oil with Israeli friends and officials outside the country (see Appendix). Otherwise, I visited all the countries mentioned above and talked with local rulers, leaders, ministers and technicians, as well as to American, European and Japanese oilmen working on the spot. I also discussed (in their own countries where possible, otherwise in their places of hiding or exile) the future of Middle East oil with Arab and Iranian opponents of the status quo. In New York, San Francisco, London, Paris and The Hague I visited the headquarters of the major oil companies, talked with their chief spokesmen, and made use of the expert knowledge and archives which they most willingly offered. I spent some time with the so-called "independents" like Jean Paul Getty, Dr. Armand Hammer, Bunker Hunt and Dr. Wendell Phillips, and chewed the fat with old-time drillers, geologists and Arabists wherever they have retired after their labors in the desert. Finally, I have had the help in my researches of the indefatigable Joan St. George Saunders, who, among other invaluable aids, lent me the unpublished history of the Anglo-Iranian Oil Company written by her late husband, Hilary St. George Saunders.

Acknowledgments

TRAVELING IN THE MIDDLE EAST in search of contacts and information is not easy. It takes time to procure visas, and in some of the Arab countries visitors are not admitted without written proof that they are either Christian, Muslim, or at least not Jewish. Also, some oil companies have had unfortunate experiences with writers whom they have helped or sponsored, for afterward they have been blamed by the rulers under whose aegis they work for what was written. Hence, getting into certain countries is like trying to join certain unions: they won't have you until you get the job, and you can't get the job until you have a union ticket. The embassy politely tells you that a visa will be granted the moment you produce a letter from an oil company, and the oil company says that once you have a visa they will give you all the letters you want.

Luckily, this is where experience as a foreign correspondent is useful. The visas or entry permits are secured—and then the real struggle begins. The oil fields of the Middle East are thousands of miles apart, and some of them are extremely awkward to reach. It is true that each country is linked by a remarkable efficient network of commercial airlines, which I used, but I would never have been able to get to several of the wellheads had it not been for the private aircraft which the oil companies (particularly in Abu Dhabi) and the state (particularly in Saudi Arabia) put at my disposal. I have tried to make sure that this has had no effect on my impartiality in this book, but I am indeed grateful to them for their help and hospitality.

Of the hundreds of experts, officials and ministers to whom I talked during the course of my research on this book, it would be impossible to give a complete list and unfair to include a partial one. I have therefore decided to give thanks by name only to those whose aid both before and during my travels went far beyond the call of duty. I must, for instance, particularly single out the continuous advice and information which was put at my disposal by Mr. Fuad W. Itayim. Mr. Itayim is editor and publisher in Beirut of the knowledgeable, authoritative and influential bulletins of the *Middle East Economic Survey*, a weekly review of news and views on Middle East oil. He placed his office and his files at my service, and he and his staff were always ready to answer my questions. Furthermore, the news editor of *MEES*, Mr. I. N. H. Seymour, whom I encountered in several remote parts of the Persian Gulf, was always ready with news and introductions to oil sheiks and oilmen, with scores of whom he is on special terms of trust and friendship. I am most grateful both to him and to Mr. Itayim.

In the United States, both in his office and in his home, Mr. Howard W. Page, that wise spokesman of the oil industry, took hours from his other occupations to retrace the critical moments in the story of Middle East oil. In Saudi Arabia, Sheik Ahmed Zaki Yamani, the oil minister, did likewise, at a moment when every ruler in the Middle East was seeking his counsel. It was also during this same period of great crisis, when the future of Western oil interests in the Middle East was in the balance, that Mr. Frank Jungers, president of Aramco, talked at length to me about his problems.

The names of a few other people whose services, for space reasons, I will not specify:

Messrs. Shafiq Ombargi, Farid Riszk, Ihsen Hijazi, Sinclair Road and Sulaiman Olayan in Beirut; Harry McDonald, Faysal al Bassam, Bill Mulligan and Pete Speers in Dhahran; Mohammad Almana in Al Khobar; Abdullah Hashim in Dammam; Ghaleb Abu al Faraj, Abdullah al Ghanim and Fahd al Sedayri in Riyadh; Dr. Tom Temperley and Mr. Abas Sindi in Jidda; Messrs. Tony Leigh-Morgan, Peter Rawlston and Richard I. L. Stephens in Kuwait; Vittorio Gismondi in Dubai; Ahmed Obaidli, Zaki Nusseibeh, Ahmed Hijji, R. A. de Freitas, Geoffrey W. G. Ball, Alan Horan, Robert Bell and John Millward in Abu Dhabi; Ronald F. Rosner and Joseph Johnston in New York; Julius Edwards and Geoffrey Keating in London (and, in the case of the latter, in several parts of the Middle East); Dr. Wendell Phillips in London and New York; Dr. Nadim Pachachi and various other members of OPEC and OAPEC in all parts of the Middle East.

Finally, there was my traveling companion, who as usual not only looked, listened and took notes, but who also made the desert smile in the blackest moments. To her my special thanks; to the rest, my grateful salaams.

L. M.

Contents

Contents

Part One

THE CONCESSIONAIRES

PRECEDING PAGE: *A 100,000 barrel-a-day gusher on fire in Bibi-eibat, Baku, in 1899.*
PHOTO: JOHN WEBB/BROMPTON STUDIO

CHAPTER ONE

The Millionaires of Baku

THE FIRST OIL CONCESSIONAIRES in the Middle East were not British, American, French or Dutch but Armenians and Tartars. A hundred years ago they exploited the great field of petroleum in the Caucasus lying beneath the western shores of the Caspian Sea, and in bringing it to the surface turned the scrublands of Azerbaijan into a black desert, polluted the rivers and the sea, demeaned and degraded their workers, and sowed the seeds of the Russian Revolution. They also kept half of Europe's lamps alight and broke the Standard Oil Company's stranglehold on the world paraffin market.

Today the great oil fields of Baku are a tightly run Soviet monopoly with virtually no contact with the great petroleum industries to the south of them. But in the nineteenth and early part of the twentieth century they were part of the Middle East free-for-all. Baku and its Azerbaijan hinterland had belonged to Persia until 1813, but even after the Russians took it over it remained Persian in spirit, and only the ruthless domination of the Cossack armies kept it under control. The area was all but awash with oil. Along the Caspian shores, and particularly on the Apsheron Peninsula, the flat scrublands covered a vast reservoir of petroleum and gas. The peninsula was dotted with flaring gas fires and small Zoroastrian temples to which pilgrims came to worship from India. If one rowed five miles into the Caspian and threw a tow of lighted rope overboard, the sea would catch fire from the gas bubbling up from the sea bed. For generations the natives had painstakingly dug wells by hand, filled leather bags with the seeping oil and transported it to market for sale as fuel or lubricant. Then in the 1870s someone brought the first steam drill to the Caucasus. It was the same kind as those which were transforming the U.S. oil industry and making a multimillionaire out of John D. Rockefeller, and a team of Pennsylvania drillers came with it. They sent the metal bit spinning into the soft, sandy soil and up came what was known locally as "a spouter." Oil had been struck at 124 feet. From that moment the rush was on.

Always on the lookout for sources of revenue, the Czarist government in St. Petersburg was quick to exploit the bonanza. The viceroy received urgent orders to seize most of the oil-producing land and to lease concessions to the highest bidders. A great auction was held in which most of the rich Armenians from Tiflis, Batum and Constantinople bid wildly against each other, and against the local Azerbaijani and Tartar khans, for the right to exploit "spouters" of their own. They bid half a million pounds for land which had previously been valued at one hundredth of that sum, and then went off excitedly to erect derricks and start drilling. Since the land which had been auctioned off was confined to those areas where hand-dug wells had previously been in operation, they did not need to be told where to start. But a young Swede named Robert Nobel, who had been in the Caspian area investigating a timber scheme, quietly acquired a strip of territory away from the coveted areas, and paid only £1,000 for it. Then he called in his three brothers—Emil, the frail but shrewd

* Better known for his invention of dynamite and for the Nobel prizes, which are named after him.

young scholar; Ludwig, the arms manufacturer; and Alfred,* the chemist and inventor—and asked them for advice and financial help. The oil field which they started in 1874 soon became the biggest and most prosperous in Baku, with refineries to purify the oil and pipelines to take it to the railhead. News of their success fired the enthusiasm of the British, who came in and began buying concessions. Only the Americans held back; they were confident that no field could ever equal their own in Pennsylvania, and were skeptical of reports that Baku sat on a veritable sea of oil. In any case, Baku crude was so sulfurous and smoky, and it cost so much to transport to Western Europe by cumbersome routes through Russia that Standard Oil was convinced that its monopoly of the market could never be broken.

Sure enough, to begin with, the great Baku oil rush seemed to be an unmanageable mess. Down went the drills and up came spouters all over the Apsheron Peninsula. It was an extraordinary sight. "When the wells reached rich oil sands eruptions of gas, oil and sand were so violent," wrote one observer, "that the fragile casings often collapsed or became deflected by earth movements. At that time there was no such thing as screwed, welded casing in 20-foot lengths, and the weakness of the linings and their large diameter prevented the attachment of devices which would impose an additional pressure on the wells." *

Almost every day a new fountain of oil and gas burst out of the ground, but few of the workers obeyed the injunctions not to smoke, and fires were an everyday occurrence. The territory began to be covered by lakes of oil. "Not a single producer had provided himself with storage facilities, or with means of regulating the flow of oil, which streamed all over the region, soaking into the calcareous sand soil and destroying the scattered patches of vegetation on the wastelands of the Balakhani and Saboonchi oil fields. The whole surface of the field was saturated with oil, and American drillers, familiar with the small wells of Pennsylvania,

* *Oil Pioneer,* by A. Beeby-Thompson.

were amazed at the marvellous and overwhelming production of the spouters of Balakhani." *

Nobody cared about pollution or the destruction of the region's vegetable and animal life. The lakes of oil were set afire, and areas were ravaged to clear terrain where new wells could be drilled. "Nobody put any value on the wasted oil," reported Henry. Sadly, the local Baku newspaper, *Bakinskiya Izvestiya,* wrote of one great gusher in its issue of October 12, 1896: "From the town the fountain [of oil] had the appearance of a colossal pillar of smoke, from the crest of which clouds of oil detached themselves and floated away a great distance without touching the ground. Owing to the prevalence of southerly winds, the oil was blown in the direction of Bailov Point, covering hill and valley with sand and oil, and drenching the houses of Bailov, a mile and a half away. The whole district of Bibi-Eilat was covered with oil, which filled up the cavities, formed a lake, and on the fifth day began pouring into the sea. On the sixth day the wind freshened and the oil spray began flying all over the town. The square in front of the town hall in Baku was drenched with petroleum . . . Altogether 14,000,000 poods [about 250,000 tons] are estimated to have come to the surface, and most of this was lost for want of storage accommodation. The oil simply poured into the Caspian Sea, and was lost forever to mankind."

During these years, caviar was scarce in St. Petersburg and Western Europe, for the spawning sturgeon, swimming out of the Volga River into the Caspian Sea were poisoned by the drifting patches of oil. The shores of the great inland sea were covered with dead and dying birds. The grapes withered on the vines under the pall of burning oil smoke. The reek of petroleum was carried on every wind.

Not that the Baku oil concessionaires cared a whit. Caviar and the wine were local products of Bibi-Eilat, and therefore easily come by. They preferred more exotic dishes, and sent to France for pâté de foie gras and choice champagne, for by the

* *Baku: An Eventful History,* by J. D. Henry.

1890s the Azerbaijani capital had become the gaudiest city in the world. Great fortunes were being made by the well owners, and they had no hesitation in flaunting their new-found wealth. Great houses began to grow up on the seashore well away from the smoke and stench of the oil fields, and their owners imported architects from Paris and Berlin and let their imaginations run riot. One of them covered the whole face of his house with gold leaf and decorated it inside with pink Italian marble lined with solid gold. Some of the new millionaires were not without a sense of humor, and an inveterate gambler among them fashioned his house in the shape of a hand of cards, with a jack of diamonds, a king of clubs, a queen of hearts and an ace of spades painted on its façade. The Tartar oil magnates indulged their fancy for ornate private baths covered with marble and lapis lazuli, and brought in virgins and small boys from the Circassian villages of Turkey to add sensual delight to the languors of the steam rooms. Baku became a place to visit for every high-priced demimondaine in Paris, Berlin and Budapest, and jewelers from the Rue de la Paix rushed to open branches in the town, complete with their most vulgar displays of diamonds and rubies.

One of the most flamboyant of the oil millionaires was an Armenian named Alexander Mantachoff, whose wild parties lasted for days and whose appetite for food, wine, women and violence was notorious. A contemporary picture of him at a meeting of the local owners' association shows him dressed in the formal morning dress of the period, topped by a black silk cloak with what appears to be white lining, and from beneath his trousers peep highly polished and spurred riding boots. He is carrying a whip, and the expression on his heavily bearded face is both arrogant and challenging. Mantachoff prided himself on being an insatiable womanizer and unbeatable with knife, whip or his fists. He had no guilt about his excessive indulgences, and his favorite remark was: "Only the weak are good, because they are not strong enough to be bad."

So far as Baku was concerned, it was true. The weak were

the native Tartar and Georgian villagers who worked for the oil concessionaires, and they lived and worked under particularly degrading conditions. Their living quarters were a series of wooden-hutted compounds in an area about ten miles from Baku known as Black Town. It was close to the wells and at the mercy of every spouter that came in, so that the streets and houses were constantly soaked with oil, and the air was a mixed stench of petroleum and excrement. Pay was low and hours long. Food was provided from primitive canteens. Employees worked under virtually the same conditions as black men in South Africa today, forbidden to have wives with them, their movements restricted. Their sexual needs were satisfied by the weekly visitation of a cartload of drabs brought in from the brothels of Tiflis and Batum, and what money they saved from their wages they were encouraged to spend in company bars on cheap wine and vodka. Any tendency to rebel against the conditions was brutally discouraged by private armies of Cossacks maintained by the well owners who rode among the insurgents and cut them down with knouts and swords. Ringleaders were publicly flogged.

From this miserable multitude of sweaty serfs, the millionaires of Baku built an oil industry capable for the first time of challenging the all-powerful Standard Oil Company of America and breaking its monopoly of European paraffin sales. At first it was the output of the Nobel oil fields which made the running, for three reasons: they controlled most of the tank cars on the railroad which took the kerosene from the Caspian Sea all the way north across Russia to the Baltic, whence it was shipped to Europe; they had built a pipeline from their field to the railhead, which enabled them to load cheaply and in bulk; and the chemical expertise of brother Robert eliminated the high sulfur content of Baku oil and made it burn as brightly as American paraffin. Then the Rothschild brothers arrived from Paris and surveyed the situation. They decided that the best way to break the Nobels' hold on Baku and Standard's monopoly in Europe was to find a cheaper way of shipping out the oil than by the long and costly route to the Baltic, so they offered to put up the money (in return for shares in the oil fields) to build a railroad from Baku to Batum across the neck of land separating the Caspian Sea from the Black Sea. From Batum, tankers could carry the oil through the Dardanelles to Europe—and in particular to a large refinery which the Rothschilds owned at Trieste, in northern Italy. Within a year of the opening of the railroad, the Nobels were forced to use this route too, and an annual supply of 10 million tons of oil began flowing out of Baku. Standard Oil was no longer number one, and the millionaires of Baku were richer than ever.

But suddenly the boom burst—in blood.

ONLY THE NOBEL BROTHERS and a few Western European newcomers among the concessionaires had attempted to ameliorate the appalling working conditions of the local populace, and between them they had built a series of more sanitary dwellings called White Town, where food was better and movements were unrestricted; some workers were even permitted to set up house with their wives and families. But these attempts to civilize conditions caused widespread resentment among the Armenian and Tartar concessionaires, particularly when their own slave laborers began to demand improvement of their animal existence. The mayor of Baku, a ruthless Tartar with the apt name of Despot-Zenovitz, encouraged by the Russian viceroy, Prince Galitzin, found a way of diverting the attention of the workers from their grievances. Rumors were spread through the compounds that the Armenians were planning a revolt in which the Tartars' mosques would be destroyed, their women kidnapped and raped, and Azerbaijan given back to Persia. To add substance to the rumors, a false—and naturally unsuccessful—assassination attempt was made on the viceroy, who was fired upon while driving through the center of Baku.

The result was a series of bloody pogroms which culminated in the appalling massacres of 1905, when hundreds of Armenians living in Baku and Azerbaijan were cruelly murdered and their women carried off by rampaging Tartars and Georgians. Cossack troops were called in to halt the killings, but their sympathies were with the Tartars and they helped rather than hindered them in their depredations. Martial law was declared because suddenly the situation got out of hand. The raging Tartars not only were killing Armenians—which the Russian authorities did not mind—but soon turned their fury upon the oil wells as well. Saboteurs spread over the fields, setting fire to the oil in the storage tanks. Quickly the whole Apsheron Peninsula once more flared into a blaze of fires, but this time it was man-made.

The Armenians streamed out of Baku, looking for any refuge they could find, and were cut down as they fled. The Armenian concessionaires took to their private yachts and cruised out into the Caspian, Mantachoff among them. Some of their fellow refugees plunged into the water and tried to swim out to them, but were ruthlessly repelled when they tried to get aboard. As Mantachoff sailed away he could see the flames and thick black smoke rising from his oil field on the Apsheron shore. He shrugged his shoulders and went below to the comfort of a luxurious cabin and the attentions of his latest mistress. He had a consignment of gold aboard with him, for just such an emergency, and there were few worries on his mind as he set sail for Bandar Pahlavi, in Persia, en route to Egypt. But several thousand other Armenians were not so lucky, and died in the streets and houses of Baku.

EVENTUALLY the great fires were extinguished, the wrecked derricks and outbuildings were rebuilt, and slowly production of oil began again. The workers were herded back into their compounds, and for the next few months there was an improvement in their conditions. Most of the big Armenian concessionaires were gone, but their wells were shared out among the Tartar landowners or resold to the Rothschilds, the Nobels and to British and

German groups, and for a time life was safer in Baku and less exotic in the villas along the Caspian.

Three years later Baku was booming again and the villas along the Caspian were bigger and more vulgar than ever, and their owners more ostentatiously extravagant. But down in the cantonments, things were stirring. Men were working there now who had read books and talked with emissaries from Moscow and St. Petersburg, and they lectured their fellow workers not to allow themselves to become bloodthirsty tools of the bosses. Among the Georgian Communists who were employed at Baku was the uncle of the man who would become Josef Stalin; another was Anastas Mikoyan, a future president of the Soviet Union.

There were strikes and riots in the years to come, and sabotage in the oil fields; bands of outlaws preyed on travelers and held them for ransom. But only minor outbreaks of racial violence took place. Under the guidance of Communist-cell workers, the serfs of Azerbaijan were counseled to bide their time until they could wipe out the concessionaires once and for all, and the Czarist government with them.

IN THE 1880s, while the Baku oil industry was still in its heyday, and Alexander Mantachoff was the swashbuckling king of the Apsheron Peninsula, a young man had arrived to visit him from Turkey. What he saw, heard and experienced was to make a lasting impression on him, and, through him, on the future of the Middle East oil industry.

His name was Calouste Sarkis Gulbenkian, and he was nineteen years old. Like his host, he was an Armenian. His father had interests in Baku oil, and sold and exported kerosene from the markets of Constantinople successfully enough to have sent his son to France to learn the language and to England for his general and technical education. This was money well spent. Through formidable drive and a precocious intelligence, Gulbenkian earned good degrees in science and engineering at King's College, London, though he was still in his teens. As a reward, his father sent him on a trip, first to Baku and the Apsheron Peninsula, and then to Mosul, in what was then Turkish-controlled Mesopotamia. In Baku the young Armenian met the Nobel brothers, toured their oil installations, and was then passed on to his fellow Armenian. The burly and mustachioed Mantachoff, a veritable giant of a man, was at first disconcerted by his thin, soft-spoken, studious-looking guest, and after one day of showing him around his wells and refineries, left it to his underlings to take Gulbenkian on the detailed tour of the outlying fields which the young man had requested. When they returned they reported that he had made little or no comment, even when confronted by the awesome sight of a fortuitous spouter, but had scribbled copious notes.

Still, when invited to Mantachoff's ornate seaside palace, the youth demonstrated that he was much less introverted than appearances indicated, for he savored all the delights the household had to offer. True, when one of his host's fabulous shindigs was in progress, he took no part in the bareback horse races, wrestling matches and trials of strength in which the big Armenian and his Tartar guests liked to engage. But he seized whatever else was offered; he ate the rich imported foods, drank the fine wines and champagne, and applauded the tumblers, magicians and belly dancers Mantachoff imported to amuse his guests. It was all a startling contrast to the austere existence which his family followed in Constantinople. Mantachoff taught Gulbenkian what luxuries money could buy and introduced him to the pleasures of the flesh.

When Calouste Sarkis Gulbenkian returned to Constantinople, he wrote a long account of his trip which was published in 1892 under the title "La Transcaucasie et la Péninsule d'Apcheron: Souvenirs d'un voyage." A straightforward, serious study of a people and an industry, it was well received, but it turned out to be his hail and farewell to oil as a tangible, smellable commodity.

Gulbenkian was to become a power in the oil world, and thanks to it, one of the richest men in the world, but the nearest he ever got to petroleum again was in the tank of his car. He never visited another oil field; he did not need to. Baku, Apsheron and Mosul had taught him what oil was like, and Mantachoff what power and wealth an oil concession could bring him. That was all he needed to know, and from the age of nineteen on, he set out to make his fortune in oil—without actually getting the greasy stuff on his fingers.

CHAPTER TWO

The Absentee Concessionaire

UNLIKE THE MILLIONAIRES OF BAKU, most of the Middle East concessionaires who came after them shared Gulbenkian's distaste and kept as far away as possible from the reek of the precious commodity which made their fortunes. There was, for instance, William Knox D'Arcy. In the official history of British Petroleum, the giant combine whose success grew out of his manipulations, Knox D'Arcy is described as "the father of the oil industry of the Middle East." * A French report of the origins of the industry gives a graphic account of his adventures in the wilderness of southwest Persia, doggedly searching for oil with a prayer on his lips and a theodolite over his shoulder. "Wizened, his face withered by the torrid sun of the Persian deserts," writes the author, "worn out but still believing in God and his own idea, more and more on his knees in front of his crucifix, D'Arcy continued his search." † Whereas, in fact, D'Arcy's feet never even touched Persian soil, and he never saw the oil fields which made him the "father" of Middle East oil. His only visit to the region

* *Adventure in Oil,* by Henry Longhurst.
† *La Guerre secrète pour le pétrole,* by Antoine Zischka.

was a two-day excursion to Cairo and the Pyramids when his ship was passing through the Suez Canal on its way from Australia to England, and this was long before he even dreamed of investing in oil. It is doubtful that he ever saw a barrel of the stuff in its crude state. He was an absentee concessionaire from start to finish.

On the other hand, it is almost certain that if the risk had not been taken by William Knox D'Arcy in the spring of 1901, the exploitation of the oil fields of Persia would have been delayed for a generation and the course of British imperial politics considerably altered. Everyone had known for a long time that there were signs of oil in southwest Persia, for the same seepages and gas flares as those found along the Caspian to the north were present. In 1872 a naturalized Englishman, Baron Julius de Reuter (who later founded the news agency which bears his name) obtained the oil concession for the whole of the shahdom of Persia, and went broke trying to exploit it. In 1897 it was offered to the Royal Dutch–Shell group. The Royal Dutch company, which relied upon wells in the Netherlands East Indies for most of its supplies, had heard rumors that its powerful American rival, Standard Oil, was about to repudiate its price-control agreements with the Baku producers and unleash a murderous price-cutting war. Royal Dutch was looking around for new sources of supply to bolster its reserves in preparation for the bitter campaign to come, but after some investigation, it turned down the Persian concession as too risky and too expensive. (The asking price was £30,000.)

The concession lapsed, and interest in Persian oil faded until two years later, when two French explorers, Jacques de Morgan, an archaeologist, and Eduard Cotte, a geologist, returned from an expedition to the mountains of the Bakhtiari tribes with vivid accounts of abundant "petroliferous evidences." They formed a syndicate to exploit the possibilities with a colorful Persian general living in Paris, Antoine Kitabgi, who had once been director of customs and excise in Teheran. He would be their front man, they decided, in the complicated bargaining which would inevitably precede the granting of a new concession.

But who would put up the money, first to buy the concession, and then to finance its operations? They were lucky enough to find that the British minister to Persia, Sir Henry Drummond Wolff, was in Paris on leave, and they asked him if he could put them in touch with anyone rich and adventurous enough to back their project.

"I know the very man," said Sir Henry. "I will go with you to London and introduce you to him." *

It so happened that a week previously Sir Henry had been to a dinner party in London at the house of an Australian named William Knox D'Arcy, and had come away much impressed with his host. D'Arcy was indeed impressive. A banker by profession, he was a huge, plump figure of a man with twinkling blue eyes, a walrus mustache, and a beautiful wife, Nina, on whom he hung precious jewelry. He had made his millions out of the fabulous Morgan gold strike in Queensland, Australia. Ironically, he had never visited a gold mine either, but had merely financed the Australian gold prospectors, the Morgan brothers, when they came to his bank for backing. It was a decision which netted him millions, and he moved to London to enjoy the fruits of his lucky investment. He soon became the talk of fashionable London, for the parties at his house in Grosvenor Square were fabulous. At one of them Nellie Melba and Enrico Caruso, the two greatest opera singers of their time, sang for their supper. Statesmen, society hostesses and aristocrats flocked to his receptions for the food, the entertainment, the sight of beautiful guests decked in all their finery, and because they found the Australian an amusing and amiable host.

D'Arcy seemed happy with his social success. But Drummond Wolff noted that at one point in their after-dinner conversation

* The syndicate had promised that the British minister "would not be forgotten" if their venture succeeded.

someone teased D'Arcy about his purchase of a new automobile, and maintained that the noisy and smelly horseless carriage would never take the place of the carriage-and-pair. To which D'Arcy replied, "I think you are mistaken. One day, I believe, the motor car will revolutionize our means of transport. There will be horseless omnibuses to carry the people, and horseless carts to transport their goods. It is already happening in America, and soon it will happen in the rest of the civilized world."

At which Drummond Wolff interpolated, "Yes, but not without the oil to propel them. And if the world is suddenly full of horseless carriages, where will the oil come from to fuel them?"

"We must find and extract more of it from the ground," D'Arcy said. "The oil is there. We must dig for it."

It was this vision of a new age of locomotion which made the Australian millionaire more than willing to meet and talk to Jacques de Morgan (the fact that his name was the same as the gold-mining brothers who were responsible for his fortune may have helped too). Drummond Wolff effected the introduction and D'Arcy was soon fired to enthusiasm by the young French archaeologist. The upshot of this and subsequent conversations was that D'Arcy agreed to back the venture. But he began cautiously by sending his secretary's cousin, Alfred Marriott, to Teheran in February 1901 to negotiate a concession ("but don't pay out any cash"), and a geologist, H. T. Burls, to make a professional report on the oil possibilities in the area De Morgan had been exploring.

Negotiations began promisingly with the grand vizier of Shah Muzaffar-ud-Din's court, who promised to report favorably to his royal master on the question of a concession. Unfortunately, it soon became apparent that no bribes were going to be passed, so the grand vizier let it be known through the bazaars that the British were seeking oil. At once the Russian minister rushed over to protest, and threatened reprisals if the shah allowed the British to operate on Persian territory, so close to its neighbor to the north, at the same time insisting that if concessions were available, only the Russians should be given them.

It looked as if D'Arcy's project had failed at the start. But in his secretary's cousin he had chosen a worthy emissary. Marriott went to see the grand vizier again, and this time made a financial offer of £10,000 down, to be handed over to the grand vizier as soon as an agreement was made, and no questions asked about its disposal.

"Ah, how I wish we could come to an arrangement," said the Persian, showing interest for the first time. "Alas, the Russians are exigent. They have insisted that any concession agreements must first be seen by them."

"In that case—" said Marriott, getting up as if to go.

"In these circumstances," the grand vizier interrupted, "we must try to arrange that the Russians see, and at the same time do not see." *

The wily Oriental and equally wily Englishman put their heads together and worked out a solution. It was that the newly appointed British minister to Persia, Mr. Arthur Hardinge, who was a Persian scholar of no mean repute, should write a letter to the Russian legation giving details of the proposed concession and asking the Russians to set forth any objections they might have to it. The grand vizier suggested that the minister should write the letter in Persian.

"He was aware," Hardinge wrote later, "that Mr. Argyropulo [the Russian minister] could not read Persian, more especially in the written or *shikaste* character, which is illegible, owing to its peculiar abbreviations, even to scholars familiar with the printed language. He also knew from his own spies that the Russian Oriental Secretary, M. Stritter, who alone could read it, was about to leave Zergendeh, the summer residence of the Russian legation, for a short sporting excursion in the neighboring hills. He therefore sent the letter to Zergendeh, where it lay several days untranslated, awaiting M. Stritter's return, and as no objection to the proposal contained in it was made by the Russian Minister,

* Quoted from an unpublished history of the Anglo-Iranian Company by Hilary St. George Saunders, lent to the author by his widow.

who could not read it, and never suspected the importance of its contents, all the Persian members of the Government supported the Grand Vizier's decision to sign the concession to Mr. D'Arcy. M. Argyropulo was far from pleased when he learned what actually happened; but the Grand Vizier could not be blamed for the accidental and temporary absence of his Legation's Persian translator, and the Russian Minister accordingly adopted the sensible course of accepting the accomplished fact." *

The deal was done, and on May 28, 1901, the shah signed over to William Knox D'Arcy a sixty-year exclusive concession to "obtain, exploit, develop, render suitable for trade, carry away and sell natural gas, petroleum, asphalt and ozocerite through the whole extent of the Persian Empire," with the exception of a crescent of territories along Persia's northern frontiers abutting Russia. As a gesture of appeasement toward the Czarist government, which had tried to get the concession for itself, these areas were reserved for Russian exploitation. Despite these exceptions, D'Arcy had been accorded a vast territory in which to search, find and develop—altogether 480,000 square miles. The Persians' demand in return was that he should form a company to exploit the concession within two years after signing the agreement. In this company the Persians would be given 20,000 £1 shares and 16 percent of the profits, plus a further £20,000 in cash. They also insisted on the appointment of an "imperial commissioner" to safeguard their interests in the new venture, and chose for the job General Antoine Kitabgi. They did not appear to be aware that Kitabgi was already a member of D'Arcy's syndicate.

A happy D'Arcy countersigned the agreement in London, bought his wife a new tiara, and celebrated the event with a huge party to which Lord Curzon, the British foreign secretary, and many other government leaders came. Then he waited for the oil and profits to flow in.

But it wasn't as easy as that.

* *A Diplomatist in the East,* by Sir Arthur H. Hardinge.

FIRST OF ALL, D'Arcy's geologists began drilling in the wrong place. Jacques de Morgan had enthused over "petroliferous evidences" in the Bakhtiari country, but the geologist D'Arcy sent out himself had recommended an area some hundreds of miles to the west. It certainly had oil, but the quality and quantity were both disappointing. Moreover, the conditions under which the field parties began their work were daunting in the extreme. As his field manager D'Arcy chose G. B. Reynolds, a hard-bitten Englishman who had drilled for oil in some remote places in the Far East; Reynolds in turn picked as his assistant a wiry American ex-driller named Hiram Rosenplaenter who had worked in Texas and Mexico. But neither of them was prepared for the dust, desolation and heat of the Persian desert. By seven in the morning the temperature was 110 degrees, and it rose steadily throughout the day. The local tribesmen, who acknowledged no authority from such remote places as Teheran, rode threateningly into the tented encampment and demanded tribute. Since he had no force with which to repel them, Reynolds took them into his employ instead as encampment "protectors" and bestowed plentiful baksheesh on their chiefs. But the guards could not be trusted, and their khans were "as full of intrigue as a nightingale's egg is pregnant with melody," Reynolds reported; and at the same time that they stole from the camp, they left behind fevers and fleas that drove the men crazy.

Rosenplaenter had used Polish drilling crews to begin with, but they stopped work so often on the grounds that another religious holiday had come around that he first cabled home for a Roman Catholic calendar, and then gradually replaced them with Canadians and Americans. He began a new well at a godforsaken spot called Chiah Surkh. "By June 1903," the official record recounts, "it had reached a depth of 1292 feet where 'fine, dark-coloured sand rock' was encountered 'which cuts all our drills to pieces.' The drillers were suffering from the effects of bad water caused by the bloated dead bodies of thousands of grasshoppers which had chosen a watery grave in the Sekhuan stream, the only source of supply. To thirst and dysentery, heat-stroke was soon added. Everyone, even the Indian doctor, suffered from it. 'We are now five Europeans suffering from a touch of the sun. Stewart was yesterday in a critical condition . . . I have built two thick-walled stone houses for the drillers but even in these the temperature rises to 120 degrees Fahrenheit and, as the nights are hot, the men get very little rest and less sleep. In spite of this we are working night and day at the well.' " *

Back in London, all D'Arcy had to worry about was money. He had spent £250,000 on getting the men and equipment out to Persia; he had had to pay off Drummond Wolff, who had got tired of waiting for his reward; and he was having trouble with his other erstwhile partners. Regretfully, he canceled the order for one of the latest and most expensive of the new automobiles to come on the market, a Rolls Royce, and gave his bank a block of Morgan gold shares against his mounting overdraft. He even contemplated giving up his private box at Ascot racecourse—except for the royal family, he was, at that time, the only race goer to have one—but reluctantly decided that perhaps this would be going too far. But he cabled Reynolds and Rosenplaenter to press on. At last, in January 1904, oil came surging out of a second well Rosenplaenter had sunk, and it looked like a winner.

"Glorious news from Persia," D'Arcy announced. "It is the greatest help to me."

But not for long. Two months later the well ran dry. Sick with dysentery and broken-hearted at the failure of his superhuman efforts, Rosenplaenter allowed himself to be shipped back to the United States. Reynolds stayed on to reconnoiter new terrain and to start drilling again. By this time D'Arcy was harassing him almost daily to find him some oil, "before disaster engulfs us all." He did not let Reynolds know that he was negotiating with an American syndicate (which eventually turned him down) and

* Saunders, *op. cit.*

with the Rothschild interests in an attempt to find outside money.

The negotiations with the Rothschilds took place in the spring sunshine along the Croisette at Cannes and aboard the Rothschild yacht floating in the harbor, over glasses of champagne and delicate snacks of caviar and pâté de foie gras, at a moment when in Persia Reynolds was working on despite a particularly weakening attack of dysentery, and grief over the loss of his pet dog, which had succumbed to heat stroke. It was the ostentatious background of D'Arcy's talks which turned the tides of fortune for him, however, for they were reported back to the British government in London. Alarmed that D'Arcy might be planning to sell out to French interests—as indeed he was—they dispatched an emissary to see him at once with instructions to appeal to his patriotism and sense of duty toward his Empire. He was asked to return to London for a meeting at which British rather than foreign aid would be discussed. D'Arcy acceded with a theatrical display of willingness to sacrifice himself in the cause of king and country, though in truth he was losing little, for the Rothschilds were driving an exceedingly hard bargain in between the glasses of champagne.

It was a lucky moment for William Knox D'Arcy. Two years earlier the Royal Navy, egged on by an unswerving believer in oil fuel, Admiral Lord ("Jackie") Fisher, had agreed to experiment with oil as the motive power for the dreadnoughts of His Majesty's fleets. It was at the height of the coal-versus-oil controversy. At Fisher's insistence a navy cruiser, the *Hannibal,* was fitted with oil tanks as well as coal bunkers, and sailed out of Portsmouth harbor on a series of sea trials. For the first two hours of her voyage the ship surged forward under coal-fired boilers, only a light smoke trickling from her three stacks. Then the order was given to change to oil firing. Almost at once, though the speed remained constant, the ship was engulfed in great clouds of black smoke.

"You see!" cried the defenders of coal, choking delightedly on the acrid fumes.

Fisher was also choking, but with anger. "All I see is *sabotage*," he replied. And he may have been right, for it turned out that the cruiser, in spite of meticulous instructions, had been fitted with old-fashioned vaporizers instead of special new burners. Naval friends of the big coal owners, who were fearful of losing one of their most important markets, were suspected of deliberately misinterpreting orders; whether or not this is true, the changeover was delayed.

But later in the year, when Fisher moved to the top of active naval direction as First Sea Lord, he crushed his opponents once and for all. An Oil Committee was set up to superintend the conversion from coal to oil for all vessels in the Royal Navy. There was, however, one danger. The tactical advantages of oil for naval use were evident by now, but strategically, maintained E. G. Pretyman, a member of Parliament who headed the Oil Committee, the British navy would be placed at a great disadvantage: "Whereas we possessed in the British Isles the best supply of the best steam coal in the world, a very small fraction of the oilfields of the world lay within the British Dominions, and even these were situated within very remote and distant regions." Therefore, if Knox D'Arcy had even a fifty-fifty chance of finding viable oil fields in Persia, they must not be allowed to fall under foreign control. The Admiralty had already given a contract to the Burmah Oil Company, a British firm, to supply them with oil from their fields in Burma and Assam. When D'Arcy arrived in London from his conferences with the Rothschilds in Cannes he was told that with the help and approval of the British government, Burmah Oil was willing to bail him out of his difficulties and henceforth pay a share of the expenses of his Persian operations. It did not occur to D'Arcy—or to any of his new directors—to go out to Persia themselves with the good news and encourage their weary and dispirited but still indefatigable field manager, G. B. Reynolds. They simply sent him even more peremptory orders to find a viable field.

But the tie-up with Burmah Oil, and through that company

with the British government, did remove one worry from Reynolds' roster of woes. Local tribesmen had been getting increasingly hostile, and had taken to sabotaging equipment and stealing supplies when the amount of baksheesh they were being paid fell below expectations. But this was 1905, the heyday of the British Empire, when one man waving a Union Jack and threatening to bring in a gunboat could cow a nation. Though Persia was an independent country, it sometimes almost came apart at the seams under the pressure for its control between Russia and Britain, a tug of war in which the Russians planted their teams in the north and the British in the south. Prodded by the Russians, the Shah Muzaffar-ud-Din complained when he heard that the terms of D'Arcy's concession was about to be altered to accommodate Burmah Oil. Immediately the new syndicate sent a note to Whitehall hoping that they could count upon the support of the government "in the event of any attempt by the Shah of Persia to cancel or interfere with the concession or on the part of Russia or any other neighbouring State to hamper the working of it."

In reply, the Foreign Office dispatched a carefully worded missive which said nothing any foreign government could object to, but suggested everything from gunboats to armies. "I am to state," the undersecretary wrote, "that it is obviously impossible that a definite statement should be made in regard to hypothetical contingencies but your clients or any other British Company formed to acquire and work the Concession can count on such support and protection as British subjects are always entitled to expect from His Majesty's Government."

That the note meant much more than it said is indicated by the fact that shortly afterward, and much to Reynolds' relief, a small contingent of troops from the British Indian Army was sent to protect his field operations from further molestation from the Bakhtiaris. The British government did not inform the shah of this decision until the troops were actually there, and then, in an explanatory note, strongly suggested that he would be foolish to object to their presence on his territory. The contingent was commanded by a young English lieutenant named Arnold Wilson, a sensitive young man of heroic good looks, whose subsequent distinguished career in British business and politics was to end on a note of gallant tragedy.* He set up his camp beside the oil company's rigs soon became an enthusiastic admirer of Reynolds, who was, he wrote in his diary, "dignified in negotiation, quick in action, and completely single-minded in his determination to find oil."

But years went by, and for all his efforts, Reynolds had still not found enough oil to justify the cost of the operations. The concession had been signed in 1901; it was now 1908. Even the Burmah Oil Company and its British government sponsors were beginning to lose heart. As for D'Arcy, he was keening over his overdraft again and canceling the option on his latest Rolls Royce. In the post-Christmas hangovers of 1908, the members of the syndicate met in London to discuss the gloomy prospects and apparently decided that they'd had enough. It was at just about this time, as the sun burst through the January rains in southwestern Persia, that Reynolds spudded in a new well among the riot of spring flowers at a spot known as Masjid-i-Sulaiman (the Mosque of Solomon), a place name that has now won a deserved place in the history of the oil business. One night when young Lieutenant Wilson went across to join Reynolds for their usual sundowner of whiskey and chlorinated water, he found the oilman in the depths of depression. He had received a cable that morning, brought by the camel courier from the telegraph office in Baghdad, saying that "funds were exhausted and the decision

* Sir Arnold Wilson (he was knighted in 1920) became an oilman himself later and a director of the Anglo-Persian Oil Company. In the years before World War II he entered Parliament as a Tory and preached an Anglo-German alliance, strongly opposing those who wanted war with Germany. But once war began he waived his parliamentary immunity and (though fifty-five) volunteered for the RAF. He was commissioned as a rear gunner in a bomber and was shot down and killed over Dunkirk by German ack-ack while bombing Nazi-held France.

reached finally and irrevocably was that he was to cease work, dismiss the staff, dismantle anything worth the cost of transporting to the coast for reshipment, and come home." *

Wilson was distressed at the emotional state of the older man, and indignant at "the short-sighted decision which may result in the cancellation of the concession. I am tired of working here for these stay-at-home business men who, in all the years they have had the concession, have never come near it," he wrote in his diary. "They have all the vices of absentee landlords. Cannot Government be moved to prevent these faint-hearted merchants, masquerading in top-hats as pioneers of Empire, from losing what may be a great asset?"

But when he went to see Reynolds the next day, the field manager had recovered his old spirit, and defiance had taken the place of despair. He was damned, he told Wilson, if he was going to take precipitate action on the strength of a cable. The drilling would go on. He sent a message to London to tell the syndicate of his decision, and received written permission to go on drilling until he reached 1,600 feet.

On May 26, 1908, oil was struck at Masjid-i-Sulaiman at 1,200 feet and a gusher came up over the top of the derrick. Joyfully, Reynolds called for a camel courier and sent off a message to the telegraph office in Baghdad to inform his head office of the good news. But young Wilson had telegraph facilities of his own to the British resident (political officer) in Bushire, and was anxious that the British government should know about the strike as urgently as possible. He realized that there would be a run on Burmah Oil's shares as a result of the news, and therefore did not want any clerk along the wire to read his message and spread the news before his masters had seen it and taken appropriate action. "I at once sent the news to Bushire," he wrote later. "As I

* Oilmen have always cast doubt on this story, pointing to the fact that no copy of the cable exists in the archives of Burmah Oil or British Petroleum. In view of what subsequently happened, someone may quietly have removed it from the files. Certainly there seems no reason to doubt Wilson, who had no motive for making up the story about the telegram.

had no Telegraph Code I wired to Lorimer [the consul]: SEE PSALM 104 VERSE 15 THIRD SENTENCE AND PSALM 114 VERSE 8 SECOND SENTENCE. This told him the news and in the circumstances was quite as effective as a cipher."

He presumed (rightly, as it turned out) that Lorimer would send it on to London.*

The strike at Masjid-i-Sulaiman proved to be the foundation of the syndicate's fortune, and no one expected to wax fatter on the glory of it all than William Knox D'Arcy. In April 1909 the Anglo-Persian Oil Company † was formed to exploit this field and the remainder of the Persian concession. Its original capital of £1 million was provided by Burmah Oil, and at the suggestion of the British government, the choice of chairman of the company was left to the civil servants in Whitehall. To D'Arcy's fury, they passed him over in favor of an eighty-eight-year-old Tory peer named Lord Strathcona, through whom they hoped to keep a tight hold on the affairs of the new company. Then they awarded Anglo-Persian a £20 million contract to supply the Royal Navy with fuel oil.

In 1914, at the urgent insistence of Admiral Lord Fisher and his close friend, Winston Churchill, who sat in the Cabinet as First Lord of the Admiralty, the government went even further. Thanks to Churchill's oratory and dire warnings of Britain's danger of "oil starvation" in the event of war with Germany, a bill was driven through Parliament giving the government the right to take over control of Anglo-Persian Oil. Just a few months before the outbreak of World War I, the government bought 51 percent of the company's shares for £2.2 million. Thus they gained control of a company which today is capitalized at £425 million and whose sales in 1970 were £2.6 billion.

As for William Knox D'Arcy, he was soon complaining that he

* He used The Book of Common Prayer (according to the use of the Church of England): "That he may bring out of the earth oil to make him a cheerful countenance . . . the flint stone into a springing well."

† The name was changed to Anglo-Iranian Oil Company in 1935, and to British Petroleum in 1954.

had been cheated out of his place in the hall of fame, and that the new controllers of Persian oil were trying to conceal the fact that he was the father of it all. He died, an unhappy man, in 1917, never having visited the concession which changed the pattern of the world oil industry.

CHAPTER THREE

Big Deal

SO FAR THE AMERICANS HAD KEPT CLEAR of the Middle East oil rush. They were too busy coping with the flow from their own fields in Pennsylvania, California, Texas and Oklahoma Territory. But when news reached the United States in the spring of 1908 that the D'Arcy concession had struck a rich petroleum field in Persia, several interests in New York decided that it was time to put a toe in the Middle East pool. Since the British and Russians between them had parceled up Persia, the obvious pool to dabble in was Mesopotamia, where it had long been known that there were petroleum manifestations in the areas around Mosul and Kirkuk. At the time the restive and reluctant Kurds and Arabs who populated these areas were under the control of the Ottoman Empire, and it was therefore from Constantinople, the Turkish capital, that any concessions must be obtained. Constantinople was ruled by the ineffable Sultan Abdul Hamid and was probably the most corrupt capital in the world.

It so happened that an American concession seeker was already in Turkey when news came of the D'Arcy oil strike. Colby M. Chester was a retired rear admiral in the U.S. Navy, a veteran of the sea battles of the Spanish-American War, and a former naval attaché at the U.S. legation in Turkey. He was not worried when other Americans appeared on the scene, nor was he concerned that other nations were haggling in the bazaars, for he was convinced that he already had the whole deal signed and sealed. His earlier service in Turkey had taught him that the way to get a bargain from the officials at Sultan Abdul Hamid's court was to keep your temper under all circumstances and be ready with the bribe. His mission had the backing of a consortium consisting of the New York Chamber of Commerce, the New York Board of Trade and the New York Transport Association, and they had provided him with ample funds with which to suborn the venal officials of the Turkish court. In addition, thanks to friends in the Navy Department, he had good relations with the U.S. government and could always rely on the aid and influence of the State Department.

With all this behind him, and thanks to a substantial bribe which he handed over to the grand vizier of the sultan's court, he was able in the spring of 1908 to emerge from Topkapi Palace brandishing not one concession but several, giving him the right to exploit oil and minerals in all parts of the Ottoman Empire. Unfortunately, a few weeks after the concessions were signed the sultan lost his throne in the coup d'état of the Young Turks,* and the privy-purse territories to which Colby Chester had been given the rights reverted to the state. The concessions were worthless, the admiral had to begin all over again, and this time he had to contend with rivals of more serious caliber and determination.

This was the period when the Germany of Kaiser Wilhelm was challenging the might of the British Empire, the greatest power of the time, by proposing to build a railroad all the way from Constantinople to Baghdad, which would poise the Germans on the threshold of British India. As part of their project,

* They were the successors of the Young Ottomans, members of a secret society which was established fifty years before. Their aim was to bring about constitutional change.

they had secured a concession from the Turks for a thousand-mile strip of territory through Turkey and Mesopotamia, and they not only had the right to construct a railroad through it but had also acquired permission to exploit all oil and minerals to be found for twenty miles on either side of the line. True, these rights, like those of Colby Chester, had been declared null and void when the Young Turks came to power, but since the new administration failed to reimburse the Germans the $100,000 they had spent surveying the route, Berlin refused to accept the cancellation and maintained its rights under international law.

Alarmed at these activities on her imperial doorstep, Britain decided that it was time for her too to interest herself in Turkish affairs. With the approval of the British government, a group of London financiers agreed to send one of their number, Sir Ernest Cassel, to Constantinople to look over the possibilities. Cassel could hold his own in the London banking world but realized that he would need a guide through the serpentine corridors of Turkish high finance. He picked Calouste Sarkis Gulbenkian as his adviser, and he could not have chosen anyone better.

Much had happened to the Armenian in the years since he had supped at Alexander Mantachoff's lavish table in the Baku oil fields in the 1890s. Gulbenkian was thirty-nine now, and had honed down his youthful enthusiasms to the pursuit of three objectives: oil concessions, money and women. There had been a time when all his energies were engaged simply in the fight to survive. In the mid-nineties, when the Turks had slaughtered the Armenians in their midst, Gulbenkian had succeeded in making his escape in a boat leaving for Egypt, and at the last moment good luck sailed with him. Milling among the panic-stricken crowds on the quayside, whom should he see but Mantachoff, who had come to Constantinople on an unheralded visit and had found himself in the middle of a Turkish pogrom (a foretaste of what was to follow in Baku years later). Gulbenkian used his influence with the ship's owners, who were related to his in-laws, to get the oil millionaire aboard and forced his own young wife, who was pregnant, to quit her cabin in order to make Mantachoff comfortable. He was ever ready to do a service for a potential benefactor, and his eager servility toward the Armenian giant paid off. First he became Mantachoff's secretary, then the representative in London of his and other Russian oil interests. He had never looked back. He had discovered his skill in complicated negotiations at a time when American oil interests (especially Standard Oil) were trying to cut-throat their way to control of the world fuel market. Gulbenkian was convinced that the only way for non-U.S. oilmen to survive was consolidation so that they could jointly meet the American competition with a united front. First he had brought the Russian oil interests into the big European organization, Royal Dutch. Then, when this rejuvenated company came under the control of a powerful Dutch operator, Henri Deterding, he had scurried back and forth as his emissary in the tortuous negotiations with the Shell Oil Company, out of which came the European consortium Royal Dutch–Shell, a combination powerful enough to stand up to any price-cutting competition from Standard Oil. As a result, Standard had called a truce.

Now Calouste Gulbenkian was well on the road to the eminence he had planned for himself. He was a naturalized Briton, the father of a boy and a girl; he had houses in fashionable Hyde Park Terrace in London and Boulevard Haussmann in Paris. He also had an apartment at the Ritz Hotel in Paris, where, for the rest of his life, he installed the regular succession of mistresses who ministered to his extramarital requirements. He had a growing bank account, and as his choice by Sir Ernest Cassel proved, he was in demand as a negotiator.

It was always Gulbenkian's policy never to let a past insult or injury deflect him from a deal with the person or persons who had inflicted it. So despite the way in which he and his fellow Armenians had been treated by the Turks, he had no hesitation in negotiating with them. He believed in an old Arab saying: "The hand you cannot bite, kiss it." He also believed—rightly, as it proved—that the Young Turk revolution had really changed nothing in Turkish administrative life, and that the skills he had learned at bribery and corruption in the old regime would stand him in good stead when he negotiated on behalf of British interests. What the British wanted to do was to set up a bank in Constantinople and infiltrate themselves into an influential position in the Turkish state through ordinary banking activities. The bank was indeed established (called the Turkish National Bank), but Gulbenkian immediately set out to convince Sir Ernest Cassel that the last thing they should embark upon was banking, an activity which in Constantinople was so crooked that no upright British financial institution could possibly compete. What they should aim for, instead, were the oil concessions in Mosul and Kirkuk.

Sir Ernest protested that his colleagues were not interested in oil, and that in any case there were already other interests in the field. Admiral Chester, for instance, had struggled back into the running, and by his imperturbable admixture of good humor and financial glad-handing had persuaded the Ministry of Public Works to restore his lost concessions to his syndicate, the Ottoman-American Development Company. Ratification of the concession rights was due to go before the Turkish parliament in 1911. Sir Ernest cited Chester's success as an insuperable obstacle, but Gulbenkian, who had heard in the bazaars that Chester was referring to him as "an Armenian trickster," merely shrugged. He could easily be taken care of, he implied.

On the other hand, it would be necessary to do something about the German interests. Even though the new Turkish government now repudiated their right to build a railroad through to Baghdad, the Germans insisted that the concession was still valid according to international law. It would be no use fighting them, Gulbenkian said; far better to compromise and bring them in. So persuasive were his arguments that Cassel and his colleagues in London soon found themselves agreeing to the formation of a company, African and Eastern Concessions,* in which the £80,000 share capital was divided up as follows: 20,000 shares to the

* The name was later changed to the Turkish Petroleum Company, and in 1928 to the Iraq Petroleum Company.

Deutsche Bank (the German interests); 28,000 to Sir Ernest Cassel and the Turkish National Bank; and 32,000 to Calouste Gulbenkian. Within a few months Gulbenkian had passed over 20,000 of his own shares to his friends in the Royal Dutch–Shell organization, keeping 12,000 shares (15 percent of the holding) for himself.

With this settled, Gulbenkian went to work on Turkish officialdom, coolly ignoring the anxious protests from Admiral Chester, now aware of the intrigues going on around him. A piquant picture of how he operated has been given by his brother-in-law, Atvarte Essayan: "Calouste Gulbenkian's success was due to his never making a mistake with the baksheesh. His inside knowledge of each man's precise position in the price chain to the top was uncanny. He also knew the exact state of each minister's health and frame of mind and the most propitious moment to see him. He found out by bribery exactly whether the minister had enjoyed a satisfactory session with the lady of his choice from the harem (or with one of his boys, who were often more prized than the women), whether the pasha had eaten well or had a stomach ache. Calouste seemed even to know when the pasha was going to belch or break wind . . . The serious business then started. What was the minister's cut on the contract to be?" *

By finessing his way through the corridors of Turkish power on the hills of Pera, Gulbenkian had made sure by 1912 that the Turkish Petroleum Company had control over the oil and mineral rights along the thousand-mile strip of territory in Mesopotamia previously in dispute between Turkey and the Germans (who were now partners in the Turkish Petroleum Company). Once this bone of contention was disposed of, he started on the trickiest part of his negotiations, in which he set out to gain the oil concessions for Kirkuk and Mosul. These were part of the area which had been promised to Admiral Chester and his Ottoman-American Development Company, and the American angrily declared

* Quoted in *Mr Five Per Cent: The Biography of Calouste Gulbenkian,* by Ralph Hewins.

that not only would he fight to retain them but he had been promised the support of the U.S. government in his battle. No one took much notice, least of all Calouste Gulbenkian, for by now he had a much more threatening crisis on his hands.

In London the two great proponents of oil-for-the-Navy were beginning the maneuvers that would culminate in the British government's purchase of control of the prospering Anglo-Persian Oil Company. Winston Churchill, the First Lord of the Admiralty, and Admiral Lord Fisher, the active head of the Royal Navy as First Sea Lord, were determined that if they did take over control of Anglo-Persian it must become the most powerful oil company not only in Britain but in the whole of Europe. Therefore, under no circumstances must the potentially invaluable oil fields of Mosul and Kirkuk be allowed to fall into any hands but those of Anglo-Persian. The British government had no qualms about Colby Chester, whose claims, no matter how valid, they were convinced could be by-passed. But they had a healthy respect for Calouste Gulbenkian's negotiating genius on behalf of himself and his partners, so the order went out that Gulbenkian must be stopped. It was decided that the best method was to eliminate the Turkish National Bank interests (Sir Ernest Cassel, et al.), the Royal Dutch–Shell group (headed by Henri Deterding) and Gulbenkian from the Turkish Petroleum Company, and that Anglo-Persian should take over their shares. Then Anglo-Persian would join forces with the Deutsche Bank to procure the Mosul-Kirkuk concessions between them.

To this end they began with a little high-level arm twisting. Emissaries came and went between the English directors of the company and the British government, and when they hesitated, Churchill played his trump card: snobbery. It was indicated to them that not only were they being unpatriotic but Buckingham Palace was also concerned about their reluctance to cede their shares. The stratagem worked. Sir Ernest Cassel told Gulbenkian that he and his colleagues had decided to place their 28,000 shares at the disposal of the British government. "The difficult position in which I found myself can easily be imagined," wrote Gulbenkian, "the more so as during these *pourparlers* I received a telegram from Mr. Alwayn Parker [a Foreign Office official, then in charge of the Near East Department] by which I was informed that it had been decided that I should transfer my shares at once. The style was somewhat peremptory. The position became very confused owing to the National Bank and Sir Ernest Cassel and their associates getting out and leaving us with the Germans, who were astonished at these most unexpected developments." *

When Henri Deterding of Royal Dutch–Shell heard of the ultimatum, his rage knew no bounds. He was a hard-driving Dutchman who was physically so tough that he took a swim in an unheated pool every morning, winter or summer, and once skated to tenth place in the Eleven Towns Race of close to fifty miles that is held every winter in northern Holland. He loved food and women as eagerly as Gulbenkian, and had given his wife, a beautiful ballet dancer, a diamond-and-emerald bracelet worth $300,000 while she was still married to a Russian general. He not only risked a challenge to a duel from the irate husband but also risked being sued by Cartier's, because he had bought the bracelet on credit and could not pay when the bill came in. "When . . . I told him what we were expected to do," wrote Gulbenkian, "Mr. Deterding got into a state of frenzy. He became wild and stated that he would never agree to it. He threatened that if I should decide to part with my shares without his consent, it would mean a definite rupture between us." †

It was not only oil that Deterding was angry about, but the knowledge that if Anglo-Persian took over it would be a triumph for that company's managing director, Sir Charles Greenway, whom he hated more than anyone else in the world. Several times he and his wife had been snubbed by the Englishman, who had let it be widely known that he considered Deterding to be a man of no breeding. Deterding replied that Greenway was

* Quoted in *ibid.*
† *Ibid.*

an incompetent who knew nothing about oil and that "if this gentleman had not had the backing of the British Government, he would certainly have not been anywhere near the head of a corporation." *

Vowing that he would never forgive or forget Gulbenkian's action if he turned his shares over to his enemy, Deterding stomped off to complain to the Dutch government and asked them to protest against this attempt to interfere with his legitimate business activities. They promised to back him and urged him to hold on to his shares. And Gulbenkian prudently decided to hold on to his too. For the moment, anyway.

It was now that Rear Admiral Colby Chester made a last attempt to reassert the validity of his own and Ottoman-American's concession, calculating that the squabbling among the British would put them out of the running once and for all, and that some high-powered pressure and some further bribes would persuade the Turks to reactivate his still-legal concessions. Accordingly, he asked the U.S. government to make an informal approach to the authorities in Constantinople expressing their deep disappointment at the unjustified difficulties which the admiral and his colleagues had encountered in their attempts to stimulate Turkish-American commercial relations. This démarche might well have produced the desired effect had not the British government intervened with a ploy which was designed to sabotage Gulbenkian and Royal Dutch–Shell, but instead eliminated Admiral Chester and the American interests from the game.

It so happened that the subjects of foreign nations living in the Ottoman Empire were allowed complete personal, religious and commercial freedom from the government's jurisdiction, so that the amount of duty on any goods they brought into Turkey, for example, was settled by arrangement between their governments and Turkey, and could not be increased by Turkey without prior agreement with the so-called protecting powers. As far as the Big Powers were concerned, it was a cozy arrangement which

* *Ibid.*

put Turkish customs revenues into the hands of the foreigners. The system, which the Big Powers were always ready to back by force, was known as the Capitulations and it lasted until shortly after World War I.

It was under the Capitulations that Turkey, always desperately short of money, now applied to the British government for permission to increase its duties on imported goods. For London, the request could not have come at a more convenient moment. The Turkish ambassador was called to the Foreign Office and quietly told that under no circumstances would His Majesty's government allow Turkey to increase her revenues so long as she continued to hand out oil concessions to all comers, particularly such syndicates as the Turkish Petroleum Company (the Royal Dutch-Shell–German–Gulbenkian group) and the Americans. On the other hand, if the concessions were to be granted to the Anglo-Persian Oil Company, then His Majesty's Government would be prepared to give favorable consideration to Turkey's application . . .

It was British diplomacy at its most ruthlessly efficient, and the amiable tone in which Earl Grey, the foreign secretary, delivered the ultimatum did not hide from the Turkish envoy the gunboats circling in the background. He thanked the foreign secretary for his "most helpful" advice and agreed that no word of their conversation should be allowed to leak to "outside interests." The British government was anxious to confront the rival syndicates with a *fait accompli* and did not wish them to have any forewarning of what was afoot.

Earl Grey, however, was ignorant of the fact that there were no Turkish secrets to which Calouste Gulbenkian was not privy. Sometime previously Gulbenkian had had himself appointed to the job of senior financial adviser to the Turkish government, and so was immediately aware of the British move and of its potential consequences to himself and his associates. He rushed over to inform Henri Deterding of what was going on, and the Dutchman was all for creating trouble. No, no, insisted Gulbenkian. He knew the impossibility of winning an open battle with an entrenched government; that was the way to lose everything. Far better to go behind the scenes, to negotiate, to propitiate and above all to compromise.

To persuade the reluctant Deterding that it was better to talk than fight, and to strengthen his hand in discussions with the British, Gulbenkian made a painful sacrifice. He had 15 percent of the shares in the Turkish Petroleum Company. Sir Ernest Cassel had already given the British government the 35 percent held by him and his associates. The German group had 25 percent, and so did Deterding's Royal Dutch–Shell group. Now Gulbenkian turned over a further 10 percent of his own holding to the Dutchman so that he would be able to negotiate on equal terms with the British. Now he had only 5 percent left, but if his gesture thwarted the British plans for a complete takeover, the sacrifice would have been worth it.

So began a series of discussions behind closed doors in London to which, at the British government's request, only "major shareholders" were invited. This excluded Gulbenkian from the discussions, and Deterding, who was now able to talk as an equal with his adversaries, suddenly became reticent with the associate who had made it possible. ("Never expect gratitude in business," the Armenian said later.) But Gulbenkian was never without his private line to the most confidential negotiations. In the spring of 1914 the Turkish government sent a minister named Hakki Pasha as its emissary to the talks. "My father," wrote Gulbenkian's son, Nubar, "often entertained Hakki Pasha to dinner at our house in Hyde Park Gardens. Hakki Pasha was a very astute man. He was small, not much more than five feet high and nearly five feet round, wore thick glasses and had a marvellous appetite to which he gave free rein. He was certainly nobody's fool. He and my father held long discussions." *

Hakki Pasha revealed to his Armenian friend that the British government was still exerting enormous pressure on the Turks to

* *Pantaraxia: The Autobiography of Nubar Gulbenkian.*

grant the whole concession to Anglo-Persian. But stiffened by the persuasions of his host and the blandishments of his table, Hakki Pasha continued to hold out for a compromise. "The final conference took place at the Foreign Office on 24 March, 1914," Gulbenkian wrote later. "The resulting agreement was signed by the German Ambassador, Sir Eyre Crowe for the British Government, and all the other representatives."

It was a stiffer compromise than Gulbenkian had hoped would come out of the talks. The Turkish Petroleum Company was reconstituted, and in the rearrangement Anglo-Persian had 50 percent, Royal Dutch–Shell 25 percent, and the German group 25 percent. All Gulbenkian got was 5 percent, 2½ each given to him as "a gesture" by Anglo-Persian and Royal Dutch–Shell from their own holdings. These were the shares which would one day make him known as Mr. Five Percent.

Convinced that he could have done much better had he been allowed into the conference room, the disappointed Armenian bitterly complained that considering the work, time and money he had put into the project—and which had been his idea in the first place—he had been treated most shabbily. Perhaps; nevertheless, his 5 percent would one day make him the richest single individual in the world.

On June 28, 1914, the Turkish grand vizier, Said Halim Pasha, wrote letters to the British and German ambassadors officially granting the oil and mineral concessions for Mesopotomia to the Turkish Petroleum Company. At once the U.S. government protested to Constantinople about the way in which its nationals had been by-passed. But it was too late; the papers had been signed. Admiral Colby Chester put his own concession—worthless once again—back into his briefcase and departed for America in disgust.

On July 10, with British permission, the Turkish government raised its tariffs from 8 to 11 percent. Not that it got much benefit from the increase, for on August 4, 1914, the Great War began, and the Turkish Petroleum Company's concession was left in abeyance until 1918.

LONG BEFORE WORLD WAR I WAS OVER, the British and French governments had decided between them what was going to be done with the old Arab dependencies of the Ottoman Empire. As early as 1916, two years before the defeat of Germany and Turkey, the two Allies signed the Sykes-Picot Agreement, in which their zones of influence within conquered Turkish colonial territory were precisely defined—so precisely, in fact, that there was considerable embarrassment when the British army entered Baghdad and Mosul in 1918 and immediately began to sink wells and build pipelines and railways, only to have it pointed out to them that Mosul came within the French sphere of influence. The difficulty was eventually smoothed out at a meeting in Prime Minister David Lloyd George's apartment in London in 1919, when the French premier, Georges Clemenceau, agreed to hand Mosul over to the British zone—in return, of course, for a quid pro quo. In any case, another clause in the agreement expressed French willingness to respect and sanction any British concessions in their prewar zones, which should have safeguarded the Turkish Petroleum Company's concession. But with the French it was always best to make sure.

During the war several loose ends had been tied up, and the Turkish Petroleum Company was now indubitably British. The German group's 25 percent holding in the company had been seized by the Custodian of Enemy Property in London at the outbreak of the war, and since the British government controlled Anglo-Persian, which owned a further 47½ percent of the shares, it now had direct control of 72½ percent of the company's holdings. In addition, Henri Deterding had become a British citizen in 1915 (he was later knighted for his "war services"), and brought Royal Dutch–Shell's 22½ percent with him to Britain. And of course Calouste Gulbenkian, with the remaining 5 percent, was already British. Hence, the British government thought it had no "foreigners" to worry about as it set about preparing the conquered Arab territories for exploitation.

However, this brisk and businesslike parceling up of conquered territory by secret agreements and old-boy deals in private clubs failed to take into account the fact that the lands wrested from the Turks were, in fact, Arab lands, and that the Arabs had fought on the side of the British and the French on the understanding that once the war was over, independence would be theirs. Just before the armistice with Germany and Turkey, the two Western Allies announced their belief in this independence:

> The object aimed at by France and Great Britain in prosecuting in the East the war let loose by German ambition is the complete and definite emancipation of the people so long oppressed by the Turks, and the establishment of national governments and administrations deriving their authority from the initiative and free choice of the indigenous governments and administrations in Syria and Mesopotamia [Iraq], now liberated by the Allies . . . Far from wishing to impose on the populations of these regions any particular institutions, they are only concerned to ensure by their support and by adequate assistance the regular working of governments and administrations freely chosen by the populations themselves.*

But this early flush of idealism was soon superseded by more self-interested policies. When the Arab leader, Faisal, who had victoriously entered Damascus with Lawrence of Arabia in October 1918, was proclaimed king of Syria there in March 1920, he was immediately sent an ultimatum by the French. Either he must acknowledge a French mandate over Syria and agree to accept their overall control, or the French army would remove him by force. Conveniently for the French, who did not wish to have an Arab king in Syria, even a puppet king, Faisal's acceptance of the ultimatum was "delayed" en route to French army headquarters outside Damascus, and the army commander, General Henri Gouraud, gave the order to march on the Syrian capital. Faisal

* Joint declaration by the Allies to Emir Faisal, quoted in *The Struggle for Arab Independence*, by Zeine N. Zeine.

fled to the provincial town of Deraa, whose inhabitants shortly afterward were showered with leaflets from a French plane. The leaflets threatened aerial bombardment of the town if the inhabitants failed to expel Faisal within the next twenty-four hours. The Arab leader therefore moved on to Haifa, in Palestine, and asked for the protection of the British. France moved her forces into position and took over the whole of Syria and Lebanon (remaining in control for the next twenty-five years, until independence came to the two Arab countries at the end of World War II).

In the meantime Britain had established herself in Palestine and Iraq. The British had made vague promises to Faisal's father (King Husain, sherif of Mecca) that the whole of Arabia from the Euphrates to the frontiers of Turkey proper would be unified after the victory, and that his son would be monarch of it all. Now that the French had thwarted part of that pledge (there were those who maintained that Britain was much relieved that they had), they offered Faisal as compensation the throne of Iraq, with his capital in Baghdad. But first he had to promise that he would always accept the office of the British high commissioner as "juridically the supreme power." *

The British and the French were having so many secret meetings together to dispose of the postwar spoils that one might be tempted to forget that there was another principal ally who had contributed much to the victory. As far as the Middle East discussions were concerned, the United States might never have taken part in the struggle at all. One of the most important of the postwar conferences took place at San Remo, in Italy, in April 1920, at which the British awarded themselves the mandates (in other words, the control) over Palestine and Mesopotamia, and agreed to French assumption of mandates over Lebanon and Syria. To complaints that it was outside the rights of the two Allies to help themselves to mandates in this way, and that only

* See *The Report of the High Commissioner on Iraq Administration, October 1920–March 1924* (H.M. Stationery Office, London).

the newly functioning League of Nations had the power to do so, the British foreign secretary coolly replied that this was a misconception. "It rests with the Powers who have conquered the territories, which it then falls to them to distribute, and it was in these circumstances that the mandate for Palestine and Mesopotamia [Iraq] was conferred upon and accepted by us, and that the mandate for Syria was conferred upon and accepted by France." *

The United States was not invited to the distribution party at San Remo, nor was it informed that during the meeting one of the decisions made had an important bearing on the future of oil exploitation in Iraq. Since France had agreed earlier to the transfer of Mosul from French to British control, at San Remo she got her reward. At Gulbenkian's suggestion, the 25 percent German shareholding in the Turkish Petroleum Company, which Britain's Custodian of Enemy Property had seized during the war, was passed over to the French. Henceforth she would have a quarter interest in the oil of Iraq, which would transform France from a petroleum-buying power (mostly from the United States) into a nation with oil resources of her own. The French were so delighted with the arrangement that they willingly agreed to build a pipeline across Syria to the Mediterranean for the Mosul oil, and waived duties and harbor charges on its shipment to the markets.

When news of this arrangement leaked out in the United States the American oil companies were outraged, and a furious campaign began in the newspapers against what one of them described as "this iniquitous carveup." Standard Oil Company of New Jersey reported that one of its geological surveying crews, operating on a previously agreed terrain in Palestine, had been pressured out of the area by the British authorities, and that others had been refused permission to proceed to Iraq. It was pointed out to the British that one of the articles of the Arab mandates guaranteed that "concessions in the nature of a general monopoly shall not be granted," and in any case, the U.S. govern-

* Lord Curzon. House of Lords debates. 5th S.XL. 1920.

ment did not recognize the prewar concession granted to the Turkish Petroleum Company.

It was a period when there was something of a panic in the United States over oil supplies. The automobile had come of age; all America was buying itself a car (by this time there were more than ten million cars and trucks riding U.S. roads), and the price of crude oil and gasoline was rocketing. Experts predicted that U.S. wells were running dry, and that unless some form of fuel rationing were introduced, the nation would have to rely on foreign oil to keep its wheels turning. The situation was not helped by cocky articles by British experts in American magazines forecasting the supersession of the United States as the chief oil power by Britain, whose Shell Oil Company had already established itself along American highways. "Britain will soon be able to do to America what Standard Oil once did to the rest of the world—hold it up to ransom!" wrote one brash commentator in the London *Daily News.*

In such a climate of anxiety and indignation, government, oil companies and public were united in their determination that America was not going to be cheated out of some participation in the hoped-for Middle East oil bonanza. An acerbic exchange of notes began between Secretary of State Robert Lansing and his British counterpart, Lord Curzon, in which America insisted on a share in the fruits of victory while Britain, though accepting the principle, maintained that it did not apply to rights which had been granted before the war.

What rights? It was Washington's contention that if the British were relying on the concession of the Turkish Petroleum Company, it was no more valid—in fact, it was less valid—than that granted by the Turks to Rear Admiral Colby Chester, also before the war.

At about this time the unsinkable Admiral Chester had just dug his concession out of the bottom drawer and sailed back to Turkey with it. His subsequent activities were to bedevil and confuse the question of the Mosul oil concession, and for a while it looked as if the U.S. government was going to back him to the full. Once more Chester persuaded the Turkish parliament to grant him oil and mineral concessions, as well as a strip of land for a railway and a pipeline, all the way from Anatolia to Mosul, and then lobbied untiringly among anti-British and anti-French congressmen and senators to secure support at home. He also found an ally in a fellow naval officer who proved to be as dogged as himself, and even tougher and more antagonistic toward his French and British Allies. This was Admiral Mark L. Bristol, who commanded the eastern Mediterranean naval detachment of the U.S. fleet while he was at sea, but became U.S. high commissioner to Turkey when he went ashore. His British and French colleagues were well aware that he preferred the Turkish ex-enemy to his wartime Allies, and they winced each time he rose to talk at one of their joint conferences. It was known that he had taken to intercepting and decoding their telegrams to Paris and London, and he relished quoting their own statements back at them. He believed that the United States should have a mandate over the whole of Turkey, including the former Turkish colonies, and that all oil and mineral rights should be given to Colby Chester. He was unabashed in his use of U.S. Navy ships to bring American oil technicians and business experts to Turkey, and when the question of the British and French mandates were discussed at the Lausanne Conference,* he lobbied alongside Colby Chester to get the Mosul enclave carved out of the British mandate over Iraq. Both at Lausanne and subsequently at the League of Nations it was decided otherwise, but it was only when both Bristol and Chester appeared to have developed embarrassingly close relations with a pro-war party in Turkey (which advocated taking back Mosul from the British and the Arabs by force) that the State Department decided to withdraw support from Chester once and for all.

An additional reason was the fact that the British had by now,

* This peace conference convened on November 20, 1922, to settle differences between Greece and Turkey, which had been at war for three years. Turkish nationalists had bitterly opposed the Allied armistice agreement, which, they felt, infringed on the sovereignty of their country.

under constant prodding by Washington, agreed to make room for the Americans in Iraq. Once more it was Calouste Gulbenkian who acted as go-between. Fearful for his 5 percent, unswerving in his belief that it was always better to compromise than fight, he used his influence with Sir William Tyrell of the Foreign Office to persuade the British to accept American shareholders on the Turkish Petroleum Company's board. After due consideration, London suggested that all the American applicants for admission to the Mosul oil concession * should band together under the name of the Near East Development Corporation, and that NEDC should be given a "proportionate" share of Turkish Petroleum Company's capital.

This did not suit Sir Henri Deterding of Royal Dutch–Shell, who was trying hard amid these tortuous negotiations to sap the predominant strength in the Turkish Petroleum Company of the Anglo-Persian Oil Company and its hated chief, Sir Charles Greenway. Calouste Gulbenkian's son, Nubar, who was involved in the negotiations, later wrote: "[Deterding] pointed out that on that basis [a "proportionate" share] the Anglo-Persian would have the largest share and could always gang up with one of the other three groups to get a majority. He insisted that all the shareholdings should be equal: one quarter to the Americans, one quarter to the French [Compagnie Française des Pétroles], one quarter to Royal Dutch–Shell, and one quarter to the Anglo-Persian, all the holdings subject of course to my father's five percent. This meant cutting down the Anglo-Persian 50 percent shareholding by half. To persuade them to accept this, Deterding suggested that they should receive in lieu a royalty on the oil produced." † Which is how it eventually worked out, but it took time.

After some heavy pressure from the British government, King Faisal I of Iraq confirmed the Turkish Petroleum Company's right to drill for oil in his kingdom by granting it a new

* They consisted of the Standard Oil Company of New Jersey, Standard Oil Company of New York, Gulf Refining Company, Atlantic Refining Company, Pan American Petroleum and Refining Company (Standard of Indiana), Sinclair Oil Company and the Texas Oil Company.

† Gulbenkian, *op. cit.*

concession in 1925, and work began at once on finding a suitable field. But it was not until 1927 that a gusher began to spurt oil at Baba Gurgur and set the company on the road to prosperity, and it was not until 1928 that the British, French and Americans finally agreed on their shareholdings in what had by then become the Iraq Petroleum Company.* The British had finally opened the door to the Middle East oil fields and let the Americans put one foot inside. From now on the struggle would be to keep them from bringing in the other.

THE FULL STORY OF THE PLOTTING, backbiting, lying and deception which preceded the agreement over the exploitation of Iraqi oil is too tortuous to detail here.† "I must say that in the oil business not even one's best friends are to be trusted," Nubar remarked at one point in the negotiations. His father had once said, "Oilmen are like cats—you can never tell from the sound of them whether they are fighting or making love." But as far as he was concerned, they were always fighting, and not by Queensbury rules. None of the interested parties (not even the French, who owed their participation to Gulbenkian) hesitated to try to cheat him out of his 5 percent. It was one Armenian against the richest corporations and smartest lawyers in the world, and only the bargaining he had learned at the courts of Abdul Hamid enabled him to emerge triumphant. From the sordid wheeling and dealing of those years one incident involving Gulbenkian should be described, for it had important repercussions on subsequent ventures into Middle East oil.

* The final outcome gave each participant equal shares, as Gulbenkian had suggested. This meant that at board meetings Royal Dutch–Shell, Anglo-Persian, the French and the Americans all had one vote each. (Gulbenkian's 5 percent was nonvoting.) They still have. Since the American share is now equally divided between Standard of New Jersey and Mobil, this means that the two companies must agree on policy; otherwise their half-vote each nullifies the other. "The knowledge sometimes brings about a salutary solution to our differences," a Standard official told the author.

† I recommend *Pantaraxia,* Nubar Gulbenkian's autobiography; the biography of his father by Ralph Hewins, which Nubar authorized; and *An International Oilman,* the autobiography of Sir Henri Deterding, for the often unsavory details.

Big Deal

When the original shareholders of the Turkish Petroleum Company came together in 1914 they had signed a pledge agreeing that in all future oil exploitations in the Ottoman Empire, they would act in concert, and that any fields found by one of the signatories would become the property of them all, but that all nonmembers of the group would be kept out. When the French were admitted to TPC in 1920 they were asked to accept this proviso as the price of membership, and they did so without a qualm. But when the Americans were told that a similar acceptance was the price of a shareholding, they balked. The U.S. government, which believed in the Open Door policy, was particularly opposed to such a self-denying ordinance.

A modus vivendi was found which enabled the American oil companies to accept TPC's restrictive policy and at the same time satisfy Washington (or at least those who did not look at the fine print in the contracts) that the Open Door policy was being observed. However, once oil was struck at Baba Gurgur and it looked as if there would be finds in other areas of Iraq, the major shareholders of what was now known as the Iraq Petroleum Company were all for striking out in other directions—but without Gulbenkian. "It was suggested that the other oil groups would endeavour to obtain concessions themselves, leaving me out," he wrote in his private memoirs, "but this was stopped by the threat of legal proceedings to enforce the provisions against competition contained in the Foreign Office agreement of 1914."

Three years of wrangling followed, but finally, in July 1928 in Ostend, Belgium, at a joint conference of all the main shareholders, Gulbenkian's contention was accepted. Not even the Americans could break the self-denying ordinance. The pledge was renewed: none of the shareholders could exploit oil in the territories of the old Ottoman Empire without the consent and participation of the others.

There remained one other point to be clarified. What were the boundaries of the old Ottoman Empire? "This question could involve untold millions," wrote Gulbenkian's biographer. "The Gulf Corporation of Pennsylvania (one of the American companies

[then] involved in the American group of the [Iraq] Petroleum Company) was already interested in Kuwait and Bahrain, although there had as yet been no oilstrike in either sheikdom. The British Government too had special interests in the Arabian Peninsula, for example treaties with a number of sheiks who had never admitted Ottoman claims to sovereignty over them." *

How were they ever going to define the limits of Ottoman-controlled territory? No one seemed to be sure. But when it looked as if the conference was going to collapse over the failure to clarify this all-important question, Gulbenkian had an inspiration. He sent out for a large map of the Middle East, laid it on the table in front of the delegates, and then drew a line around the central area with a red pencil. "That was the Ottoman Empire which I knew in 1914," he said. "And I ought to know. I was born in it, lived in it, served it. If anyone knows better, carry on."

The delegates leaned over the map and peered at the territories cut through by the thick red line. The British were satisfied, because they saw that Bahrain, Qatar, Saudi Arabia and the Persian Gulf sheikdoms were all inside the line. The Americans saw that Kuwait, where they wanted to operate, was outside the line and that they were free to go ahead. The others looked and raised no objections.

This became the famous Red Line Agreement which shaped the pattern of oil development in the Middle East for the next twenty years. By agreeing to Gulbenkian's definition, the delegates —though they did not realize it at the time—had ensured the wily Armenian of an income of $50 million a year, for that was his 5 percent share of the oil profits which would be made within the Red Line area in the decades to come. The map with the Red Line was initialed and incorporated in the agreement, which the delegates signed before trooping off to a euphoric celebratory lunch.

It was another secret agreement, of course, and no one told the Arabs anything about it.

* See Hewins, *op. cit.*

CHAPTER FOUR

The Wells of Ibn Saud

HE WAS AN AMERICAN CITIZEN who lived in New York, and he was proud of being fluent in Bowery and Brooklyn slang. But thanks to a childhood spent in Lebanon, in the shadow of Mount Hermon, he spoke several Arabic dialects equally well. He was a practicing Christian but admired the teachings of Mohammed and the tenets of Islam. In Western clothes, clean-shaven save for a trim mustache, he looked like a bright young graduate from one of the fashionable East Coast universities (as indeed he was), but in flowing Arab robes with a short beard added to his chin for desert wear he could easily be, and often was, mistaken for a sheik. His name was Ameen Rihani, and he had come to the Arabian peninsula in 1922 not to look for oil, but to see and explore the Middle East and to get to know more about the Arabs with whom he had spent his boyhood. Nevertheless, he did get mixed up with oil, and not very successfully. But his fellow Americans should be grateful to him for his well-meaning intervention. Had it not been for the fact that he backed the wrong man and persuaded an Arab king to do likewise, today the richest oil fields in the world would be in British rather than American hands.

The Arabian peninsula, that great, sack-shaped slab of mountain and desert stretching from the Red Sea to the Persian Gulf, and from the Gulf of Aqaba to the Indian Ocean, is today three-fourths Saudi Arabian territory under the rule of King Faisal, son of the great desert warrior Abdul Aziz ibn Saud.* Its capital is Riyadh, but the two most cherished jewels in its crown are the holy cities of Islam, Mecca and Medina.

When Ameen Rihani came there in 1922, however, Mecca and Medina and the whole of the Red Sea coast and its hinterland were under the shaky control of King Husain, the British puppet who had been dubbed sherif of Mecca after the Turks were driven out of the peninsula in World War I. At that time, Abdul Aziz ibn Saud, the future lord of Arabia, and his Bedouin armies had conquered only the Persian Gulf coastal area, the Great Oasis of Al Hasa Province, and the desert fortress of Riyadh, which he had named as his capital and there proclaimed himself sultan. There would be many bloody battles with rival claimants to the sherifdom before Ibn Saud and his wild tribesmen swept westward across mountain and desert into Mecca and sent King Husain fleeing into exile from his Red Sea capital of Jidda.

In one sense, Ibn Saud was a British puppet himself, having been given a yearly subsidy of £60,000 during the war to fight on the British side against the Turks. Since he would have fought against the Turks, anyway, he accepted the bribe and went on accepting it. As far as he was concerned, it would make no difference to his plans for the conquest of Arabia. The British, on the other hand, believed that with £60,000 they had bought his political loyalty and safeguarded their plans to keep Husain as sherif of Mecca and themselves in control of the peninsula. It was to prove a costly error.

Why the British ever backed the weak, avaricious and arrogant Husain in preference to Ibn Saud is one of the mysteries of Middle East history. In this scorched land of wandering Bedouins, the man who wielded his sword most mightily cut and chopped his

* The other quarter is divided between Yemen, South Yemen, the Federation of Arabian Emirates and Oman.

way to the top, and the man who then showed mercy to his enemies but prudently kept them under his eye stayed there. Husain was one who paid others to do his fighting for him, and then got rid of them. No man trusted his word; no man feared his sword. On the other hand, Ibn Saud's weapon was feared and his word was accepted even by his bitterest enemies. One day he showed off one of his most valued swords, of chased gold and a silver scabbard. With this sword, the sultan told Rihani, he had dispatched one of his bitterest foes in a desert battle. "I struck him first on the leg and disabled him," he said. "Quickly after that I struck at the neck—the head fell to one side—the blood spurted up like a fountain. The third blow at the heart. I saw the heart, which was cut in two, palpitate like that." He illustrated with a shiver of his hand. "It was a joyous moment. I kissed the sword." *

This was in the great battle against his hated Bedouin rivals, the House of Rashid, but after the fighting was over, Ibn Saud took all the remaining members of the Rashid family back to Riyadh, where they lived out the rest of their lives at his expense. He married the widow of the head of the house, and adopted his children of his own. And since he was shrewd as well as magnanimous, he bound the successor to the House of Rashid to him by taking his daughter as yet another wife.

By dhow from Bahrain † to the tiny fishing port of Oqair, and from there by camel for four days across the desert toward the oasis of Al Hofuf, Rihani came for a rendezvous with the mighty warrior. It was the culminating moment of six months' preparation and correspondence with the great Arab warrior. It was a meeting that was to stay in his mind for the rest of his life, for Rihani had expected to meet the sultan in Al Hofuf, but Ibn Saud arrived unexpectedly across the desert by night at the place where the American was encamped nursing a groin made painfully sore

* *Ibn Saoud of Arabia: His People and His Land,* by Ameen Rihani.

† Bahrain is an island about twenty miles off the Saudi Arabian coast in the Persian Gulf. Like most other sheikdoms and emirates in the Gulf, it was under British control.

by his camel saddle. He heard the royal caravan approaching from far off across the desert.

> Soon the heights on which we were encamped reverberated with the cavalcade of the Sultan. More than 200 camels guggled and growled as they were crouching, while the *ikh, ikh* of the riders and the sound of their bamboos on the necks of their mounts were like the patter of rain in a grove of palms. Soon after the tents were pitched, the fires were lighted, and the tintinabulations of the mortars in the coffee pestles were heard. We hastened forth to meet the great guest, but he was quicker in coming towards us, followed by two of his suite . . . We first met on the sand, under the stars, and in the light of many bonfires that blazed all around. A tall majestic figure in white and brown, overshadowing, overwhelming—that was my first impression . . . and the thing that dominates in him is his magnetic smile.*

Ibn Saud took Rihani's hand in his, and there was an instant rapport between them. From that first meeting came an admiration on the American's part and a trust and respect on that of the desert king which were to have a significant result in the days to come. Together, Rihani's tiny caravan and the sultan's army of retainers traveled down to Oqair, where Ibn Saud was to meet Sir Percy Cox, the British high commissioner in Baghdad. The sultan hated the heat and humidity of the coast, and he disliked even more being absent from his wives and concubines, but it was from Sir Percy that he received his annual subsidy, which he needed to buy arms and supplies for his Bedouin fighters. Therefore he was willing to talk to the high commissioner about the future of Arabia, though he was resolutely determined to make no promises about his own role in the development of it.

When Sir Percy and his party arrived in Oqair—weeks later than they had indicated, for it was difficult to keep specific dates at that time—they found that Arab hospitality had gone to great lengths to make them comfortable, though Rihani suspected that they did not appreciate it. Chairs and tables and real camp beds had been found with which to furnish their tents, together with bowls of fresh fruit, and even bottles of Perrier water. There was a

* Rihani, *op. cit.*

great banquet at which a whole camel was roasted. At the dinner given by the Englishmen for the sultan, Sir Percy and his aides appeared in tuxedos. Ibn Saud left early, for he knew that the visitors were anxious to drink brandy and smoke cigars, and that they could not do so in his presence.

From the start the discussions went badly. The main subject of the talk, other than the subsidy, was to define the northern borders of Ibn Saud's conquered territory with the new kingdom of Iraq. But on the day after his arrival, Cox sent a note over to the sultan, roughly scribbled in pencil, in which he asked Ibn Saud to give him written assurances of his friendship toward Britain. The terms of the letter the sultan was asked to write were set out in precise detail in an accompanying note. If His Highness returned him the required letter forthwith, Cox added, he "could then be assured of His Majesty's Government's protection" in the territory he now held.

Ibn Saud was furious that anyone would dare to send him a scribbled note, he whose own communications were masterpieces of verbal and calligraphic splendor. He was even more outraged that the Englishman should suggest that he was not capable of protecting himself. "We are only afraid of Allah!" he roared.

It was into this curdled atmosphere that a strange figure waded ashore from a dhow which had brought him, his interpreter and a Somali servant across from Bahrain. He was a European, but over his ordinary clothes he wore "a thin aba [cloak] which concealed nothing, and over his cork helmet a red kerchief and ighal [cord] which made his head seem colossal. But in this attempt to combine good Arab form with comfort and hygiene he certainly looked funny." *

Frank Holmes was a New Zealander who had served in World War I in the British navy (Royal Marines), but before that he had worked as an engineer in the Far East and counted Herbert Hoover, subsequently President of the United States, among his colleagues and friends. Now he represented a London syndicate and was

* *Ibid.*

looking for an oil concession which he and his associates could prepare for exploitation and then sell off to the highest bidders, preferably the Americans. Rihani, not overly enamored with the haughty British visitors himself, took immediately to the New Zealander, who shared his growing admiration for the great Ibn Saud. So when the sultan sent along a clumsily written Arabic document for Rihani's examination and opinion in which Holmes, through his translator, made his application for an oil concession, the American set to work at once to reword it, and at the same time hinted to Ibn Saud that the New Zealander was a man worth dealing with.

When Sir Percy Cox heard that there was a stranger in the camp, and one, moreover, seeking an oil concession, he at once called in Holmes and said, "Go slow about the concession. The time is not yet ripe for it. The British Government cannot afford your company any protection." *

But Holmes was no fool, and he knew that Sir Percy had cabled London and Baghdad to inform his friends in the government-owned Anglo-Persian Oil Company that a rival entrepreneur was trespassing on what he considered to be Anglo-Persian territory. "Yesterday, Abdul Latif Pasha [Ibn Saud's adviser] showed me a letter," Rihani wrote in his diary on November 30, 1922, "which he has just received from his friend Sir Arnold Wilson (President of the Anglo-Persian Oil Company at Abadan †) in which he says he is coming soon to see the sultan and 'maybe we can strike a deal about oil.' " ‡ He confirmed that Wilson had been alerted by Sir Percy Cox.

As Rihani remarked, though Cox had told Holmes his application was inopportune, "evidently it is not untimely for the APO [Anglo-Persian Oil] to negotiate for a concession."

Not for the first time, a British diplomat had decided that he

* *Ibid.*

† He was in fact resident director. This is the same Arnold Wilson who, as a young army lieutenant, had seen the first oil well gush on the D'Arcy concession in Persia in 1908.

‡ Rihani, *op. cit.*

would squeeze out the national of whom he did not approve in favor of a national of his own class. Major Holmes may have been a British subject, but he was also a brash colonial, whereas Sir Arnold Wilson was an old friend and former colleague. This, plus Anglo-Persian's interest in the Saudi concession, made Cox determined to frustrate Holmes's efforts. To the New Zealander's despair and Rihani's fury, the British high commissioner used the future of Ibn Saud's subsidy as a means of pressuring him against Holmes's application.

> Sir Percy Cox has asked the Sultan to write a letter to Major Holmes saying that he cannot give his decision till he has made certain inquiries of the British Government and consulted them about the matter. He sent a copy of the letter to be written, with a lead-pencil note to Dr. Abdullah [the sultan's secretary] saying: "Will the Sultan please write letter in the above terms to Major Holmes and send me a copy of it" . . . Sayed Hashem said that three times the Sultan had refused and three times the High Commissioner insisted.*

But Ibn Saud wanted his £60,000, and the letter was written. It might have ended there, with the defeat of Holmes and the entry of Anglo-Persian into the oil fields of Saudi Arabia. Ameen Rihani, however, was not only angry at the treatment the New Zealander had received, but also ashamed of Ibn Saud for knuckling under to the high-handed British diplomat. In fact, so far as the sultan was concerned, it did not matter who got the concession so long as he received a rental payment in return—until, that is, Rihani appealed to his sense of pride by writing him a letter, on December 10, in which he said:

> Your Highness is a sovereign in your own land, and you have a right to give a concession to whatever company you please, so long as it is English. Your pact with the British Government does not bind you to accept the company they prefer. Here are two English companies, one of them practically owned by the British Government, while the other has nothing to do apparently with politics, is free from all government influence, and you have a right to your own choice in the matter . . . The least of politics with capital the better for Arabia.

* *Ibid.*

Concessions on a purely business basis and with a purely business motive, without any political tags to them, or any lead-pencil suggestions from British officials concerning them—these are best for the Arabs and for the English.*

The letter hit home. If Rihani considered that Ibn Saud had knuckled under, then the sultan must demonstrate that he had not. But by this time Major Holmes had left the Gulf and installed himself disconsolately in Baghdad. There, some weeks later, Rihani saw him again. The gangling New Zealander told him that he was having tea the next day with Lady Cox, in order to say good-bye to her before he left for England. When Rihani asked him why he did not go back and ask for his concession, Holmes shrugged his shoulders. "My own government is against me," he said.

"I will give you a letter to [Ibn Saud] and I am certain you will get the concession," Rihani replied. "Never mind what Sir Percy says . . . By all means accept the invitation of Lady Cox to tea and tell her you are going back home . . . Say goodbye too to Sir Percy. For if he suspects you are going back to Al Hasa he might get ahead of you to the Sultan with one or two of those lead-pencil notes . . . Goodbye and good luck." †

Major Holmes returned to Ibn Saud's primitive sultanate once more a few weeks later, and the following spring he and his syndicate were granted the oil concession for 60,000 square miles of the desert territory stretching from the shores of the Gulf to the edge of the fertile oasis of Al Hofuf. The rental agreed to between the syndicate and Ibn Saud was £2,500 a year for the right to survey and drill over what has since become one of the richest oil fields in the world. "And everyone seemed satisfied with the deal," wrote Ibn Saud's friend, Harry St. John Philby, later.‡

Unfortunately, though the oil was there, Holmes failed to find it. The syndicate did not have much money and its surveys were

* *Ibid.*
† *Ibid.*
‡ In *Forty Years in the Wilderness.*

superficial. Nor, as they had hoped, could Holmes's company persuade any of the big American companies to take over the concession. From their point of view, Saudi Arabia was too remote, there were no adequate roads or harbors, and fighting was still going on between Ibn Saud and his rivals. After 1927, Holmes stopped paying the annual rental to the sultan and allowed the concession to lapse.

Thus the Holmes concession would have been nothing more than a footnote in the story of Saudi Arabian oil were it not for the fact that, failure though it proved to be, it thwarted Anglo-Persian Oil at a moment when that company was ready to take up the concession. Had it done so, almost certainly its experts would have found the oil which had eluded the New Zealander's syndicate, and the world's most profitable oil field would have passed into British hands. The high-handedness of Sir Percy Cox, and Ameen Rihani's antipathy toward him, saved it for the Americans.

CHAPTER FIVE

35,000 Gold Sovereigns

BY 1930 ABDUL AZIZ IBN SAUD HAD KILLED, captured or put to flight all his enemies, and the land he ruled as king stretched all the way from the Red Sea to the Persian Gulf.* Mecca had given up without a fight. Jidda, the great Red Sea port, had surrendered after the British puppet, Husain, had fled into exile, and became Saud's administrative capital. But there had been bloody battles against rival claimants to lordship of Arabia, and Ibn Saud owed his victory not only to his own strategy and tactical skill on the battlefield, but also to the fanatical bravery of his Wahhabi Bedouin troops and their spearheads, the *ikhwans*. Like many a Christian general, Ibn Saud had found that a mixture of the sword and the Holy Book worked wonders on the battlefield, and the *ikhwans*, who believed in every word of the Koran, who considered any Muslim who did not an infidel, and who died joyously in battle because they were going straight to paradise, had shed their blood most efficaciously in his behalf.

Now, as their reward, they were allowed to impose their puritanical regime on the new kingdom, and they did it with a

* It comprised Nejd and Hejaz, a dual kingdom formed in 1926 by Ibn Saud; it became a single kingdom in 1932 and was renamed Saudi Arabia.

pious passion that was to leave its mark on Saudi Arabia to the present day. Unlike Moses, they did not raise tablets upon which the Commandments of the Prophet were written, for anything suggesting a graven image was henceforth forbidden. But the Commandments were no less exigent because they were spoken. Pictures were torn down from the walls. Statues and monuments were destroyed. *Thou shalt not smoke.* A man caught with a cigarette in his mouth or the smell of tobacco on his clothes was taken into the bazaar square, spread-eagled on the ground by burly slaves, and subjected to forty lashes. *Thou shalt not drink spirituous liquor.* An offender received sixty lashes, or had his hand chopped off at the wrist. *Thou shalt not commit adultery—at least during daylight.* The *ikhwans* and their *ulema* (priests) believed that man was not ordained to have sexual relations unless darkness had fallen, not even with one of his wives or concubines, and transgression was subject to the lash. Adultery during Ramadan, the religious time of fasting, was punished by a ceremonial execution, and all Jidda or Riyadh or Al Hufuf was summoned to watch a black slave swordsman lop off the offenders' head. (The guilty female did not merit such a ceremony and was stoned to death.)

Women were banned from the streets. A man who wore silk robes or walked with a swing of the shoulders was reprimanded for arrogance. It was forbidden to laugh out loud or sing or otherwise make music, and a ceremonial beating awaited anyone who failed to stop whatever he was doing five times a day to hurry to the statutory prayers at his mosque. Allah be praised, it was a saturnine and sanctimonious regime which had descended upon the Arabian peninsula.

Not that the new king was affected by it, for his own particular pleasures did not suffer under the round of prohibitions. He had four to which he was particularly attached. True, for the moment he was not indulging in one of them, the joy of fighting hand to hand with an enemy, but this gave him more time for the other three. The first of them was simply talking, gossiping with his courtiers about old enemies he had killed or captured, discussing world politics with his friend and adviser, the Englishman Harry St. John Philby, or generally reminiscing about the women he had loved and the permutations and techniques of sexual relations. "Women and world politics continue to divide the honours as prime subjects of the king's conversation at private sessions," Philby reported in his diary in 1930.

The second of his pleasures was hunting. For long he had gone out on hawking expeditions into the desert after gazelle, buzzard and game birds, but now the motorcar had arrived in Saudi Arabia and had changed everything. Instead of a hawk to peck out the eyes of the hunted animals and bring them within range of the guns, the automobile now enabled the hunters to chase the quarry and outdistance it. With the help of his new car and a fast-driving Italian chauffeur, Ibn Saud and his companions managed to thin game in the Arabian desert down to extinction levels. The ostriches which roamed the foothills until 1935 were wiped out, and the last of rare breeds of oryx and gazelle were driven into the mountains. Philby once saw the king wipe out a herd of rare gazelles in fifteen minutes, and there was no doubt of the great joy he got from the process of extermination.

But the third and undoubtedly the most savored of Ibn Saud's pleasures was sex, and since he agreed with the *ulema* that it should never be indulged in during daylight hours, when he considered it indecent, he encountered no disapproval from them of his activities with the opposite sex, marathon though they were. From the age of eleven until a few weeks before his death in 1953 at the age of seventy-two, he had sexual relations with a different woman every night, save during battles and while the *hajj*, the pilgrimage to Mecca, was in progress, when all good Muslims are pledged to eschew sex. Each night at nine-thirty he would glance at his watch, say adieu to his guests and depart for his harem. He had married for the first time at the age of fourteen, and it is likely that he loved his young bride, for he grieved deeply when she died in an epidemic. But thereafter he gave himself over to what Philby euphemistically called "his marked tendency to uxuriousness." He is said to have told the ruler of Kuwait that the pleasure he found "most worth living for" was to "put his lips on the woman's lips, his body on her body and his feet on her feet." After this remark became known, there was some comment that, if taken literally, this must severely have limited his choice of females in Arabia, since he was six feet six in height. That all he asked for from a woman was sexual relations —and children, of course—was indicated by his remark that he had never had other than small talk with a member of the female sex, and until American oilmen in Arabia began to bring their wives, and were sometimes entertained by the king, he had never seen a woman eat or drink, not even his mother.

The king never exceeded the quota of four wives allowed him by Koranic law, but since he sometimes found it politic to marry the widow of a defeated chief or the daughter of a tribal leader whose allegiance he was seeking, the quadrumvirate of wives remained constant but their identities changed frequently as he solemnly pronounced one divorced before wedding the next. He had sexual relations with all of them, though he confessed that with some he never even bothered to take off their veils.

In the months following his conquest of Arabia, Ibn Saud was in his sexual prime, and though it seemed to make no untoward demands on his physical reserves, it certainly depleted his treasury. "Each of his four wives having a house of her own with a full complement of servants, slaves and attendants, in which they received his visits in rotation," wrote Philby, "he also had a house of his own . . . which was run for him by four favourite concubines, enjoying a status indistinguishable from that of wedded wives. And in addition he had four favoured slave girls to complete his matrimonial team of a dozen, to say nothing of his right to pick and choose from other numerous damsels at his disposal." *

In 1930 Ibn Saud confessed to Philby that he had married

* From *Arabian Jubilee.*

135 virgins, to say nothing of "about a hundred others" up to that time. But he was finding them a desperate drain on his financial resources, especially as he also had to keep their servants and any children they produced. He had resolved to limit himself to two new wives a year from then on (since it was wives who were most expensive), but even this would not solve the monetary troubles in which he and the new kingdom found themselves.

Harry St. John Philby was an Arabist scholar and explorer who had once worked for Sir Percy Cox, but disenchanted by Anglo-French treatment of the Arabs, he had resigned from the colonial service. He had won the friendship of the desert king and become the only man at court who dared to argue with him, or proffer advice unasked and have it taken. He had even turned Muslim and built himself homes in Mecca and Jidda. To help keep his wife and children in England, he had also secured for himself the lucrative Ford agency for Saudi Arabia and sent Model A's out onto the desert tracks where before only human and animal feet had touched.

Philby knew that no economy in wives would save the kingdom from bankruptcy, and that in any case the royal resolution to abstain would be abandoned the moment a particularly pleasing concubine or slave began to plead for marriage. Saudi Arabia's financial problems were much more serious than that. The main income of the country was garnered from the pilgrims who flocked each year to Mecca, and because of the unrest in Arabia and the depression abroad only small numbers had come in 1930 and even fewer were expected in 1931.

One day in the fall of 1930 Philby accompanied the king on a desert drive. He soon realized that Ibn Saud was worried and asked what the matter was. The king replied that the financial situation was so bad that the government could see no way to make ends meet, and was at the same time confronted by difficulties that it could not hope to cope with. "I replied, as cheerfully as possible in the circumstances," Philby wrote later, "that he and his Government were like folk asleep on the site of buried treasure, but too lazy or too frightened to dig in search of it. Challenged to make my meaning clear, I said I had no doubt whatever that his enormous country contained rich mineral resources, though they were of little use to him or to anybody else in the bowels of the earth. Their existence could only be proved by expert prospection, while their ultimate exploitation for the benefit of the country necessarily involved the cooperation of foreign technicians and capital. Yet the Government seemed to have set its face against the development of its potential wealth by foreign agencies."

Instead of moaning about poverty, Philby went on, Ibn Saud should think about the Koranic saying: "God changeth not that which is in people unless they change that which is in themselves."

"Oh, Philby," exclaimed the king, "if anyone would offer me a million pounds, I would give him all the concessions he wanted."

"Well," replied Philby, "it won't be as bad as that. But nobody will give you a million pounds without a preliminary investigation of the potential resources of your country. Yet there is a man in Egypt now—an American whom you did not take the trouble to meet when he visited Jiddah a few years ago. He has recently visited the Yemen and done a good deal to help that country in various ways. He could help you too, for he is a very rich man with important contacts in the American industrial community. I suggest you should meet him as soon as possible." *

The man to whom Philby was referring was a U.S. philanthropist named Charles Crane, whose passionate love affair with the Arab people had survived snubs and setbacks and even armed attacks (during one of which a companion traveling with him through the desert was shot dead). His affection for its people left him surprisingly ignorant of some things about Arabia, however. He arrived in Jidda the following February and had several discussions with the king and his ministers. Just before his departure, Ibn Saud made him a gift of two magnificent Arab stallions.

* *Ibid.*

Crane's return gift was, of all things in a land whose oases grew nothing else, a box of dates (Californian). But he did give useful advice about the exploitation of mineral resources, and within six weeks, had dispatched to Jidda a Vermont mining engineer named Karl Twitchell who, at Crane's expense, would prospect the desert for oil possibilities.

Karl Twitchell, in the words of the author of the official history of Aramco (the Arabian-American Oil Company), "went up and down Arabia sniffing for oil seeps and quartz outcrops," and his reports in 1932 "so encouraged Ibn Saud about oil and gold prospects that he commissioned Twitchell to communicate to oil or mining companies in the United States Arabia's willingness to discuss concessions." *

What neither Twitchell nor King ibn Saud knew was that Standard Oil of California officials had been trying to contact the king for months in the hopes of discussing concessions. But they had been stalled and hindered in their efforts by none other than the dashing New Zealander, Major Frank Holmes, whose image during these events becomes rather less burnished than it was when Ameen Rihani had recommended him to Ibn Saud in 1922.

Having failed to find a buyer for the Al Hasa concession Ibn Saud had granted him, and lacking the financial resources to explore the terrain adequately himself, Holmes had simply let the concession lapse, without informing the king or paying the modest annual rental of £2,500 which was due. All communications from Arabia were unanswered, and it was assumed that Holmes had gone back either to England or to New Zealand. He was, in fact, only twenty miles away, on the island of Bahrain, just off the Arabian coast. There in 1925 he was granted an oil concession by the ruling sheik, with the approval of the British agent,† and the concession was renewed in 1927. In that year Holmes sold an

* *Discovery!*, by Wallace Stegner.

† In the Persian Gulf a political resident, the British government's representative, acted for all areas and was represented by political agents at the three local centers of Manama, Masqat and Al Kuwait, capitals of Bahrain, Oman and Kuwait, respectively.

option on it to the American company Eastern Gulf Oil, a subsidiary of the Gulf Oil Company. The U.S. oilmen sent a geologist, Ralph Rhoades, to look the territory over; they were delighted when Rhoades reported that Bahrain had a perfect geological structure and recommended that drilling for oil should begin immediately.

But there was a snag. Bahrain was within the red line drawn on the map by Calouste Gulbenkian in his definition of the confines of the Ottoman Empire, and the Red Line Agreement stipulated that no member of the Iraq Petroleum Company could exploit a concession within the area independently of the others. At that time Gulf was one of the consortium of American companies which had a share in IPC, so Gulf offered the Bahrain concession to IPC, which it promptly turned down on the grounds that its geologists considered the oil prospects poor.

It was the first serious indication to the Americans that Gulbenkian's red line was going to be a halter around their necks, and shortly afterward Gulf bowed out of IPC in order to regain independence of action, as did several other U.S. companies. But it was too late to save Bahrain. Gulf sold out their option to Standard Oil of California (Socal), which almost immediately ran into trouble with it.

No one had told Socal that in the British-controlled areas of the Middle East only British firms were allowed to operate, and it took them two years to arrange a way around the regulations. First a company had to be registered in Canada, then an office established in Britain run by a British subject. It was stipulated that an active director of the company must be not only British but officially approved by the government. Further, the majority of the employees must be either British or Bahraini, and the company must undertake to appoint in Bahrain itself a chief local representative, vetted and approved by the British government, and this representative would never approach the ruler of Bahrain except through the offices of the British agent.

It was not until August 1, 1930, that Socal formally assigned its concession to its Canadian subsidiary, Bahrain Petroleum Company (Bapco); at the same time it announced the appointment of Major Frank Holmes as its chief local representative.

Bahrain's first well came in, confounding IPC's geologists, on October 14, 1931, but long before this, Socal's visiting oilmen knew that oil was there. More important than that, their soundings told them that if there was an oil pool under Bahrain there was probably an even bigger one under the terrain they could see vaguely shimmering in the heat, twenty miles away—in Saudi Arabia.

A Socal geologist named Fred A. Davies, a huge fullback from Pennsylvania, was so convinced that Saudi Arabia was where most of the oil was waiting that he galvanized his home office in San Francisco into agreeing to go after the concession. It was decided to approach Ibn Saud, and the man who could take them to him, or so they believed, was obviously Major Frank Holmes, who was now on their payroll. Holmes was told to set up a meeting and he agreed at once to do so. What he does not seem to have told them is that he had failed to pay the rental on his own lapsed Al Hasa concession, and was therefore not exactly anxious to meet his creditors. While Socal officials waited in the steamy heat of Bahrain a short tantalizing distance away from the potentials of Al Hasa, Holmes fobbed them off with excuses. First he told them that King Ibn Saud was fighting a campaign, then that he was mourning a favorite wife. Eventually, tired of waiting for a rendezvous that looked more and more like a mirage, Davies and his fellow Socal officials sailed for home.

In the meantime the company's adviser on foreign relations, Francis B. Loomis (who had once been an undersecretary of state under Theodore Roosevelt), had heard of Harry St. John Philby's influence with the Saudi monarch and had written him a letter asking him to arrange a meeting for Socal representatives with the king. A few days afterward, at a luncheon party in Washington, Loomis found himself talking to Karl Twitchell, back in the United States for a vacation. He mentioned Socal's interest in a

Saudi concession. "Then why haven't you been in touch with the king?" asked Twitchell. "Yours are just the people he wants to meet."

Loomis wasted no time telling him about Holmes's procrastinations,* but at once asked Twitchell to return to Saudi Arabia with a Socal negotiator and act as go-between in the discussions with the king. Socal chose a plump, forty-year-old lawyer and land-lease expert from California named Lloyd N. Hamilton. The two men arrived in Jidda aboard the S.S. *Burmah* on February 20, 1933, and they brought their wives, Airy Hamilton and Nona Twitchell, along with them. Hamilton had the slightly glazed look of an innocent American abroad, but appearances were deceptive, as his Arab opposite numbers and rival oil companies would discover. It was a fateful arrival. "Change," as Aramco's official historian put it, "entered in a suit of wrinkled whites and a sun helmet." †

Meanwhile, urgent conferences were being held in London over the whole question of oil concessions in the Persian Gulf. To say that the Anglo-Persian Oil Company was concerned over U.S. incursions on an area hitherto considered a British bailiwick is to put it too politely. The new chairman of the company, Sir John Cadman, was sternly asking his British associates in the Iraq Petroleum Company why they had let the Bahrain concession go to the Americans, when they had the prior right to take it up themselves. It was explained to him that IPC geologists had not believed it possible that oil would be found in porous limestone of Cretaceous age (the deposits in Iraq and Persia being in much older rock formations). He is said to have replied, "In that case, I suggest you sack our geologists and hire Americans."

With the oil of Bahrain, it would have been easy to get the concession. Since the ruler was under British control, all IPC had

* Holmes was only too right in his hesitations, as far as he personally was concerned. He arrived in Jidda during subsequent Socal negotiations and was at once told to pay up his concession rentals or get out within forty-eight hours. He got out.

† Stegner, *op. cit.*

to do, under the Red Line Agreement, was take over. It had missed its chance. But with Saudi Arabia the situation would be much more difficult. The time was past when King ibn Saud could be bullied by the British. He would sell to the highest bidder. The Saudi concession would now have to be fought for with an American rival already successfully established just across the water in Bahrain, and one, moreover, which had signed no prior agreements with anyone and was not bound by Red Line or any other agreements. Cadman frigidly indicated to his subordinates that IPC should not be dilatory or lacking in combativeness on this occasion. From his point of view, the situation was doubly dangerous, commercially and politically. If Socal succeeded in Saudi Arabia, it would not only bring an American oil rival into an area where British petroleum interests had so far been dominant but since the flag followed trade, it would also bring U.S. influence into the Persian Gulf, a zone which Britain considered vital to her imperial interests and was determined to keep walled off from any other foreign interference.

If the Anglo-Persian chief was aware of the important strategical as well as tactical advantages of securing the Saudi concession for Britain, however, no one in IPC seems to have explained them to the British negotiator the company sent to Jidda, Stephen Hemsley Longrigg. Longrigg is a most able and experienced veteran of the Middle East oil industry, and he has written an authoritative history of how it evolved, but in that account he devotes only one paragraph to his own and IPC's part in the negotiations for Saudi Arabia's oil. His reticence is understandable, for the company blundered badly. All he says is:

The Iraq Petroleum Company had this time decided to contest the issue. Their representative (the present writer) arrived at Jiddah to find negotiations in progress between Hamilton [of Socal] and the Saudi ministers and was invited to make his own offers. Both negotiators interviewed the King, both advanced their proposals, each was assured that his company and his nationality would, all things being equal, be more acceptable to the Saudi King. But the IPC directors were slow and cautious in their offers and would speak only of rupees when gold was demanded. Their negotiator, so handicapped, could do little; and agreement was reached without difficulty between Hamilton and Shaikh Abdullah Sulaiman on 29 May 1933 for a 60-year concession for the Al Hasa province. *

But there was much more to it than that. IPC might have sent its negotiator to Jidda even later in the day had not Harry St. John Philby informed Anglo-Persian in London that Socal representatives were about to arrive to make a deal. This was just after he had received a note from King ibn Saud (dated February 23, 1933) saying, after a friendly preamble:

> I am confident that you will protect our interests, both economic and political, just as you would protect your own personal interests. So I shall expect your assistance in this matter and also I shall expect you to give me the benefit of your personal advice, which will be treated as confidential and with all consideration.†

It was to secure a better bargain for the king that Philby urged Anglo-Persian to prod IPC into sending a negotiator, on the correct assumption that competition for the concession would force up the price that was offered for it. But by the time Longrigg arrived, the situation had radically changed. In his conversations with Philby, the Socal representative, Lloyd Hamilton, had quickly divined that the Englishman, like Ibn Saud himself, had a money problem—though on a humbler scale, to be sure. The Ford agency was not bringing in as much profit as he had hoped, and there were fees to be paid for his son at Cambridge and his three daughters at school.‡ Hamilton therefore made a discreet proposition. If Philby would help Socal to secure the concession by pressing the American case with the king and his ministers, and by keeping Hamilton informed of any rival offers, he would be paid a fee of £1,000 a month for a minimum of six months, added to which there would be bonuses when the concession was signed and when oil was struck.

* *Oil in the Middle East,* by Stephen H. Longrigg.

† *Arabian Oil Ventures,* by Harry St. John Philby.

‡ The son, Kim, grew up to become a departmental chief of British intelligence in World War II, and in 1963 defected to the Soviet Union, where he is now an officer of the KGB.

Philby accepted, apparently without hesitation. The only stipulation he made was that no one should be told of the arrangement. Hamilton did not even tell his own wife or Twitchell, and Philby did not inform the king or his finance minister, Abdullah Sulaiman. Longrigg was so completely in the dark about what was going on that when he arrived he asked Philby if he would like to take over the negotiations for IPC. Philby refused on the grounds that it would affect his "impartiality."

Despite the fact that he was now on the Americans' payroll, Philby might have kept the IPC representative in the game had not Longrigg decided at one of their meetings to take his fellow countryman into his confidence. The two men had once worked together in the colonial service, and they had sat on the same side of the table at many an Arabian conference; Longrigg had no reason not to trust his compatriot. He even appears to have felt that for all his declarations of "impartiality," Philby would do his best for him and for Britain in the negotiations. Now Longrigg confessed that IPC was not really interested in Arabian oil at all. The company did not care whether any was found or not; all it was anxious to do was prevent the concession from falling into the hands of the Americans. Petroleum had slumped to 10 cents a barrel in the world markets, there was a trade depression everywhere, and IPC had all the oil it could handle from its wells in Iraq.

This was jolting news to Philby. For Ibn Saud's sake, he wanted to see Arabia's oil potential exploited as urgently as possible. It reinforced the suspicion he had always had that IPC was not the right company for the concession, and Longrigg's next revelation confirmed it. All he had been authorized to offer for a concession was £5,000. Philby told him that he had better go away at once and get the stake raised, and make sure that whatever sum he came back with was payable in gold. Longrigg hurried off to the British legation to have a cipher telegram dispatched to London, and reappeared subsequently with the news that he could go up to £6,000 but that there would be no question of payment being made in gold. Britain had gone off the gold standard and the Bank of England would never allow it, for it would be tantamount to suggesting that the pound sterling was unstable. The king would have to accept sterling, or the equivalent in sterling-backed Indian rupees.

Longrigg stayed around in Jidda for some time thereafter, but Philby eliminated him from the negotiations. So far as he was concerned, it was now a straight contest between Abdullah Sulaiman, the Saudi finance minister, and Lloyd Hamilton of Socal, with Philby acting as the go-between. Years later the Saudi minister told Philby that "if he had known at the time that I was actually supporting the negotiations of the California Standard Company for the oil concession in 1953, he would have done his best to block the transaction." * In fact, he was doing Philby an injustice, for the moment he realized that IPC was not a contender worth being seriously considered for the concession, Philby did his best, while strongly supporting Socal's case, to get a really generous sum of money out of the Americans. But he was no oilman, and he did not realize the huge potentials involved. He ferried himself back and forth in his brand-new Model A between the Grand Hotel, where the Lloyd Hamiltons were installed in a huge room with a brass bedstead and an airless bathroom with cockroaches but no running water, and the king's palace beyond Jidda's city walls. For all his shrewdness, Hamilton had expected the negotiations to take only a week or two, but there were endless rounds of conversations and innumerable cups of coffee consumed during which no word about oil was spoken. Finally, at Philby's suggestion, Socal was asked for an immediate payment of £100,000, and he was considerably shaken when Hamilton told him that all Socal was prepared to offer was $50,000 against future royalties. Weeks went by and it was not until April 20, when the humid heat of Jidda was boiling up to the temperature of a Turkish bath, that Hamilton met Abdullah Sulaiman and presented his final offer of an immediate loan in gold of £30,000, plus a further

* Philby, *Arabian Oil Ventures.*

loan of £20,000 gold eighteen months later, and a yearly rent of £5,000 gold. Philby knew from his conversations with Hamilton that this was indeed the final offer and had conveyed that fact to Abdullah Sulaiman, so the Saudis were therefore ready to accept it when a Reuters telegram reached them. As the historian of Aramco remarks, it "threw the whole tangled negotiations into a snarl again," for April 20, 1933, was the day on which the United States announced an embargo on gold. "Within a few days Lombardi [a Socal vice-president] was cabling from London to hold everything, and the Netherlands Bank in Jidda had raised the value of the gold pound from $4.87 to $5.60. Within another few days it climbed to above $6—carrying with it the whole price of Hamilton's carefully negotiated agreements." *

New negotiations began, as Hamilton tried to turn the Saudis away from gold pounds to paper dollars. They were adamant. It was not until May 25, 1933, that Abdullah Sulaiman came into the king's presence and began reading the long and complicated agreement which he had made with Lloyd Hamilton. It took two days to read, and Ibn Saud, puzzled by its convolutions, went to sleep during most of the presentation. But toward the end he woke up with a start, fixed Philby, sitting on the ground near him, with his one good eye † and asked him what he thought of the agreement. Philby assured him that it was the best Saudi Arabia could obtain.

"Very well," said the king to the finance minister, "put your trust in God and sign."

On May 29, Abdullah Sulaiman for the Saudis and Lloyd Hamilton for Socal signed the concession agreement that was to bring American capital, brains and technical skill into the Persian Gulf and change the course of Arabian history.‡ The door of

* Stegner, *op. cit.*

† Like many another Arab, he suffered from trachoma.

‡ The company which was established in 1933 as a result of these negotiations was called California-Arabian Standard Oil and was wholly owned by Standard Oil of California. In 1936 the Texas Oil Company became a half-partner.

Although the company's name was not changed to the Arabian-American Oil Company until the end of 1943, the present-day title (or Aramco for short) has been used throughout for clarity.

the Middle East had opened wide enough for the United States, with one foot planted in Bahrain, to get a second onto the sands of the mainland. It did not happen immediately, but it was the beginning of the end of British control of the Persian Gulf. Britain had not only let the Americans into a cherished sphere of influence; she had also let slip the biggest Middle East oil field of them all.

FOR HIS PART IN THE NEGOTIATIONS, Harry St. John Philby received the warm thanks of King ibn Saud and the gift of a house, known as the Green Palace, with a garden not far from the king's palace, beyond the walls of old Jidda. This enabled him to sell his old house in the city, the Beit Bagdadi, to Lloyd Hamilton as temporary headquarters for the company's Saudi operations. He had done well by Socal, and he had earned his fees and bonuses. Everyone seemed satisfied, and it was only later, when the gushers began to come in, that the Saudis began to claim that they had been cheated. In fact, it was only by a stroke of luck that they received their first payment at all. Hamilton had pledged that the initial payment in gold would be made to the Saudi finance minister within three months of the signing of the agreement. Under the new currency regulations, however, all gold exports from the United States had to be authorized by the Treasury Department. Socal at once made application for a permit to export $170,327.50 in gold to Saudi Arabia. Weeks passed and there was no word from Washington. Hamilton, who had been on vacation in France and Italy, arrived in London in July to learn of the ominous silence, and decided that drastic measures must be taken.

On July 26 he had Socal officials go into the London gold market and buy 35,000 gold sovereigns from the Guaranty Trust Company. It was an illegal purchase for an American company under the new U.S. regulations, but the future of the concession was at stake. He was more than ever convinced that he was right to have taken the risk when a reply was at last received, on July

28, from the Treasury Department, signed by the Undersecretary, a certain Dean Acheson, formally turning down Socal's application.

The 35,000 black-market sovereigns were boxed and put aboard a P and O liner which sailed for the Persian Gulf on August 4. On August 25, three days short of the deadline, Karl Twitchell sat at a table in the Jidda branch of the Netherlands Bank and counted out 35,000 sovereigns one by one under the wary eyes of Abdullah Sulaiman. When he got a receipt, the Americans were in business in Saudi Arabia at last.

CHAPTER SIX

And Then There Was Kuwait

TWO HUNDRED AND FIFTY MILES NORTH of Bahrain and Al Hasa, at the neck of the Persian Gulf, a wedge of coastline and desert shaped like the bisected half of a six-pointed star has been inserted into the borders between Saudi Arabia and Iraq. This is the sheikdom of Kuwait. Today it is a car-clogged land of millionaires, of skyscraper hotels and ornate government buildings, and it sits on a vast, if not inexhaustible, pool of high-quality oil. But in the 1930s it was a sleepy, pearl-fishing village. As to oil, all that the geologists knew was that there were bitumen seepages at Burgan, in the desert away from the sea, and that underneath, large quantities of oil probably did exist. Still, no one could say for certain until they had been drilled for.

Into Kuwait in 1930 came a character who has figured several times before in this story, none other than the bouncy New Zealander, Major Frank Holmes, in search of yet another oil concession. He already had a customer for it. When he sold out his Bahrain concession to Eastern Gulf Oil Company he had agreed, as part of the contract, to obtain a concession in Kuwait should Gulf ever ask for it. In the intervening period Gulf, because of its commitment to the Red Line Agreement (and their anger at this forced sale was doubly compounded when Socal struck oil in Bahrain in May 1931), had been forced to sell off Bahrain to Standard Oil of California. The Gulf people told Holmes they were exercising their option under their contract with him; they wanted the Kuwait concession. They wanted Kuwait not only because they were confident that oil was there but because they believed that in Kuwait they would have a free hand. When Gulbenkian drew the red line defining the boundaries of the Ottoman Empire, he had left Kuwait on the free side of the line.

This time Holmes needed quick action, for he was short of money, and he arrived in the sheikdom bearing gifts. For his first serious talk with the ruler of Kuwait, Sheik Ahmad al Jabir as-Sabah, he arrived in a brand-new Sunbeam automobile, driven by his associate, T. E. Ward, and when the sheik came out to admire its sleek English lines the New Zealander said with a flourish, "It is yours, Excellency." Slipping the knot of the green gauze veil which he wore to protect his face from the sand, and doffing his white cork sun helmet, he followed the ruler into the palace. All seemed to be well. "It is interesting to observe the greedy attitude the Sheik of Kuwait exhibited when I hinted that the Company I was representing had an American tang to it," he remarked to Ward afterward.*

In the next months the greed was nurtured by gifts which Ward later enumerated in the account he drew up for William J. Wallace of the Gulf Oil Company. Four members of the state council received 4,000 rupees, 45,000 rupees were reserved for certain high officials, and 20,000 rupees went into the ruler's private purse.

The concession was as good as granted—or so Sheik Ahmad indicated. But soon rumors began to circulate around the bazaars in Kuwait that the British political agent, Colonel Harold Dickson,

* *Negotiations for Oil Concessions in Bahrain, Al Hasa, the Neutral Zone, Qatar and Kuwait,* by T. E. Ward.

had been to see the ruler several times, and that the sheik had been reminded of certain "obligations." Suddenly there were no more conversations between the sheik and the New Zealander, and further baksheesh was politely turned away.

The Gulf people were growing increasingly impatient at the sudden halt in the negotiations, especially now that they were aware of the oil strike in Bahrain. But it was not until well into 1931 that they discovered the reason for the delay. As it had done in the case of Bahrain, the British government let it be known that it was invoking an old treaty with the sheikdom to prevent an American or any other foreign company from operating in the territory; and this time, it implied, the difficulty would not be overcome by the registration of a U.S. company in Canada to give a false coat of Britishness to its colors. It would have to be a British company or nothing.*

At which point a genuine British company appeared on the scene right on cue, in the person of Archibald H. T. Chisholm, representing the Anglo-Persian Oil Company. Just as Gulf was not restricted by the Red Line Agreement in Kuwait, neither was its partner in the Iraq Petroleum Company, Anglo-Persian; and APO had the British government and a sheikly treaty behind it in the campaign which it began to oust Gulf. Chisholm later admitted that his company was not primarily interested in finding oil in Kuwait (as mentioned earlier, these were the days of the great world depression, and petroleum sales were drastically down).

* To reinforce their argument that Kuwait's oil was reserved for Britain, the government reproduced a letter written by the then ruler to the British resident in Bushire in October 1913, in which, in return for the promise of British protection from the Turks, he wrote:

> We are agreeable to everything which you regard as advantageous and if the Admiral honours our country [with a visit] we will associate with him one of our sons to be in his service, to show the place of bitumen in Burgan and elsewhere, and if in their view there seems hope of obtaining oil therefrom we shall never give a concession in this matter to anyone except a person appointed from the British Government.
>
> (*Treaties, Engagements and Sanads relating to India and Neighbouring Countries.*)

"But we *had* to get the concession so as to protect our huge investment in Persia," he said. "For Kuwait is nearer [world] markets and, if competitors found oil there, the Persian business would have been undercut and ruined." *

Chisholm set to work to stop the American incursion. The wife of the political agent in Kuwait, Mrs. Violet Dickson, paints a cozy picture of two friendly rivals dabbling in oil concessions between social rounds, while her husband sat with Sheik Ahmad at the parleys and gave him impartial advice. "Chisholm in particular," she writes, "with his youthful charm, his gaiety and his skill at tennis, was an asset to all our social gatherings." And Holmes was "forceful and often rude, but could be good company nevertheless." And, of course, "Sheik Ahmad had grown to trust [Harold] as a friend." †

The situation was somewhat tougher than that. The two men were complete contrasts in types. Chisholm, thirty years old, Anglo-Irish, son of a former editor of the *Times* of London, with an honors degree from Oxford, a year on the staff of the *Wall Street Journal,* a family connection with Sir John Cadman, the Anglo-Persian chief, dressed like a character out of a Noel Coward play, sported a monocle, and larded his conversation with the latest Mayfair slang. But behind the foppish front was a cold, calculating mind that would make him one of the most successful military intelligence officers in the Middle East in World War II.

Frank Holmes was over fifty, and when he walked through Kuwait in his ridiculous solar topee and green face veil, small boys would walk behind him, laughing at the white sunshade he carried and singing:

> "*Ingaresi bu taila,*
> *Assa yi mutun hal laila!*"
> (Englishman with hat on your head,
> May you die tonight in bed.)

* Quoted in *The Golden Dream,* by Ralph Hewins.
† *Forty Years in Kuwait,* by Violet Dickson.

But here again, the eccentric outer garments and florid front were a cover for a tough man who knew exactly what he wanted and was determined to get it. So far as Holmes was concerned, the Kuwait oil concession was a matter of life or death. All Chisholm would get if he failed was a slap on the wrist from his relative on APO's board, but if Holmes failed he knew Gulf would sue and squeeze him dry.

"Holmes and I were at daggers drawn for a year," Chisholm said later. "Month after month we lived as neighbours, meeting with invariable courtesy and no rancour, each making calls on the shaikh, each in turn suggesting some new clause and knowing that the other would act accordingly. Every Sunday we met in church." *

By this time the ruler of Kuwait, Sheik Ahmad, had begun to like the situation. So long as Holmes and his Americans remained in the competition, he had everything to gain. Even if British pressure did eventually force him, as he suspected it would, to favor the cool, bemonocled Anglo-Irishman over his New Zealand rival, the preliminary struggles would drive the price up; and since Sheik Ahmad had expensive and exotic tastes, especially in women, he needed all the money he could get.

Not that Holmes considered that he and his friends were beaten yet, despite the British declaration. Nor did Gulf. They knew they had a high card in their hand, and on November 27, 1931, they played it. On that date they formally called the attention of the State Department "to the fact that the so-called British Colonial Office was insisting on the so-called 'nationality clause' in the Kuwait concession. This clause in effect prevented anyone except a British subject or firm from obtaining a concession in Kuwait." †

The State Department's reaction was prompt. On December 3, 1931, "instructions were sent to our Embassy in London

* Quoted in Hewins, *The Golden Dream.*

† Hearings on American Petroleum Interests in Foreign Countries, memorandum submitted by Charles Rayner, petroleum adviser to the State Department. 79th Cong., 1st sess., June 27–28, 1945.

to make representations to securing equal treatment for American firms." As a State Department memorandum later pointed out:

> These negotiations were long, and complicated at a later date by the fact that the British-controlled Anglo-Persian, which had previously expressed its disinterest in Kuwait, suddenly endeavored to secure a concession from the Shaikh of Kuwait. Here again the Department insisted on the "open door" policy, and our Embassy in London was assiduous in its endeavor to expedite a settlement, and continuously and frequently pressured the British authorities for action.*

Frank Holmes and his sponsors knew only too well why the U.S. ambassador in London was so "assiduous" on their behalf. Andrew Mellon, the Pennsylvania millionaire, who was American envoy to Britain at the time, had until his appointment been an active director of Gulf Oil, and the company was controlled by the Mellon family interests. But though he "continuously and frequently pressured" the British government to let Gulf take over Kuwait, as long as Anglo-Persian insisted on the concession, Whitehall resisted all U.S. demands. And for a while Anglo-Persian was inflexible; they even brought rigging crews over to Kuwait from Persia, ready to start work. As the competition dragged on, Sheik Ahmad delightedly rejected each company's bid in turn and upped his demands: for more money down, for more rental, and—worst of all, in the eyes of both rivals—for a Kuwaiti director on the board of the winning company.

Then one Sunday morning in 1932, halfway through the service in the stuffily hot little Anglican church in Kuwait, his sensitive musical ear cringing under the blast of Frank Holmes's foghorn baritone thundering out the hymns behind him, Archie Chisholm thought of Solomon. Why not do what the Biblical king suggested, and cut the baby in two?

Almost simultaneously, T. E. Ward, Frank Holmes's legal adviser and a fellow director of his syndicate, had the same idea.

* *Ibid.*

The time was right for compromise. He had just read a speech made by Sir John Cadman of Anglo-Persian to a convention at the American Petroleum Institute in Houston, Texas, in which Cadman complained of a "glut" of petroleum and cutthroat competition among its producers. "Consumption everywhere has decreased," Sir John had said, "competition is too keen, and prices contain no margin for gain. One possible step towards rehabilitation of trade is evidently the readjustment of supply to demand and the prevention of excessive competition by allotting to each country a quota which it will undertake not to exceed." *

It was the first public admission that the oil companies were moving toward the establishment of an international cartel to control petroleum supplies and prices. To T. E. Ward the words used by Cadman "were not only a keynote for the convention but also for the Anglo-Persian and Gulf Company directors on the subject of equal partnership in Kuwait." †

So almost at the same time Chisholm and Holmes arrived at the conclusion that it was useless to go on fighting for a monopoly of Kuwait's oil. There is no record that they ever discussed the question of a compromise together, but both of them went to work on their home offices.

Suddenly it was all over. Sheik Ahmad discovered that not only were the two negotiators no longer bidding against each other, they were actually demanding that he scale down his demands. First they forced him to reduce the royalty on the oil they hoped to extract. Then they rejected outright his demand that a Kuwaiti join the board of directors of the new company. As the price dropped at each meeting, it became obvious to Sheik Ahmad that he had best settle for what he could get, for his dreams of a vast subvention were fast fading.

His fears were confirmed on December 22, 1934, when his trusted friend Colonel Dickson, the political agent, strongly advised him to sign the agreement which had been drawn up by

* Ward, *op. cit.* Quoted in *Middle East Oil,* by George W. Stocking.
† *Ibid.*

the two companies. It had already been approved by the British government. It gave him a down payment of $170,000 instead of the $500,000 he had expected, and a yearly rental of $35,000 instead of $100,000. The payments were, moreover, to be in Indian rupees and were not to be backed by gold, as were those to Saudi Arabia, Iraq and Persia. The concession was to be for seventy-five years, and the price stipulated as royalty per barrel of oil extracted was by far the lowest granted to any petroleum-producing country in the Middle East—20 percent less than the royalty rate for Persia and Saudi Arabia.

It was a salutary lesson to the Arabs of what happens when a country lacks independence and the companies competing for its petroleum decide to work together. But Sheik Ahmad had no option, and he signed the next day. Chisholm signed on behalf of Kuwait Oil Company, the new company which had been formed to exploit the new field, and Holmes on behalf of its associate and equal partner, Gulf Oil. Trusted friend Colonel Dickson witnessed the signatures. It was a happy outcome for the two negotiators, and on Christmas Day, as Chisholm and Holmes filed into their pews in Kuwait's Anglican church, they actually smiled at each other, and Chisholm's patrician face failed to show its customary wincing expression as the New Zealander's plangent voice was raised in song. It was the season of good will, and none knew it better than these two.

THE PRE–WORLD WAR II GAME of concession grabbing was almost over in the Middle East, and along the Persian Gulf there was only one likely area left where oil prospects were promising enough to bring the protagonists back to the table again. This was Qatar, a small sheikdom just across the water from both Bahrain and Saudi Arabia, and therefore probably concealing beneath its rolling sand dunes the same rich reservoirs of petroleum.

Stephen Longrigg, who was at the time a member of Anglo-

Persian Oil Company's field staff, writes: "Meanwhile the Qatar prospect could not but be followed up, and a negotiator was sent to Doha [the Qatar capital] to act nominally on behalf of Anglo-Persian so as to avoid confusing the old ruler with apparent rivalries, but in fact to acquire rights for the I[raq] P[etroleum] C[ompany]." *

In other words, the British did not want another of those auctions which were so apt to delight the sheiks by driving up the price of the concession, and when Anglo-Persian heard that both Standard Oil of California and Major Holmes and his friends were planning to bid, they speedily asked the Colonial Office to take action—for of course Qatar was yet another British dependency. "The Political Resident at Bushire felt it well to advice the Shaikh [of Qatar] to make the award to a British company and discouraged accordingly a tentative Californian démarche; he was aware also that Holmes and his Syndicate, who appeared with offers for the Qatar concession, were middlemen scarcely to be preferred to a fully competent operating company." †

But though everything seemed to be in Anglo-Persian's favor, the negotiations took time. The wily old ruler of Qatar refused to be rushed, and exasperated officials in London changed a polite and patient negotiator for one with less hesitation about invoking the name of the British government. It was not until May 1935 that the ruler, whom Longrigg describes as "senile and suspicious," consented to award the concession to Anglo-Persian. They at once assigned it to the Iraq Petroleum Company, which, still surfeited with oil, was in no hurry to exploit it. Not until 1938 did the company spud in its first well.

NOW THE MIDDLE EAST WAS DIVIDED UP between the giants. Persia's oil was in British hands, as was Qatar and half of Kuwait. The Americans had moved over from Bahrain into the oases of

* Longrigg, *Oil in the Middle East.*
† *Ibid.*

medieval Saudi Arabia. A suspicious consortium of British, Dutch, French and Americans controlled the wells of Iraq.* There would be new rivals by the end of World War II, and new oil fields to conquer: in the deserts of Cyrenaica and Tripolitania, in the exotic emirates of Abu Dhabi, Dubai, Muscat and Oman, and Dhofar, and beneath the waters of the Gulf. But for the moment all the productive as well as the most promising territories had been parceled up. The rulers of the leased lands had had little to say about how it had been done, and their people nothing. There was not much they could have done about it, anyway; the leaseholders had powerful governments behind them to back up the terms of their tenancy. In any case, the Arabs had no technical knowledge or skills to exploit their riches themselves. It was a time to take the money (in the case of the rulers), what jobs were available (in the case of the people), and to remember the Arab saying, "The hand you cannot bite, kiss it."

* Baku had now been walled off from the Middle East, and was still recuperating from the depredations of the Russian Revolution.

Part Two

THE OPERATORS

PRECEDING PAGE: *Calouste Gulbenkian.*
PHOTO: IRAQ PETROLEUM CO. LTD.

CHAPTER SEVEN

Gusher

IT IS HARDLY SURPRISING that the Anglo-Persian Oil Company resented the intrusion of the Americans into the oil fields of the Middle East, for the arrival of the newcomers upset careful arrangements with their counterparts across the Atlantic to limit and control the world's supplies of petroleum. Great corporations always prefer to combine rather than compete, and while the antitrust laws in the United States precluded the establishment of a legal tie-up of the markets, the three giants of the oil industry, Standard Oil of New Jersey, Royal Dutch–Shell and Anglo-Persian, had got around that by means of what came to be known as the Achnacarry (or As-Is) Agreement. In September 1928, it "so happened" that Walter Teagle, president of Standard Oil of New Jersey, Sir John Cadman, chairman of Anglo-Persian, and their two wives were invited to shoot grouse on the estate at Achnacarry, Scotland, of Sir Henri and Lady Deterding. Deterding, of course, was chairman of Royal Dutch–Shell. As Teagle put it later: "While the game was the primary object of the visit, the problem of the world's petroleum industry naturally came in for a great deal of discussion." *

* *Oil and Gas Journal* (September 20, 1928).

From these discussions emerged a document which was called "Pool Association" by its compilers, but was dubbed the As-Is Agreement because its obvious purpose (though its authors denied this) was to divide the world oil markets between the three companies by eliminating "excessive competition." It was not true, the document complained, that the petroleum industry operated under a policy of "greed" or that it was given to "wanton extravagance," as the public was led to believe by certain newspapers. (One can imagine the three executives, gathered in the study of Achnacarry in the Scottish gloaming, revived by their first whiskey after a hard day in the butts on the grouse moors, choosing the words more in sorrow than in anger, and lamenting the ignorance of the world outside.) Their only aim, the document went on, was to cut out waste, eliminate duplication of markets, and halt the ruinous price war that was depressing shares as well as petroleum prices.

Whatever public-spirited design lay behind the As-Is Agreement,* its effect was to designate operating and trading zones for the Big Three and discourage competition from outsiders and new oil ventures likely to upset the equilibrium. Anglo-Persian, the colossus of the Persian Gulf, could only view the arrival of Standard of California in Bahrain and Saudi Arabia, and of Gulf Oil in Kuwait (neither of which was a signatory to the As-Is Agreement) as brash upstarts, dangerous disturbers of the status quo, and their arrangements and operations were watched with a mixture of stiff-lipped apprehension and scarcely concealed contempt.

For by the time the Americans brought in their first well on Bahrain in 1931, Anglo-Persian had been going for more than twenty years and had more than two hundred wells operating farther up the Gulf in the Bakhtiari hills of Persia. Well F7, the fount of the company's fortunes, had been sealed off at Masjid-

* The document was subsequently subpoenaed by the Federal Trade Commission and was a prime exhibit in the hearings on The International Petroleum Cartel, U.S. Senate, 82nd Cong. (Government Printing Office, Washington, D. C., 1952).

i-Sulaiman in 1926 after flowing continuously since 1911 and producing 7 million tons of oil. It was not that the well had gone dry, but simply that the wellhead fitting had not been designed for unusually high pressures, and had been sealed up for safety because gases were building up underneath. F7 was closed out with a small ceremony at which a trumpeter blew the Last Post, and a plaque was erected to demonstrate the company's gratitude.

By this time Anglo-Persian ran a tightly knit, efficient and enlightened operation in southwest Persia whose success was acknowledged by the other big companies and envied by the newcomers. At one period it had grown dangerously arrogant and tended to regard the bony mountain country where it operated and the ports from which its tankers sailed as a country within a country. Field managers made treaties with the Bakhtiari chieftains without reference to Teheran, the Persian capital, and the great refinery base at Abadan and the pipelines through the hills were in operation long before the Persian authorities had approved the plans for them. But now the days of British gunboats were past and a new shah, Reza Khan, had taken over in Teheran, a rough-riding and rough-spoken dictator determined to demonstrate to his people that he was master of his country and that not even the British could overawe him. When they tried to, in 1932, he simply cancelled the concession. It was a ploy, of course; at that moment there was no one to take Anglo-Persian's place, and the Persians lacked the skills to run the great oil complex themselves. The quarrel reached the floor of the League of Nations, where that supine organization promptly told the two contending parties to start talking again. And an arrangement was made, and a new concession worked out with better terms for Persia. (The company, which changed its name to Anglo-Iranian in 1935, went back to the wellhead and started the oil flowing again—with its concession lease extended until 1993. But it paid more royalties, it agreed to "Iranianize" its operations and promote more natives to the higher echelons and technical sections of the operation. More important, the British stopped acting as if they owned and ruled the land and the people around them, and the change, if not physically very visible—for the hospitals were still run by the company, the roads still built by it, the Bakhtiaris paid by it, the transport maintained by it—was psychologically stressed by the symbolic presence everywhere of the new ruler, Reza Khan, his picture on the walls of all the office buildings, the Iranian flag flying from the mastheads of all the company's properties, and the presence inside every company branch of an Iranian government official. He could be bribed, of course, when hard-pressed officials wanted to cut corners or red tape, but he was there.)

In its twenty-odd years of operations Anglo-Persian had pioneered many things: new methods of drilling, for instance,* new ideas for saving oil wastage,† and elaborate welfare systems not only for employees and their families but for all the inhabitants of the region who came in search of social or medical help. It employed a cosmopolitan conglomeration: Polish, British, American and Canadian drillers, Muslim and Hindu clerks, Sikh mechanics, and hordes of Iranians. In the past, when the Indian clerks went on strike, they and their families were shipped en masse back to India on the next boat. But now the policy was talk, bend and compromise. As one of their managers, a red-faced, blue-eyed Scot named John Jameson, said, "When you've been through what most of us veterans have experienced out here, you get to learn to be tolerant . . ."

Certainly, practically everything disastrous that could happen in the Middle East had happened to Anglo-Persian long before the Americans came on the scene: fire, flood, disease, rebellion, death and destruction. These were pre–air-conditioning days, and they had endured temperatures of over 120 degrees with no ice and no fans. Their buildings and derricks had been swept away

* The company was the first to drill beyond 9,000 feet with an electric drill.

† Anglo-Persian is credited with inventing "recycling." In the days when oil was used for industry far less than today, the fuel elements of crude oil were often burned as waste. The company pioneered the method of reinjecting millions of tons of fuel oil and distillates under pressure back into the pores and fissures of limestone, for later use.

by floods, their bridges and roads destroyed by avalanches of mud, the flowers and vegetables they carefully nurtured had been eaten by plagues of locusts, and they had skidded and waded through a stenching morass of dead locusts for weeks.

So they smiled with contemptuous superiority as the Americans came first into Iraq and then into the Persian Gulf itself. *They will know nothing of the hardships we had to endure,* they told each other smugly.

Perhaps not. But as the Americans were to find out, the Persian Gulf had not exactly been tamed.

THE FIRST CHALLENGE to the supremacy which the Anglo-Persian Oil Company had established in the Persian Gulf came not from the Americans but from a company in whose policy and destinies Anglo-Persian had a considerable influence. The Iraq Petroleum Company was a quarter-owned each by Royal Dutch–Shell, Anglo-Persian, Compagnie Française des Pétroles and a consortium of U.S. oil interests. The company had been in operation in the new kingdom of Iraq long before King Faisal confirmed its concession in 1925, but as late as 1927 it had not yet come in with a discovery capable of even whispering a challenge at the British giant across the border in Persia. Most of the shareholders did not seem to be much concerned about that. Royal Dutch and Anglo-Persian were quite well satisfied with their operations elsewhere, as was Standard of New Jersey, one of the principal members of IPC's American faction. So long as no one else was able to exploit Iraq's oil, why waste money by hurrying? When the time was ripe and the markets were more favorable, that was the moment to begin spending on an accelerated program.

However, the other American members of IPC's shareholding (except for Standard of New Jersey) were restive and eager to find oil. This was their pilot venture in the Middle East and they needed to make a success of it if future undertakings were to be sanctioned by the directors back home. As far as they were concerned, by the spring of 1927 the news from Iraq was black, and not with oil. In April that year a well had ceremonially been spudded in, with King Faisal invited to watch it, at Palkhana, in the Mosul district, but after drilling had gone down to 2,000 feet at this and one other well, work was suspended. The limestone in which oil is trapped had not been reached.

Work then began at Tarjil, near Kirkuk, but all that came up was salt water. In May the crews moved on to Jabal Hamrin, "where a well at Khashm al-Ahmar was abandoned after encountering high pressure gas and heaving shales, and another at Injana was for similar reasons not taken below 3,500 feet. A well at Jambur, south of Kirkuk, met with minor shows of oil and gas. Results at Qaiyara, west of the [river] Tigris, were more significant." *

Such failures were disappointing and expensive. As the meager results came in, members of the IPC board held an emergency meeting in London, and someone pointed out that 2,500 Iraqi workers, 50 British and 20 foreign drillers were now operating in Iraq, and maintaining residential and office buildings, shops, stores, garages and vehicles. In a time of economic difficulties, this was monstrously costly. A majority of the board seemed to be in favor of cutting back in staff and expenditure, and at least for the moment, calling it a day. On the other hand, the minority American shareholders, the French and Calouste Gulbenkian (whose fortunes depended at this time on the finding of oil in Iraq) were desperately eager to go on. It was finally agreed that there would be one last attempt to find oil at the spot where the main drilling crews had established their headquarters. It was a dried-up river bed (or wadi) in the desert near Kirkuk called Baba Gurgur, and most members of the board were skeptical of the chances of anything worthwhile being found there. The company's chief geologist, Professor Hans de Bockh, had reported that there was "an absence of the lagoonal phase in the tertiary sedi-

* Longrigg, *Oil in the Middle East.*

ment, which is the source bed of Asmari oil," * meaning that a viable field was unlikely. But the professor's opinion was strongly contested by the Americans in the syndicate, for they had sent out their own geologist, Dr. Richard Trowbridge, and he had reported enthusiastically on Baba Gurgur's prospects. So the board reluctantly consented to one last try. On June 31, 1927, American drillers spudded in the well, and drilling commenced in the sickening heat of an Iraqi summer.

It so happened that there was something special about Baba Gurgur. About a mile and a half away—between Baba Gurgur and the Kurdish city of Kirkuk, five miles farther on—a ring of flames about six yards wide and four feet high rose from fissures in the parched, stony black desert. By day the flames were invisible, and all that could be seen were trails of black smoke rising into the sky. But at night the flames could both be seen and heard, and the locals said they "licked like hyenas' tongues and hissed like vipers." They were known—and still are—as the Eternal Fires, and they are said to have been burning continuously for five thousand years, since Shadrach, Meshach and Abednego marched through the "fiery furnace" and the idolatrous King Nebuchadnezzar cringed before them in mortal fear of the wrath of God. Not unnaturally, the drilling crews considered the Eternal Fires a symbol and an augury of the success of their latest well.

Drilling went on all summer, but as Beeby-Thompson, a visiting oil engineer, wrote afterward: "So little interest was taken in this test [well] that the drillers were left much to their own devices, and although they repeatedly reported encouraging showings of oil in the mud returns, little notice was taken until cores showed that the basement bed [of rock] had been reached, and it was considered useless to proceed further. Orders were therefore given to cement the well for exclusion of upper water and to make a test bailing; but on completion of that operation the onus of testing was left to the American drillers." †

* Quoted in IPC archives.
† Beeby-Thompson, *Oil Pioneer.*

Shortly after midnight on October 14, 1927, a veteran Texas driller named Harry W. Winger left the field compound and joined his team of two other American drillers and a crew of Iraqi helpers at the wellhead. The drill had now gone down 1,500 feet. Though he was not looking for miracles from the final test, which he and his fellow drillers had now decided to make, Winger was too old a hand not to take precautions against the unexpected. The rotary drill had already been exchanged for a percussion drill, to prevent unnecessary heat. Now he ordered the Iraqis to bale out the casing, but told them to leave 500 feet of mud at the base as a cushion against any oil which *might* be surging around 1,500 feet below. Beside him the steam engine which drove the percussion drill began puffing away like an old-time shunting locomotive as it worked the probing bit down into the rock below. It was a cool night and a light breeze blew across the desert from the direction of the Eternal Fires, bringing with it the pungent smell of petroleum gas, a smell Winger had grown so accustomed to that he thought the air smelled dead whenever he moved out of the oil fields. He looked across at the fires and thought what false prophets they were; if this well was a poor giver, as everyone seemed to believe, then he would have to start thinking about a new job.*

After a while he gave the order to withdraw the bit and clean the well of the cuttings which had been made, and it was then that it happened. There was a noise, faint at first, deep in the bowels of the earth, which reminded Winger of a fast train approaching through a tunnel. It grew louder and louder, and at the same time mud and oil began to slop and then gush onto the floor of the derrick.

Winger didn't wait to see more. He clanged the alarm bell and dashed across to the steam engine to douse the boiler, at the same time screaming to his Iraqi foreman, Sayed Nur Ali, to switch off all current at the electric switchboard.

Just before the lights went out, startled day-shift workers,

* This account is based on documents in IPC archives.

alarmed by the bell, dashed from their bunks and were in time to see a shining black column of oil burst through the floor of the derrick like a genie out of a bottle and soar into the sky. Then all went black. They stood there, listening to the noise of the escaping oil, which was now an ear-piercing howl. They began to choke on the powerful stench of escaping gas, and then, from the blackness overhead, they felt rain falling on their heads and realized, when they touched it, that it was oil pouring down on them.

It took a little time for them to sense that the darkness was not total, and then they understood why. Over a humpy hill a mile away, the Eternal Fires still flickered against the blackness of the night. If the wind were to turn and blow the erupting pillar of gas and oil that way, every one of them would be blown to bits —and perhaps the nearby city of Kirkuk with them.

It was then that Henry Hammick, the field manager, decided to do what no man had done for five thousand years. He ordered Alec Kinch, his English labor boss, to gather together men and all the spades they could lay their hands on. There was nothing they could do about the spouting well until morning light came, but in the next seven hours, with rocks and heaps of sand and rubble, they smothered out the flames of the Eternal Fires. Then, in the smoky, sulfurous light of dawn, they wiped the sand and oil out of their eyes and went back to the job of getting the runaway under control.

By now the column of oil looked like a gigantic cypress tree bending in the wind against the pink morning sky. It was already 200 feet high and growing. For ten miles around the sky rained oil, and in the camp itself it was already ankle-deep and flowing down the wadi in an ever-growing stream.

FIVE MEN DIED in the fight to control the well at Baba Gurgur. Two American drillers wandered into a pocket of gas in a hollow near the camp and succumbed, as did three Iraqis who went in to try to drag them out. The reports in the records of the

Iraq Petroleum Company give a salutary picture of the appalling conditions under which the struggle was waged. The English manager of the company who had traveled up by train and car from Baghdad as soon as he heard news of the emergency was confronted by a nightmare scene. The oil was belching out of the well at the rate of 12,500 tons a day and the river of black crude flowing down the wadi was now 100 feet wide.

"Sand and oil, coupled with the whistling wind, and the roar of the escaping oil," wrote Hilary Bull, the manager, "made one wonder if it was not beyond man's resources to cope with the situation. No man could work for long under such conditions and ten minutes seemed to be the average endurance . . . With gas masks on, and oil dripping from their hats, the drillers worked feverishly in the cellar [of the derrick]. The man actually at the bottom of the cellar fixing the tie rods under the flange had a rope tied round him so that he could be hauled up if overcome by gas. All day long men were collapsing and being hauled up unconscious to be revived outside in the comparatively fresh air. Some of the men were gassed two or three times a day, and yet staggered back to their jobs." *

No one had believed enough in the potentialities of Baba Gurgur to make provision in the event of a gusher, so makeshift bunds (embankments) had to be built in the sand to contain the spurting oil as the drillers wrestled to contain it at the wellhead. The first bund filled up in one afternoon, and there was a frantic race to get a second and bigger one dug before it spilled over. Six bunds, each larger than the last, were erected. Although in those days oil companies did not think much about pollution, everyone knew that it would be a disaster of the first magnitude if the flow of oil reached the nearby river Tigris. Finally a site was chosen seventeen miles from the wellhead, and 800 men from a local tribe were hired to make a reservoir there. It was to be six feet high and seven hundred feet long, and it would hold 200,000 tons of oil. The menacing black river, slopping over from one

* Report in IPC archives.

bund to the other, was only three miles away from this last ditch when the drillers finally got what is known as a Christmas tree (a new wellhead valve) into place and brought the filthy roaring menace under control at last.

It was 3 P.M. on October 23, 1927, nine days after the oil and gas had first burst free, and the sudden silence was uncanny. Exhausted men flung themselves down on the oily ground. Someone dared to light a fire and make some tea. Bottles of whiskey were broken out of the compound's emergency store. And as they sipped through oily lips, Hilary Bull saw a car approaching the compound from the direction of Kirkuk. It was a messenger from the local provincial governor demanding the manager's presence at once.

"And now," the governor said some hours later, his tone icy with fury, "now that you have your oil, will you please relight the Eternal Fires as quickly as possible? For nine days my people have been without them, and it has made them uneasy." *

A week later the flames were flickering over Baba Gurgur again. The Eternal Fires were relit, and the Iraq Petroleum Company was launched on a veritable sea of oil.

* IPC archives.

CHAPTER EIGHT

Domes for Sale

IN THE SUMMER OF 1934 Max Steineke waded ashore at Jubail, on the Gulf coast of Saudi Arabia, from a small Arab dhow which had brought him from Bahrain on the last leg of his journey from the United States. He had been appointed geologist-in-charge of the new American concession in Al Hasa, and on his way over he had read closely the reports of the pioneer party of ten Americans who had been in the area for over a year. But their sparse prose had hardly prepared him for the realities of the situation. The heat was blinding, and there was no relief from it even wading waist-deep in the waters of the Gulf, for the temperature of the sea was hot almost to the point of simmering. Nor was his mind made easier on these final few hundred yards of his journey by the sight of a long sea snake, its mouth open and its wicked-looking teeth bared in what appeared to be a snarl, undulating toward him. Like a frightened horse he shied violently and dropped his typewriter in the water, whence it was retrieved by an Arab crewman who almost immediately thrust him brusquely to one side, so that he fell forward, ducking his head in the water. A small school of pink jellyfish floated by, their delicate tentacles spinning like ballet dancers' arms a few inches from his nose. It was only when he reached the shore, angry and soaked, that it was explained to him why the Arab bearer had been so rough. In this part of the world you ignored sea snakes, which, no matter how fierce they looked, apparently leave human beings alone, but you took immediate evasive action when confronted by pretty pink Portuguese men-of-war, for the lightest caress of their delicate fronds could give you a rash that would agonize for weeks.*

The world was one third of the way into the twentieth century, but one would never have guessed it from this part of Arabia. In some ways it was almost like the Old West of a century before, except that the wandering tribes were Bedouin instead of Indians, and the herds they tended were camels instead of buffalo. In villages like Jubail and Oqair, and in the nearby city of Al Hufuf, the center of the Al Hasa oasis, the local emirs received their writs from King ibn Saud and ruled like feudal princes. The Americans were tolerated because they would find the oil that would make all of Arabia rich; still, they were viewed with intense suspicion by the emirs as infidel intruders come to corrupt and suborn the tribesmen. No latitude was allowed them. On the beach at Jubail, beside Steineke's luggage, lay crates of canned food which had been sent from America via Bahrain as rations for the American crews; but the emir had decided that they were liable to heavy import duties, and though the Americans pointed out that this was contrary to the terms of the concession agreement, the crates simmered in the sun, and the crews subsisted on a diet of rice and teeth-breaking camel, goat or occasional mutton. In consequence, most of them were suffering from stomach pains, boils and spots-before-the-eyes, which may have been psychosomatic—or due to the water, which tasted of camel urine. All of the visitors walked and moved their limbs with difficulty, for due to the heat, the sand and the rigors of the country they were martyrs to prickly heat in the most unfortunate places.

* This description, and other material in this chapter concerning Steineke, is from the notes he was writing, and left unfinished, when he died in 1952.

In the circumstances, Steineke found the advance party a surprisingly cheerful and optimistic team, which suited him, for his temperament was similar. Since he left Stanford University he had worked in all sorts of difficult places—South America, Australasia, Alaska—and had learned to put up with physical hardships, and though he disliked the heat, flies and the stench of his new location, the people and the landscape appealed to him. He was a bluff, laughing man with a rich vocabulary of swearwords, and ordinary folk (if not emirs) soon succumbed to his friendly approaches. As for the desert and its hills and formations, it looked promising for both work and pleasure. The formations of the *jabals,* or hillocks, suggested likely petroliferous indications, and both the sea and desert were still rich with enough fish, birds and other fauna to satisfy his passion for hunting, for he was both a crack shot and an expert big-game fisherman.

In 1934 the Aramco team lived in a *barrasti* (a sort of bunkhouse made from woven palm leaves) between the tiny pearl-fishing village of Jubail and the main town of the area, Dammam. They had a couple of trucks in which they bounced over the desert with their supplies (there were no roads) but otherwise got around on donkeyback, like the Arabs. They also had a plane, and this had caused more trouble than anything else. One of the clauses in the concession agreement said that the company could use up to two planes for area mapping and reconnaissance, so a Fairchild monoplane had been shipped to Egypt, and flown in from there by its pilot, Dick Kerr, via a complicated route over Palestine, Iraq and Persia. Unfortunately, someone had forgotten to inform King ibn Saud that it was coming. It was a time when the king was in the midst of one of his recurring quarrels with the imam of Yemen, and suspicious that his adversary was receiving secret aid from Persia. Kerr chose this moment to fly in his Fairchild, and when a Saudi radio operator heard him talking to an air-control station at Bushire, Persia, before coming in to land at Jubail, he signaled the alarm to the king in Jidda.

It was the job of the oil-company representative in Jidda, Bill

Lenahan, to straighten out these difficulties, and he had managed to convince the king that Aramco was not secretly aiding an Iranian invasion. The plane had been released from impoundment but its flights were restricted to coastal reconnaissance, and the emir of Al Hufuf had given orders to his irregular troops to open fire if any plane flew over the oasis.

The pioneer team had already staked out a number of likely sites for test oil drillings on and down the coast, and particularly between the coastal town of Dammam and a series of rocky hillocks or "structures" known as Jabal Bari and Jabal Dhahran. Steineke told one of his assistants, Tom Koch, that it looked as if he had an easy task ahead of him. The theory was that the "structures" out of which Bahrain's oil had come continued under the sea and on to the Saudi mainland. "The problem seemed merely one of cruising about the desert and finding jabals—an easy matter," wrote Steineke later. "In the lights of what we now know about the geology of Arabia, these early endeavours to bring to light the secrets locked under the sand seem very amusing. But they were done in deadly seriousness in those days. Krug Henry and Bert Miller [two oil company employees] had found many such jabals during their trip to the west in the spring of 1934. Muhamet Taweel, the emir of the Hasa Province, became so eager to see the oil produced that he caused many anxious moments with his trying to promote the drilling of wells on each jabal, so sure was he that each one would prove to be a field. At the first opportunity, therefore, we drove out to investigate Jabal Bari in detail."

Petroleum is usually found in so-called sedimentary rocks which were once the bed of ancient seas, and in the 1930s what a geologist looked for was evidence of fossilized shells in the rock formations. These discovered, he went on to search beneath them for the "cap rocks" in which oil, formed from fatty organic matter masticated on the sea bed by bacteria, is trapped by the movement of water and the pressure of the earth. Jabal Bari seemed promising in that it was a troughlike fold in the rock formation, but when Steineke chipped out a few hunks of rock he discovered small, round, buttonlike fossils that belonged to the Miocene period, only 19 million years old, too young for his purposes. He wanted Eocene beds like those at Bahrain, at least 55 million years old.

After a series of probes which proved both disappointing and confusing, Steineke moved westward about twenty-five miles to a spot which the pioneers had named Button Bed Hill. He had lumps of rock in his truck which had fossilized oysters, sea urchins and other small echinoids embedded in them—signposts of Eocene—but each time he dug down, the formations were thin. He even came across one outcrop of Cretaceous rock, filled with clam skeletons and giving evidence of an age of more than 70 million years. But after weeks of weary test drilling, each subsequent probe, at Button Bed Hill, at Jabal Dhahran and over a number of promising domes or "caps" which had been mapped out from the air, proved to be sterile or unpromising. Some wag put a cartoon up on the mess wall picturing Steineke weeping tears on the *jabals* and a notice over it saying, "Domes for Sale," but nobody laughed very much. The wildcatters under Steineke's command and in a desert camp outside Dammam had the worst job in the world, working, soaked with sweat, in a temperature that rose to 120 by noon, often in shrieking sandstorms, and aided by Bedouin workmen whose language they did not speak and whose minds were not attuned to this kind of labor and these newfangled tools.

"Now the bets started pro and con as to whether or not even Dhahran would be worth fifty cents as an oil field," wrote Steineke. "The year before, Buck Miller, Krug Henry and Soak Hoover had mapped Dhahran as a bona fide structure similar to Bahrain which was of course a producing field. But now the theories we had constructed based on these facts seemed to be going more and more awry. There was something very unorthodox about the whole geological set-up . . . The whole situation was causing us sleepless nights when we argued and discussed all these weighty problems trying to solve them in the light of our misconceptions."

Not that there was despair—except in Jidda, perhaps, where King ibn Saud was waiting for oil and his money—because the company's home office in San Francisco showed no signs of backing out of the project; in fact, it was continuing to add to its staff and materials in Saudi Arabia. Aramco may have committed some serious errors in the course of its career, but unlike other companies, it has never operated as an absentee concessionaire. It has always put its senior men in the hot spots and drawn its top directors from the field. Working beside Steineke were Fred A. Davies, who was to become Aramco's president, and Floyd W. Ohliger and Tom Barger, who retired as executive vice-presidents, all three of them geologists and engineers. San Francisco gave every evidence of backing the Saudi project or going bust, and as far as Steineke and his crews were concerned, go bust they would not.

But things got edgy at times. There was a crisis when one of the drillers struck a Saudi workman for stupidity and insolence, and was ordered out of the country by the furious king. The drillers said they would all go if he went, so he stayed, but relations with the local emir and his officials were not improved. Henceforth the Americans were roughed up by the police for the slightest infraction of local rules and customs, and were finally confined to their compound or camps. At the end of 1935 Max Steineke had decided to stake his all on the so-called Dammam Dome, and King ibn Saud's eldest son, Saud, came across from Al Hufuf to turn the wheel on Dammam Number One, which had gone down to 3,200 feet and was producing oil. But the royal hand worked no miracles and the well whimpered away after a promising start. As a compensation the company lent the king £15,000 because his wives and other extravagances had put him heavily in debt again.

"Still going for bust," as Steineke put it, the company ordered a whole series of test wells to be drilled on the Dammam Dome, and the geologist realized that the crucial phase of his professional career was now looming before him. If the tests failed, not only would Aramco go bust, but he would be too.

It was now 1936. After a good start, Dammam Number Two suddenly vomited salt water. Number Three brought up a sluggish sulfurous oil so viscoid that all it could be used for was as covering for the desert roads the men were building. Number Four went down to 2,318 feet and stopped there, on November 18, dry. So did Number Five, at 2,067 feet. Number Six was never even spudded in, but was simply abandoned with a derrick erected over a probing hole.

It was left there because Max Steineke had decided to put all his money, all his prospects and those of his colleagues and his company, on one final well, Dammam Number Seven. He was so confident of success (or pretended to be) that after seeing it spudded in, on December 7, 1936, he departed with an engineer, Floyd Meeker, on a trip across the Arabian Desert to Jidda and back. The company was encouraged by his optimistic predictions to such an extent that they gave the go-ahead to a program for sending wives to join their husbands in the Gulf, and ordered a married-quarters compound started.

Fortunately for all concerned, seven is a lucky number in the oil business. On December 31, 1937, Dammam Number Seven "blew out" and sent rigging and drill high in the air. By then the drill was down to 4,535 feet, and New Year's Day was spent pouring great loads of mud into the well to "kill" a jet of gas which was escaping at the rate of 30 million cubic feet every twenty-four hours. It was gas only, no oil with it, and at that time no one was interested in gas, only frightened of it.

For the first time, the home office was frightened too. Had Steineke guessed wrong? Should they try somewhere else? Back from his transdesert trip, the geologist was confronted with these problems and asked for an immediate answer. He agonized through a sleepless night and then told his crew to press on with Dammam Number Seven.

On March 4, 1938, oil began to flow out of the well at the rate of 1,585 barrels a day. Three days later the flow had increased to 3,690 barrels. By the end of April the well had produced 100,000 barrels, and Steineke predicted that it would go on producing. They were over an oil cap and a commercial field.*

But no one had told King ibn Saud yet.

BY THE TIME Dammam Number Seven put the Americans in the oil business in Saudi Arabia, a change had come over the Aramco encampments; the pioneering phase was over. By mid-1937 the first five American wives, and three of their children, had arrived and were installed in California-type wooden bungalows which had been erected amid the scrub and wastes of sand under the slopes of Jabal Dhahran. No one imagined then that these were the nucleus of what, thirty-five years later, would be the green garden-city of Dhahran, capital of the most prosperous oil consortium in the world.

Oil-company wives had been living in Persia, Iraq and Bahrain for years—the first British wife was at Masjid-i-Sulaiman, in Persia, as early as 1910—but at that time all these territories were under British control or influence, and years of colonial rule had accustomed the locals to the presence of European women. Saudi Arabia was different. It was a closed society where a camel was more valuable than a woman, where the king did not even bother to count up how many daughters he had sired, and where Christian females were despised even more than their Muslim sisters.

Ameen Rihani, a devout Christian, was shocked by a story told to him by one of Ibn Saud's ministers about a beautiful young Armenian Christian girl who had been bought as a slave by the king in the 1920s. He then gave her away to his minister "after he

* I am grateful to Bill Mulligan and Pete Speers, in charge of the Aramco archives in Dhahran, and to Wallace Stegner's official history, *Discovery!*, for helping me to detail this story.

had entered her—one night only." The minister went to her that night in the room to which she had been assigned. "She held down her head after she had seen me," the minister said. "And I—never in my life have I seen or heard of such beauty, such august beauty. Her skin? White as alabaster. Her hair? Like cataracts of melted gold. Her lips? Red as a pomegranate seed. Her forehead? Lofty and glowing like the dawn. And those honey-coloured eyes, so soft, so demure, so appealing to the honour of man . . . I sat before that image of beauty like a child—I tell you, like a child—and I felt the flame of shame upon me. I was ashamed to touch her, or even to speak to her. I got up and walked out of the room."

So what *did* he do? Rihani asked the minister.

"On the following day," he replied, "I sold her to a man from Kuwait for four hundred riyals. Aye, wallah, only four hundred riyals." *

Faced with such a rigidly misogynistic society, it is not surprising that someone suggested that the American women should consider wearing veils. Max Steineke's wife, Florence, settled this by replying, "I'm damned if I'm going to cover up my face for any man in the world. It may not be the face that launched a thousand ships, but those Arabs are going to have to look at it."

The women did decide, however, not to wear the slacks or shorts they had brought with them from the States, and they reluctantly packed away their bathing suits. Then they trooped out into the desert and posed under the derrick on Dammam Number Seven for the official photograph of the first six Dhahran wives and offspring: Marilyn Witherspoon, Florence Steineke, Nellie Carpenter, Patsy Jones and Edna Brown, the Steineke's two daughters, Marian and Maxine, and Mitzi Henry, all with chins up, grinning defiantly into the sunlight.

The first few months must have reminded them of those stories from American folklore of life on the plains, for they were under orders not to leave camp, and they sweltered in a patch of desert

* Rihani, *Ibn Saoud of Arabia.*

not much more than two miles by one. Appalled by the blazing heat and the driving dust, they wondered whether by refusing to wear veils they had saved their pride at the expense of their complexions. But when Dammam Number Seven came in, everyone, even the Saudi authorities, seemed to relax, and conditions got better. As proof that they recognized the permanence of the operation, the home office shipped in air-conditioning equipment for the family bungalows. Shortly afterward Max Steineke took his wife and Nellie Carpenter on one of his forays into the desert. They dressed in Arab clothes and visited two Bedouin encampments, where they spent much time struggling through sign language and gales of giggles to make themselves understood by the Bedouin women. They came back enlivened by what they had seen, and soon all the wives had Bedouin robes and were traveling in the desert in them. Thereafter they took ointments, surgical spirits and bandages with them, to tend to the sores and wounds they found among the Bedouin women and children.

The more they saw of the life around them, the more excited and also the more appalled the Americans became. It was the beginning of a love-hate relationship with Arabia which has endured to this day. They were entranced by the vast, moody beauty of the desert, its rolling seas of golden dunes, its flat, black, calm infinities of shale and cinders, its sudden, rearing mountain peaks and ranges shaped like battleships and toadstools and fairy castles. The women organized their husbands into expeditions to explore the caves and blood-red fissures of Al Mubarraz, the hot springs and old Turkish forts of Al Hufuf. Their children played with the Arab children and exchanged words of each other's language. They went out with the pearl fishers of Darin Island and gaped at the divers who went down to the sea bed with goatskins to bring up fresh water from the springs which bubbled at the bottom of the Gulf.

But the backwardness and suspicion and superstition depressed them, and the cruelty made them indignant. How could they keep silent when they saw an emaciated child, covered head to foot in sores, each scab clotted with flies, crawling in the dung-heap of an Arab village? Florence Steineke penetrated into the black women's tent of a Bedouin tribe and found a young girl with great open angry sores across her chest, into which an older woman was rubbing a loathsome fluid. It turned out that the girl had been cauterized with a branding iron against the tuberculosis from which she was suffering, and now the wounds were being treated with the urine of a female camel. Hot irons, on the face or the heel, were favorite remedies for stomach ache, and needles were used on the retinas of eyes with cataracts or trachoma.

Warned by their husbands not to "upset the apple cart" by making too much fuss, the American women imported medicines, and once nurses and doctors began to appear, set up clinics to provide medical treatment and surgery for the pathetic victims of Arabian disease and ignorance. Sometimes it seemed that every child and every woman they saw had a badly set limb, a diseased eye, yaws on the face, or was suffering from malaria, tuberculosis or diarrhea. They did what they could to alleviate the suffering, and in fumbling but increasingly fluent Arabic, offered the unfortunate females of Arabia what practical advice was possible under the social and physical conditions of this medieval land. But when they saw a pregnant mother after she had given birth, and asking her about the baby, heard her say, "It was nothing," they cried. It meant that the woman—who was often a girl of twelve or thirteen—had borne a girl, and it seemed monstrously unjust and unfair.

What also made them flinch was the implacable brutality of Saudi Arabia justice, which they could do nothing about. The first time that an American wife, driven to distraction by the petty pilfering of her household goods and cashbox, reported the thefts to a local official, she lived to regret it. The culprit, a Saudi houseboy, was speedily found guilty by the imam at Dammam. It was more than a month before she saw him again, and during that time no male member of the camp would tell her what had happened to him. Then one day she saw him on the outskirts of the camp; on the end of his arm where the hand had been was a festering stump. Thieves had their hands ceremonially chopped off in public, and only because Floyd Ohliger, the general manager, had interceded, had she not been summoned to watch it. They all began to realize that they lived in a Biblical, eye-for-an-eye society, where the relatives of a man, woman or child killed by an automobile could demand the life of the driver in return, could appoint the executioner and decide the method of dispatch. They could be bought off, of course, and persuaded to demand minor punishment, but this meant that it was only the poor who suffered.

Since the emir of Al Hufuf was a stern upholder of the Koranic law, he was rigid and ruthless in his punishment of sinners. Word of his brutality spread through the province, and he was feared far and wide. Phil McConnell, a production man, returned from a visit to Al Hufuf and told of his unwitting attendance at the public execution of a poor wretch who had been caught committing adultery not only during the day, but in Ramadan, the time of religious abstinence.

"They had shaved off his hair and his beard," McConnell said, "and they brought him on to this platform with his hands tied behind his back. The poor guy looked surprisingly calm and resigned. There were two executioners, big Negro slaves with chests like barrels. They made the man kneel down and one of the Negroes, he had a huge curved sword in his hand, and so did the other, went and stood in front of him. And suddenly he started dancing! He waved his sword above his head. He spun around in his sandals, and the crowd cheered him. As for the poor guy on his knees, he was watching, fascinated. Sayed [his interpreter] whispered to me that that was the idea. While his attention was distracted, the real executioner was getting ready behind his back. Suddenly he moved his own sword, quick as a flash, and I saw the end of it prick the back of the man's neck. His head jerked back—Sayed said that was to stiffen his muscles

—and then so fast all you could see was a blur of silver, the sword came down and the head was off and rolling on to the platform, and blood was gushing out of the neck." *

It was no comfort to have it pointed out, as Harry St. John Philby seldom failed to, that this stern form of justice kept Saudi Arabia free of thieves and murderers at a time when elsewhere in the Middle East no one was safe from robbery and violence. Faced by the horrid thought that the thief would lose his hand, or even his arm, if reported to the authorities, the Americans began to overlook petty thefts, and thus encouraged their increase. At one moment they even persuaded Floyd Ohliger, who was now on close terms with the king, to ask him to allow the culprit's severed stump to be treated, immediately after the ceremonial severing, against the flies and infections to which it was exposed. But Ibn Saud would neither allow this nor the smoking of keef against the pain. His attitude was: if a man is being punished for a crime, he should feel his punishment. And that was the end of it.

IT WAS NOT UNTIL OCTOBER 1938 that King ibn Saud was officially informed that a viable oil field had been found at Dhahran,† and by that time Dammam Number Seven was not the only well among the *jabals*. Others were being spudded in all around the original bonanza. For Ibn Saud it was a welcome relief, for of course he and his administration were heavily in debt once more. He was so thankful at the news that he informed Floyd Ohliger and Bill Lenahan, the two Aramco representatives, that he refused to grant new concessions to certain applicants who had come hurrying to Jidda jingling gold the moment the news of the Aramco strike leaked out. One of them was Stephen Longrigg of IPC, who retired voluntarily once he heard that Aramco had an option over the concession he was seeking.

* From a document in the archives at Dhahran.
† The U.S. oil township was officially named Dhahran in February 1939.

Germany's special envoy to the Middle East, Dr. Fritz Grobba, came from Baghdad to put out feelers on behalf of German and Italian interests.* So did the Japanese. All of them, the king informed the Americans, he had turned down, even the Japanese, who had made him a "colossal" offer.

But later the Americans had another meeting, this time with Ibn Saud's smooth and expert finance minister, Abdullah Sulaiman, and he made it politely clear that what had been refused the Japanese and Germans he expected Aramco to take over, at a price above what the others had offered. It was obviously going to be a much greater sum of money than the company had paid for the original concession.

For the moment, however, the bargaining over new concessions was postponed for a ceremony. The king announced that he would journey to the Gulf coast to pay his first visit to the company's oil fields and installations. He arrived on April 30, 1939, after a desert trek of a thousand miles with five hundred cars and two thousand slaves, servants, wives and concubines. On the morrow he would declare open a new Aramco port. Miraculously a city of Bedouin tents rose beside the California-style bungalows of Dhahran, and at sunset the Americans watched in awe as the voice of the muezzin rose from a nearby mosque. The giant monarch marched ahead of his followers into the desert and then the whole concourse fell on their knees in the sand and bowed their heads westward to Mecca for the last prayers of the day.

By now this part of the Gulf coast was a vastly different place from the bare, barren coastline Max Steineke had first seen in 1934. A pier had been built for ships to unload at the fishing hamlet of Al Khobar. Out on the isthmus, through the pale-blue shallows, a sea passage had been dredged for ocean-going tankers. A tank farm had been erected, and underwater pipelines led out to floating anchorages. The first tanker to tie up at the pump

* There were rumors that Standard of New Jersey, which had considerable interests in Germany, was involved in these negotiations.

head, the *D. G. Schofield,* was in position and waiting. The loading port was called Ras Tanura, and the Americans were proud of it. Telegrams of congratulation were read out loud from William Berg, president of Standard Oil of California, and Torkild Rieber, chairman of the Texas Oil Company.* Ibn Saud and Abdullah Sulaiman were presented with new automobiles (to the king a Cadillac, to Abdullah a Chrysler). In return, Floyd Ohliger and Bill Lenahan received gold watches from the king, and two golden daggers elaborately carved by the craftsmen of Al Hufuf. Then, from a gilded chair, Ibn Saud turned the wheel that set the Ras Tanura pumps in motion, and the *D. G. Schofield* began to fill with crude oil.

Ten days later, back from an official visit to the ruler of Bahrain, whose oil revenues he had always envied but whose good fortune he could now match, the mighty king threw a great banquet for four thousand people on the sands of Dhahran. More than a thousand sheep were slaughtered, and few of the three hundred Americans who now made up the company's complement escaped having to swallow sheeps' eyes—the delicacy offered them by their hosts. The party continued into the small hours, lit up by the gas flares from Dammam Numbers Seven to Twelve now busily pumping oil into the tank farm of Ras Tanura and gold into the pockets of Aramco and King ibn Saud.

After all the setbacks and disappointments, it seemed too good to be true. And it was. Four months later, on September 3, 1939, Britain and France declared war on Germany, and World War II began. The effect on the fortunes of Aramco, on Ibn Saud and on Saudi Arabia was disastrous.

* The two American companies which at that time controlled Aramco.

CHAPTER NINE

Skulduggery in Kuwait

THE SHEIKDOM OF KUWAIT, that wedge of desert at the head of the Persian Gulf between Saudi Arabia and Iraq, is only seventy miles broad and eighty miles long, but beneath its sands is a veritable sea of oil. There are so many indications of its presence that it would seem difficult to sink a well and not find it. But when Anglo-Iranian put down its first well in 1936, on its own behalf and that of its U.S. partner, Gulf, it missed the oil by miles.

Was it deliberate? At this writing the independent government of Kuwait is hoping to confirm or confound its suspicions on this point by commissioning an official history of the Kuwait concession and its subsequent operation by the Kuwait Oil Company which is being prepared for them by Mr. Archibald H. T. Chisholm from his place of retirement in Ireland. The fact that Chisholm was Anglo-Persian's chief negotiator in Kuwait should give him access to material which is unavailable to other historians.

The agreement on the concession which Chisholm and Frank Holmes (for Gulf) had extracted from Sheik Ahmad was a hard bargain for the Arab ruler and his people, and one they came to resent. But the deal which Anglo-Persian had made with Gulf in

1933, once they decided to share Kuwait's oil, was also something in the nature of a triumph for the British company. It will be remembered that Anglo-Persian did not really want to see any more petroleum produced in the Gulf, or anywhere else in the Middle East, for each new oil source prejudiced the agreement to control the world market which had been made at Achnacarry by Standard of New Jersey, Royal Dutch–Shell and Anglo-Persian. These companies would have preferred to allow Iraq and Persia, where they owned wells, to monopolize the Middle East's share of the world fuel market. The discovery of oil by Standard of California (a company outside the charmed circle) at Bahrain had been a jolt to their plans, and they also feared the effect of Saudi Arabian operations. So when Anglo-Persian forced Gulf to accept them as partners in Kuwait, the terms the British wrote into the agreement with the Americans were harsh and restrictive. Both parties pledged that Kuwait oil would not be used to interfere with or damage the world market, and that they would keep in constant consultation over where the oil was sold. There was also this clause:

> The parties have it in mind that it might from time to time suit both parties for Anglo-Persian to supply Gulf's requirements from Persia and/or Iraq in lieu of Gulf requiring the company to produce oil or additional oil in Kuwait. Provided Anglo-Persian is in a position conveniently to furnish such alternative supply, of which Anglo-Persian shall be the sole judge, it will supply Gulf from such other sources with any quantity of crude thus required by Gulf, provided the quantity thus demanded does not exceed the quantity which in the absence of such alternative supply Gulf might have required the company to produce in Kuwait . . .*

This gave Anglo-Persian a position of power over its U.S. partner both going and coming, since it was the British company which would provide the manpower and decide where the Ku-

* Quoted in a memorandum by Charles Rayner, petroleum adviser to the State Department, during hearings on American Petroleum Interests in Foreign Countries, before a U.S. Senate committee inquiry on petroleum resources, 79th Cong., 1st sess., June 27–28, 1945.

wait wells should be drilled, and it was the British company which would have the choice ("of which Anglo-Persian shall be the sole judge") of deciding how much oil should come out of them once they were in operation. Anxious as the company was to keep up the flow of oil from its main source of supply, in Persia, it is not surprising that Anglo-Persian was in no hurry to bring Kuwait into the game. The mystery remains how and why Gulf allowed them this power, eager as the Americans were for the stake in Kuwait oil.

There was one obvious place to drill for oil in the territory—at Burgan. That was where fires had burned and from which bitumen had seeped for generations. It was the wadi in the desert of which the ruling sheik, Mubarrak al Sabah, had written to the British as long before as 1913, promising to show a visiting Royal Navy admiral "the place of bitumen in Burgan and elsewhere," and pledging himself, if oil was obtainable therefrom, to make the concession for it a British one. Frank Holmes, the Gulf negotiator, had always envisaged Burgan as the spot where he would commence drilling operations.

The British geologists decided otherwise, and picked a remote spot named Bahra where there was, it is true, evidence of bitumen seepage, but nothing to compare with Burgan. Any doubts expressed about the site were demolished by self-confident company experts; and the ruler, Sheik Ahmad, the new British political agent, Colonel A. C. Galloway, and a host of dignitaries were invited to the ceremonial spudding in of the first Bahra well in May 1936. It was an augury for the future that the ceremony took place in one of the sandstorms for which Kuwait is notorious; as guests, hot and sticky with sand, peered through the brown, swirling mist at the shadowy figures spudding in the new well, only the thought of the oil to come made it bearable. But weeks, then months, went by and no oil strike was reported, though the drill bit more and more deeply into the rock. It was 1937 before the drill was pulled out at 8,000 feet and the glum announcement was made that Bahra was dry.

Now everyone presumed that the drilling crews would move to Burgan at last. Sheik Ahmad, short of money to pay for his latest cabaret girl from Beirut, behind with his doctors' bills (he was said to be suffering from syphilis), angrily demanded of Colonel Galloway to know the reason for the delays. It was rumored that the oil company was deliberately stalling, and agitators whipped up enough anti-British feeling for a demonstration, which was quickly suppressed. But Dr. Fritz Grobba, the German envoy in Baghdad, busily subsidizing anti-British movements, was interested enough to send a courier down with money for the dissidents.

Still the British company avoided Burgan. Instead it announced that the whole of Kuwait territory was to be thoroughly mapped and geologically surveyed by specialists, and an elaborate—some said time-wasting—operation began which lasted for nearly a year. Only at the end of it did the British announce that after all was said and done, they had come to the conclusion that there was one spot which was more likely than any other in Kuwait to produce viable quantities of oil: Burgan.

It was there that a well was spudded in on October 16, 1937, this time with no ceremonials. Promising oil traces were reported when the drills reached 3,400 feet, and in April 1938 the bit cracked through a cap of rock and plugged into a vast swirling cavern of high-pressure gas, sand and oil. The clocks on the gauges whirled like spinning tops, and for a while it looked as if the unleashed elements were going to blow drill and earth crust sky-high; only by desperate efforts was the pressure (greater per square inch than the strongest city firehose) brought under control.

Anglo-Iranian's joy was muted, but two other wells were drilled at once, and by mid-1938 the two partners and Sheik Ahmad knew that they were in possession of a phenomenal oil field. It lay twenty-eight miles south of the capital, Al Kuwait, and fourteen miles into the desert from the Gulf. The field was pear-shaped, with the stalk end pointing northward, and was at least fifteen miles long. No one was aware yet that Kuwait was, in fact, sitting on one of the largest reservoirs of oil the world has ever known, but everyone realized that the country's petroleum resources would henceforth alter the balance of oil production in the Middle East.

Once it went into production, that is. Unfortunately, Anglo-Iranian was not yet ready for that. First the Munich crisis in September 1938 and then the eve-of-war atmosphere of 1939 were given as excuses for holding back. There were more urgent priorities, it was pointed out, and the material was just not there to outfit a new oil field. While Aramco, farther south in Saudi Arabia, was building Ras Tanura and began ferrying tankers to the Bahrain refinery with regular loads of oil, Kuwait, with three wells ready to go, stagnated.

Sheik Ahmad, short of money, aware of restiveness in his territory—for it was the worst of times in Kuwait, its pearl-fishing industry coming to an end, its petroleum wealth unexploited—pleaded with Anglo-Iranian to get going. There were more riots from the volatile Kuwaitis to back him up, and the fact that they had been subsidized by the emissaries of Dr. Grobba did not affect their validity.

It was a difficult moment for Anglo-Iranian, but when Britain declared war on Germany, the company was off the hook. Nothing more could be done, they told the sheik, until Germany and her allies had been defeated. Until then, the oil of Kuwait must stay in the ground.

"We have been cheated!" cried one Kuwait Arabic newspaper. "We should never have allowed Kuwait's lifeblood to fall into British hands. It would have been better if we had handed it to Germany to exploit."

The issue of the newspaper was suppressed, but the irritation remained in the minds of most Kuwaitis, and their hostility would create difficulties for the oil companies and for Britain in the years to come.

Part Three

THE ARABS HIT BACK

PRECEDING PAGE: *Howard Page of Standard Oil of New Jersey and Dr. Ali Amin, the Finance Minister of Iran, during negotiations in 1954.*

CHAPTER TEN

Someone Else's War

IN OCTOBER 1940 a small fleet of Italian bombers carried out the not inconsiderable feat of bombing the oil installations of Bahrain, after a flight across the Mediterranean, the Red Sea and the Rub al Khali (the Empty Quarter of Arabia), and then back to their base in Eritrea. The results the strike achieved were more psychological than physical, for though the bombs missed their target, Bahrain's inability to react against the raiders rubbed every Arab's nose in the cold facts of the situation. Like her onetime ally, France, now crushed by her German and Italian enemies, Britain was crumbling helplessly before the inexorable might of Nazi Germany.

It would be untrue to say that this prospect alarmed most Arabs and Iranians, disturbed though they might be by the results of it. Britain had her Arab friends in the Middle East, of course, but even among those who admired or respected her, there were few who loved her. In each of the countries where Arab kings, emirs or sheiks were the titular heads, but where the real power lay with the British envoys, high commissioners or political agents behind them, there were active cells of dissident officers or politicians dreaming of the day when the English could be driven out. There was the young officers' group in Egypt led by Gamal Abdel Nasser and Anwar al Sadat. In Iraq there was the fiercely anti-British army clique of colonels known as the Golden Square. In Kuwait there was a militant political revolutionary group called Al Shabiba (the Followers). The writ of the British Colonial Office did not run as far as Iran, but the oilmen of Anglo-Iranian had long ago turned the southwest of the country into a British enclave, and the ruthless and authoritative shah-in-shah in Teheran, Reza Khan, made no secret of his resentment of them, and of his contempt for the failure of the British military in the war. Each night the hatreds and antipathies of the masses were fanned by the exiled Mufti of Jerusalem, broadcasting anti-British diatribes from Berlin, and by the Arab emissaries of Dr. Grobba, who had now moved to neutral Turkey but kept open his lines to Baghdad.*

If trouble came, the Western oilmen had no doubt that they would be the first targets of any revolutionary uprising, for they represented the wealth, power and privilege of the British behind the scenes. They were the "capitalist bloodsuckers," as the Mufti of Jerusalem called them, who were bleeding away Arabia's life and strength. In any revolt, it was almost inevitable that the oil installations would be seized and taken over as a symbol of Arabian possession of its own riches. That the flow of oil would be interrupted would not matter; in any case, it was only feeding the British war machine, and why should Arabia's oil be used in a war that was not theirs, and paid for in a currency which might soon be worthless?

It was in Kuwait that the British faced their first confrontation with Arab nationalism. The group of local freedom fighters, the Followers, misreading a message passed on to them by Dr. Grobba,† took the Bahrain raid as a signal, rose up in revolt, quickly overcame the local levies and occupied the Kuwait arsenal. Since there was no oil industry to seize, thanks to Anglo-

* He promised his pro-German Iraqi friends that he would be back by May 10, 1941, and was only four days late.

† He had told them to strike "when the air raids begin." He meant German air raids on Iraq. No one had told him about Italy's raid on Bahrain.

Iranian's procrastinations, they contented themselves by hoisting national flags over the installations at Burgan and taking over the Kuwait Oil Company's office in the town. When Colonel Galloway rushed to see Sheik Ahmad and asked him to order the rebels to hand over their arms and surrender, he found the ruler exasperatingly dilatory. Why shouldn't he be? He had no reason to be beholden to the British, who had failed to provide him with an oil income. He had nothing to fear from the rebels, even though they were loudly proclaiming that they were for his removal and for a union with Iraq. They did not know that he had already been in touch with both the Germans and the Italians, and had assurances from both that his powers would be confirmed once victory was theirs.

As the Followers held on to the arsenal and defied all attempts to get them out (they were waiting for the Germans to arrive, six months too early), panic spread through the European community of Kuwait. There were curfews and shots in the night, and urgent plans for evacuating the women and children. Then King ibn Saud, disturbed by these alarming events so close to his own oil installations, ordered troops to his northern frontiers with Kuwait and threatened to occupy the no man's land (the so-called Neutral Zone *) between the two countries. Since Sheik Ahmad was anxious not to lose any rights in this area, which was potentially rich in oil, he stirred from his lethargy. The rebels were given an ultimatum: death if they defied it, a free passage to Iraq if they gave in. They accepted the terms and departed for the nearby Iraqi city of Basra.

THE MOST SERIOUS THREAT to Britain's oil and influence in the Middle East during World War II occurred in Iraq in May 1941, when the Golden Square—anti-British, pro-German, but

* After World War I, there had been serious clashes between rival tribes claiming pasture rights on Saudi Arabia's borders with Iraq and Kuwait. To bring about order in these areas, two neutral zones were established in 1922. The Neutral Zone between Kuwait and Saudi Arabia, jointly controlled by the two countries, became the most important, due to its rich oil deposits.

even more fervidly nationalistic—organized a coup d'état and established its leader, Rashid Ali al Gailani, as premier in place of a complaisant, pro-British puppet named Nuri Said. At the time, Rashid Ali was portrayed in the British and pro-British press as an archtraitor who used Britain's current misfortunes in the Mediterranean theater as an opportunity to stab her in the back, and this reputation has been largely accepted by history. In fact, he was no more wicked or treacherous than any other national leader striving to get an outside power off his country's back, and all he can be faulted for—aside from his failure to bring it off—was his naïve belief that the revolt he inspired would bring Iraqi independence with it; for it was obvious that had he succeeded, he would merely have exchanged a British collar for a Nazi yoke.

Rashid Ali had long since made his plans with Dr. Grobba before the German envoy had departed for neutral Turkey, and as soon as he came to power, the Iraqi leader summoned the local manager of IPC and told him to assemble two large dumps of gasoline (about a million gallons in all) in 4-gallon cans at Iraqi army headquarters in Baghdad and at a post ten miles to the west. By the terms of its contract, IPC was forced to obey, though everyone knew that the dumps were being established for the arrival of the German army. At the same time the new premier reopened the telephone line with Turkey, which had been cut, so that he could talk with Grobba direct, and he re-established contact with Syria, which had a common border with Iraq and had been under the control of the pro-German Vichy regime since the fall of France. When the moment came, Syria would be used as a staging post for the German Luftwaffe's assault on the British in Iraq.

That moment was, in fact, scheduled for May 10, 1941, but the British forced Rashid Ali to jump the gun by announcing that a brigade of the Indian army was being sent to Iraq to augment British-controlled forces there. Knowing that this would enable the British to quell any Iraqi rising, the Iraqi dictator had to move fast. In any case, he felt that he could hardly lose, for elsewhere in the Mediterranean and the Middle East, the British were on the run. They had been driven out of Greece and Crete, and Rommel was advancing across the Western Desert toward Egypt. It was just a matter of time.

First Rashid Ali made what seemed to the British ambassador, Sir Kinahan Cornwallis, to be a "very gentlemanly gesture." Tensions were mounting and the Iraqi people were getting nervous, the premier pointed out, and there could be awkward incidents in Baghdad. Why not get British women and children out of the capital so that discussions could proceed without any anxiety about their safety? He suggested that they should be taken, under safe escort, to the training camp run by the Royal Air Force at Habbaniyah, fifty miles west of Baghdad. It was thence that they were accordingly sent—whereupon almost immediately Rashid Ali ordered his army to surround the encampment, and let it be known that he proposed to hold them for ransom. If the British gave in, and canceled the arrival of the Indian brigade, he would spare the women and children; if they did not, he would open fire on them. In the meantime, while awaiting the British ambassador's answer to this nasty ultimatum, he telephoned his friend Grobba and told him that he and the German air force should get a move-on.

It was a grim situation, and no one realized it more than the British oilmen, who were soon feeling the breath of Rashid Ali's minions on their necks. The man in charge of the IPC refinery at Khanaqin was a portly and genial Yorkshireman named Wellington Dix who was awakened one morning in early May by the barrel of an Iraqi officer's revolver digging into his throat. He was told that henceforth the plant was under the Iraqi army's control, and that he and his staff would continue to operate it for the government. When he picked up the telephone to speak to the head office in Baghdad, it was smashed out of his hand; it would not have helped in any case, for the line had been cut. He went out into the yard and noticed that every soldier in sight was smoking a cigarette. He immediately ordered the refinery closed down on the grounds of safety, and flatly refused all orders to start it up again.

Dix and his British staff were promptly arrested and transported first to the jail at Khanaqin and then, after a few hours' journey by train, to the town of Baquba, where they were shut up in the main rooms of the town's principal coffeehouse. Their worst sufferings for the next seven days were the blaring of a propaganda radio, the blazing heat of the rooms, and the menacing shouts of a crowd in the street outside. Otherwise the Iraqis were, as Dix afterward put it, "frightfully British." They served tea and cigarettes, and were even polite when uttering threats, such as the one made by a fierce-looking army major. "I am looking forward to the privilege of personally cutting your throat, sir," he said one evening to Dix.

Three hundred miles to the southeast, John Grafton, in charge of the Rafidain refinery in Basra, was playing a dangerous game. Once trouble had begun, he had the opportunity to slip across the frontier with his staff to Abadan, in neutral Iran, but refused because he could not take all of his Indian staff with him and he was unwilling to let them face the situation alone. Also, the Royal Air Force had a small station up the railway line at Maqil, and if he closed down the refinery it would mean that their planes would be cut off from aviation fuel. Moreover, the British reinforcement's cars and trucks, whose imminent arrival was now signaled, would be deprived of gasoline.

When Grafton's refinery was taken over by the Iraqi army, he immediately obeyed the order from the commander to put all aviation fuel aboard a series of special trains and send it north to the Rashid Ali authorities in Baghdad. The cans of fuel were loaded onto the cars, the waybills were made out and checked, and the trains duly departed. But they had to pass through Maqil on their way north, and there, warned in advance by Grafton, the cars were unloaded by the RAF and the trains sent on their way. "When complaints were made," Grafton said later, "I blandly produced the waybills which the Iraqis themselves had receipted." *

However, these were minor setbacks for the Iraqis. The major one occurred at Habbaniyah, where, as Rashid Ali himself remarked later, the British reacted "in a fashion that was unfortunately typically British." †

To the Iraqi troops which surrounded it, Habbaniyah seemed indefensible, and all its inhabitants, including the evacuated British women and children, appeared to be in their power. The Iraqi soldiers had brought up their artillery so that the landing field of the RAF encampment was within point-blank range from a quarter of a mile away, making it (as the author discovered when he landed there at the height of the siege) a highly dangerous place on which to land.‡ Except for rifles, Habbaniyah had no offensive weapons, for all the RAF aircraft at the station were small monoplane or biplane training models, and in any case, their airfield was too hazardous to use.

There were hours of uneasy calm while Rashid Ali waited for word from his German allies, and for his ultimatum to the British to expire. Then, suddenly made nervous by news that British troops were arriving in Basra, he ordered his commander at Habbaniyah to give the British a taste of what might be coming to them. Accordingly, the Iraqi gunners put a shell through the screen of the open-air cinema and one through the steeple of the Anglican church, and went on to shell targets sporadically all through the day.

"What I had not reckoned with," said Rashid Ali later, "was the British art of improvisation." The improvisation consisted of adapting the camp's polo field as an airstrip, loading the training planes with 28-pound bombs and sending them up to plaster the Iraqis in the sand dunes with shrapnel on as many as twenty-eight sorties per plane per day. While this was going on and the Iraqi artillery was distracted, a small passenger plane with a

* From an unpublished history of the Anglo-Iranian (British Petroleum) Company by the late Hilary St. George Saunders.

† In a conversation with the author in Ankara, Turkey, in 1943.

‡ A shell blew away the tail of the plane in which I landed.

Hindu pilot of Indian Airways managed to put down on the main strip. He had flown in from Karachi and brought in a number of cannon balls, which were promptly fed into the four polished Indian Mutiny cannons outside air headquarters and fired at the enemy causing little damage but much panic. On its return journey the plane took off with British women and children on board. A regular ferry service followed, and all of them were taken out of harm's way.

Even so, the revolution might still have succeeded. Though Rashid Ali did not know it, a British rescue column trying to get through from Transjordan had been halted by a mutiny among its Jordanian levies. Indian troops were landing in Basra, but they were hundreds of miles away from Baghdad.* If only the Germans had kept to their promised schedule, everything might still have worked out. Unfortunately for the Iraqis, however, the forces with which the Germans hoped to take over the country had suffered heavy casualties in their conquest of Crete, particularly in airborne troops. They came late—and not in force.

From her refuge in the British legation, Freya Stark saw a Junkers 88 passenger plane put down on the airfield and a flight of Messerschmitts sweep over Baghdad, to be greeted by a burst of enthusiastic gunfire from the Iraqi troops on the field. It was May 14, 1941, and from the Junkers stepped Dr. Fritz Grobba, four days late for his promised appointment with the rebels. Rashid Ali and his Cabinet were there to meet him, the idea being that there would be a formal banquet at which Iraqi-German unity would be announced and the German army invited to take over the campaign against the British.

It did not happen quite that way. The banquet was canceled when it was discovered that one of the Messerschmitts, piloted by Colonel von Blomberg, who was to take over operations, had crash-landed because one of the bullets fired by a rejoicing Iraqi

* For what was happening in Baghdad, *The Arab Island*, by Freya Stark, who was a refugee in the British legation, is recommended reading. She pays a warm tribute to the steps taken by the U.S. minister, Paul Knabenshue, to protect British and British Indian citizens left in the capital from the wrath of the Iraqi crowds.

soldier had pierced him through the eye and killed him instantly. It then became obvious that Dr. Grobba was by no means clear about the military situation, and thought the whole country was in Rashid Ali's hands. When he discovered that this was not the case, he hastily flew back to Syria en route to Turkey.

Wellington Dix of IPC knew it was all over when his captors transferred him and his companions from their fetid quarters in the coffeehouse at Baquba to an air-conditioned room and offered a meal of curried bully beef and Australian beer. Grafton knew it when the provincial council of embarrassed Iraqis came to the office in Basra where he was confined and announced that he had unanimously been named military governor, and would he please henceforth act as an official link with the British forces?

There had been some damage done to IPC's properties and rough handling of its Indian personnel, and one river tanker had been sunk when its crew tried to make for the sea. Otherwise damage was light, and in a report published shortly afterward, IPC recognized that most Iraqi personnel went on working during the rebellion "and it was largely due to their efforts . . . and, it must be admitted, to the reasonable attitude adopted by the [Petroleum] Ministry's officials that no dislocation of the company's activities and records occurred."

To which Rashid Ali said later, "Of course there was no interference. That was by my express orders. Why should we damage property which was ours, and which one day will come into our rightful control?"

That, however, was still a long way off. At the end of May 1941, as the British regained control of Iraq, Rashid Ali and forty of his followers (naturally dubbed "Ali and the Forty Thieves" by the British) fled to Iran, and the first major Iraqi rebellion was over.

IN JUNE 1941 the Germans launched their attack on Russia, and Winston Churchill's promise that he would help to keep the USSR supplied with any materials she needed gave Britain the opportunity to straighten out what Whitehall called "some little local difficulties" in Iran. For there were only two ways of getting supplies through to the Soviet Union: by sea from Britain along a route harassed every murderous mile by German submarines or planes, or from the Middle East by way of Iran and the Russian Caucasus.

But Iran was neutral, and as Anglo-Iranian had been finding recently, irritatingly determined to make that clear beyond all doubt. Customs officials who had hitherto been liberal in their interpretation of regulations were suddenly meticulously and painstakingly following the complicated letter of Iranian import laws. Red crosses on ambulances and on the hospital at Abadan refineries were ordered to be painted out. Tankers arriving from the war zones had to remove the breechblocks of their guns. The shah's finance minister demanded from the company an immediate loan, free of interest, of $9 million and a guarantee, whether or not it was earned, of $16 million in oil royalty payments.

There was little doubt that the shah had decided that Germany was going to win the war, and it was a prospect which did not displease him. He hated his neighbor to the north, and with good reason, for Iran had long been the victim of Russian depredations and threats. Save for the fact that the British used friendlier tones when making their demands and that their representatives were rather more personable, he regarded them as little better than the Communists. When negotiating with Iran, the British always made it clear that they had a gunboat up their sleeve.

To give himself a counterbalance against Russian political power to the north and British economic power (through its control of the oil fields) in the south, the shah had begun relying on German experts even before World War II began. He was anxious for the independent industrialization of his country, and he knew of no better technicians than the Germans to help him attain that goal. From his point of view it was a welcome bonus when it turned out, in 1941, that they also belonged to a nation on the point of conquering both Russia and Great Britain.

By this time there were around 3,200 German "experts" of one kind or another in Iran, and though most of them were indeed genuine engineers, architects, scientists and surveyors, it was inevitable that among them was a hard core of Nazi agents and shock troops. The fact that they were sitting on the threshold of Britain's lifeline to India and the Far East and that they were moving around freely in the area of her biggest oil field caused shudders in Downing Street. But until Germany attacked Russia there was little that Britain could do about it, and British remonstrances only made the shah more determined, the Iranian army more hostile, and the gendarmerie in Abadan and other Anglo-Iranian oil centers more rigidly unpleasant in their controls.

The shah's mistake was to persist in his unswerving and overt pro-German policy even after Russia was attacked. So sure was he of an imminent Nazi victory that he waved them aside when the Russians added their protests to those of the British about the German fifth column he was harboring. To underline his contempt for Britain, he immediately granted sanctuary to the Iraqi rebels when they came across the border after the collapse of their rebellion, and he publicly welcomed their leader, Rashid Ali. It was a gesture that helped to aggravate the tension and lose him his throne. On August 25, 1941, after secret consultations with the Kremlin, British and Russian troops occupied the country. The Russians came in from the Caucasus and took Azerbaijan; the British made for the oil fields in the south and took no chances. In the half-light of dawn, the Royal Navy cruiser *Shoreham* crept up the Shatt al Arab River toward the refineries at Abadan and the city of Khorramshahr just across the river. Without warning the British commander opened fire on the Iranian sloop *Palang*, in case she might be tempted to shell the tank farm and refineries behind her. In five minutes the Iranian war vessel was a blazing wreck. In similar fashion, H.M.S. *Yarra* was sending another Iranian sloop, *Babr*, to the bottom off Khorramshahr.

But then things went wrong for the British. A landing party sent to take over the refinery went aground on a mud bank, having misread the river channels on their charts. By the time a second wave of Indian troops and their British officers were ready, the Iranians had reacted. Since they had been given no orders from Teheran, they decided to fight. Not having been briefed for first-wave operations, the second wave of Indian troops got into difficulties struggling over barges lying between themselves and the refinery waterfront, and came under murderous machine-gun fire from the Iranians. Within fifteen minutes they lost their colonel and two other British officers. It took them some time before they managed to clear the dockside and force the Iranian garrison back into the general office of the Anglo-Iranian Company, a building outside the iron rail surrounding the refinery.

It was the first time these particular Indian troops had ever been in action. At their briefing they had been told that they were going into Iran to clear the country of Germans, so they believed they would be fighting the Nazis. Hence, what they were looking for was any European or anyone wearing a uniform—these must be the enemy. So when they came into the main entrance of the general office building and saw a group of men in blue uniforms, they opened fire at once—and slew eight messengers in the blue livery of the Anglo-Iranian Company. One of the company's general staff, George Wheeler, saw the slaughter and at once telephoned the company hospital for an ambulance. It was ambushed by the Indians as it came whizzing to a halt, for the red crosses on its sides had been painted out on the shah's orders. The driver and attendants were killed.

Wheeler and two of his colleagues, now coming under fire from both Indians and Iranians, took refuge in a corridor and were mowed down by an Indian machine gunner. Wheeler, badly wounded, managed to shout in Hindustani that he was British, but by then it was too late; his two companions were dead. However, he did manage to convey to the Indians that their targets were Iranians, not Europeans, and by noon they had the refinery under control. The main body of the Iranian army now began to retreat across the river, and set off across the blazing desert toward the city of Ahwaz. As hostages they took four British oil-company representatives with them. It was a ghastly journey and en route fifty-seven Iranian troops died of thirst. The British survived because the Iranian soldiers shared their rations of water with them.

At Masjid-i-Sulaiman, where Anglo-Iranian's fortunes had begun, the British staff convinced the local Iranian army garrison that to stop work would mean that "if a well got out of control, the first persons to be gassed would be the police and troops on the maidan [market place] and in the bazaar. Furthermore, were the pumping station to be shut down, everyone, beleaguers and beleaguered, would perish of thirst. These representations had the necessary effect and no harm was caused to the wells or the pumps." *

Out in the fields around the Iranian-Kurdish city of Kermanshah, the company's local manager, A. W. M. Robertson, an amiable Scot, was personally informed of the Anglo-Russian occupation by the Iranian commanding officer, General Hassan Muqaddam. Over a drink they discussed how best to protect the refinery "from possible looting by Kurds in search of food, for these tribesmen were on the edge of starvation," a member of Robertson's staff wrote later.†

"Work went on as usual," he continued, "and each morning the Iranian general reported the progress of the war to the manager. On Thursday August 28, which is the normal half holiday, the British staff assembled at the Club and arranged to play a fourball match on the golf course which they had built. They set out but without Robertson, who had been hastily sent for by General Muqaddam and asked to take himself and his staff away from Kermanshah at once. It transpired that the Shah had telegraphed an order [to the general] to surrender [to the British troops approaching the town] but the British commander, unaware of this, and fearing for the safety of the Britons of the oil company, had also informed [General Muqaddam] that unless the staff

* Records of the Anglo-Iranian Company.
† *Ibid.*

were assembled intact at a spot outside Kermanshah by 2:30 the garrison in Kermanshah would be bombed." *

Robertson told the agitated Iranian general that it was not possible for his staff to leave Kermanshah at that moment. "He explained that he could not break up a game of golf but would go himself. This he duly did, setting out with the general's chief of staff and a white flag flying from the bonnet of his car." †

Later, at a meeting between the Iranian general and Major General (later Field Marshal) W. J. Slim, the British commander of the operation, Slim remarked, "The war between us would never have occurred if the Shah had permitted us to come in and take the German fifth columnists."

But they had already gone underground or crossed the border into Turkey. However, this was not really the object of the exercise. What the British and Russians wanted was not a genuinely neutral Iran, as they claimed, but a definitely pro-Allied one, and now they had it. Reza Khan, resentful and embittered, was removed from his throne and the crown handed down to his son, Mohammed Reza Pahlevi, the present monarch. The old man was taken into guarded exile, first on the island of Mauritius and later in South Africa, where he died, his enemies unforgiven, on July 26, 1944.

His remains were not returned to his own country until 1950, but when they were, the chickens also came home to roost, as far as Britain's oil stake in Iran was concerned.

* *Ibid.*
† *Ibid.*

CHAPTER ELEVEN

Shutdown in Arabia

WHEN ITALIAN PLANES CARRIED OUT their surprise attack on Bahrain in October 1940 they also dropped a spray of light bombs on the American oil field at Dhahran in Saudi Arabia, no doubt by mistake, since Saudi Arabia was not involved in the war and the United States was still neutral. The bombs made a lot of noise but fortunately did almost no damage; nevertheless, there was a high degree of alarm in the Aramco camp while they were exploding. It was not long since Dhahran had come through the most alarming moment of its existence, when one of its wells caught fire, and everyone shuddered at the thought that it might happen again.

Fire is the worst disaster that can befall an oil field. It can blow everything to bits on the surface around it, it can cause havoc to the reservoir of oil underneath the ground, and it is hellishly difficult to put out.

It was in the stupefying heat of a July afternoon in 1939, at a moment when a cable reading "Good news. Dammam Number Twelve is on its way" had just been sent to the home office, that this same well exploded. No one has ever discovered why. The well had been test-probed to 4,565 feet and oil found in promising quantities, and now a crew was in place, ready to take the bit down a stage farther into the petroliferous strata. No one expected any trouble. The driller, Bill Eisler, had his perforating gun ready to make the downward stab; a Saudi workman was down in the "cellar" under the floor of the derrick with his hand ready to open the equalizer valve; three other crewmen stood by the main valve controls; and Monte Hawkins, a second American driller, was by the drill hoist.

Suddenly shock waves rolled like great invisible breakers through the heavy air, followed by an enormous low grumbling thunder that seemed to feed on itself and grow and grow. Dammam Number Twelve spouted yellow and purple flames which in an instant shot three hundred feet in the air. Knocked flat, Hawkins turned and saw Eisler falling off the derrick, surrounded by flames, shambling to try to get clear. Hawkins rushed forward to grab him by his cindered hands, and managed, with help from others, to get him away, but he died not long afterward in the hospital. The Saudi workman in the "cellar" had somehow managed to scramble out, just in time to miss being felled by the derrick crashing to the ground as its steel girders melted at the roots. The fire was roaring away, wailing like a banshee, a red twister of flame so hot that it blistered the paint off trucks parked two hundred yards away.

Soon everybody in the oil world knew that Dhahran had a super conflagration on its hands. Experts from Bapco, the U.S. sister company in Bahrain, announced that volunteers were on their way to help. From Abadan, Anglo-Iranian cabled that it was dispatching a team of fire specialists. Charley Potter, the company's drilling superintendent, on leave in the United States, rushed to New York to confer with Myron Kinley, a Texan with the reputation of being the world's greatest oil-fire fighter. Kinley declared himself ready to come over with his team on a chartered plane, a much more difficult journey in those days than it would be now.

But all this would take time, and fire fighters and special

equipment were needed immediately: gas masks, and asbestos screens and suits, and extra fire hoses. Aramco's London office was alerted to buy and fill a chartered plane with the desperately needed supplies. But Britain was on the eve of World War II, and expecting conflagrations of its own, she refused at first to release the equipment. When at last it was procured, the chartered plane was only allowed to fly as far as Rome, where the equipment was off-loaded and put on an Italian plane. This in turn was grounded in Basra by the British, who were determined that no Italians should fly over their oil installations, and brought on in an RAF plane to Bahrain and thence to Dhahran. Meanwhile, the fire roared on.

"Because No. 12 was a good distance from the other wells," writes Wallace Stegner, "there was no serious danger of the fire's spreading. The real danger was that the master valve and the connections on the main casing would be destroyed, which would probably destroy the well, and might also spray the entire camp with burning oil. Besides, if the well ran wild, it might seriously deplete the whole oil field by releasing gas pressure and possibly channeling water into the oil zone." *

It took ten days to douse the whirling fire and bring Dammam Number Twelve under control. King ibn Saud, convinced that his oil wealth was going up in flame, demanded hourly communiqués and talked pessimistically about the wrath of God. One particularly religious mullah announced that he proposed to immolate himself on the pyre in the hope of appeasing the malevolent furnace, but only if his sacrifice was followed by a purge of the evil and impious who were gaining influence in Saudi Arabia, and by the expulsion of the infidels who were encouraging them. No doubt by arrangement, his friends dissuaded him at the last moment but he was to be found among the crowd of gawking locals, hectoring them to repent.

Meanwhile the fight to kill the fire went on. It took long conferences and much preparation even to plan how to deal with it.

* *Discovery!*

The men who were fighting it were in no sense professional fire fighters, but it looked as if the professionals would never get there before the well blew up completely, so they had to improvise—something which oilmen seem to do well—and rely on cabled advice from Charley Potter and Myron Kinley in New York, as well as from oil offices all over the world. Finally they had managed to position asbestos screens, enclosing the wellhead, and had found enough water—which had been scarce to begin with—in a salt-water well that gave them all they needed for the hoses. Three men in asbestos suits volunteered to go to the head of the well and see what they could do there.* They had studied the plans and knew there were two large panel gates over the well; if they could get these closed down it might help. Sprayed with water from the hoses to help keep them cool, they went through a gap in the asbestos screens and by extraordinary tenacity ("it scorched your eyeballs," Eltiste said) managed to reach the wreck of the wellhead, and for thirty seconds tried to get the control wheels turning. One went around twice and then stuck; the other wouldn't budge at all. The next day they went in again, and before they were dragged clear, managed to get one panel half down. It slightly diminished the flames, and enabled Eltiste to see that the fire was coming through a split nipple in the control valve.

It was then that the home office in San Francisco came up with an idea. Don't go for the well direct, they suggested; instead, dig a tunnel to it, put a tapping machine on the base of the main valve, and start pumping mud into the hole to smother the flaming oil. This was easier said than done, but on July 18, ten days after the first mighty explosion, a tunnel had been built, and thanks to the half-closed panel gate, they had managed to get a "hot tap" ready to slip onto the valve. Two men, working night and day in great heat and peril behind shields, head-first in the cellar of the well, had moved the connecting line into place.† Now

* The volunteers were Bill Eltiste, Herb Fritzie and Walt Sims, all Aramco drillers or ex-drillers.

† Cal Ross and Ed Braun, two more drillers.

all that remained was for the bit to drive a hole in the main valve and for the mud to be pumped in before the flames from the well could spurt out through the tunnel. They played it by feel rather than by ear, though the fire was only a subdued roar now, and when they felt the bit shudder through, Eltiste raised his hand and the pump began gushing gallons of thick mud into the fiery hole. Like a damp candle, the flame simply guttered away.

FIRE EXTINGUISHED. HOLE FULL OF MUD. PROFESSIONAL FIRE FIGHTERS NOT NEEDED, cabled Floyd Ohliger, the general manager, to New York, where Myron Kinley was still waiting for a plane.

There was no expression of pride and joy in the message, but it was implicit between the lines. "In celebration of the way they had handled themselves," wrote the official historian, "the [Saudi] Government—for this one occasion—relaxed its prohibition law, and the first beer that was ever in Al Hasa came across from Bahrain, and the firemen really tied one on." *

THE GREAT DHAHRAN FIRE cost the lives of one American and one Saudi, did $3 million worth of damage, and deprived King ibn Saud of $500,000 in oil royalties. But more important, it made the men who ran the oil field feel vulnerable. When the Italian bombs dropped, there was panic, although there was no reason why the Axis powers should breach the neutrality of a kingdom whose ruler both Germany and Italy were anxious to placate. The immediate measures which were taken after the raid were perhaps well justified, though messy and complicated: roads and buildings, for instance, were painted with oil to take the sheen off them, and the gas flares which flicker over every oil field were strained and treated so that they burned a dim blue instead of a fiery red, to make them less visible from the air. More drastic was the decision, made after a pregnant wife became hysterical, to evacuate all American women and children, plus all male employees who wished to go. At the beginning of 1940, Aramco's

* Stegner, *op. cit.*

Arabian complement of U.S. citizens had been 371 employees, 38 wives and 16 children; a year later it was down to 180 males and no women and children. As a result, the whole operation was slowly running down.

Thanks to Max Steineke and his dogged fellow geologist Tom Barger, two new likely fields had been discovered. In one of them, Abu Hadriya, lying deep within the Rub al Khali (the Empty Quarter), oil had been struck at 10,115 feet in March 1940, and at about the time of the bombing, in October of the same year, a well had gone down at Abqaiq, about forty miles into the desert from Dhahran. Both areas obviously had enormous potential,* and everyone was bubbling with optimism. But immediately after the raid, Abu Hadriya was ordered closed down, and the following spring Abqaiq followed suit. Because of the evacuation of personnel, there was not enough manpower to handle them. By the beginning of 1941 only Dammam was operating, and that only at the rate of about 15,000 barrels a day, a mere trickle compared with the field's capabilities. The crude oil was ferried by barge across to the refinery at Bahrain, and Aramco's own new installations at Ras Tanura, its huge tank farm, its small refinery and its port facilities for tankers lapsed into disuse.

A gloomy hush fell over what a few months before had seemed the most promising new oil field in the Persian Gulf, and as word spread, the benevolent vultures in other parts of the Middle East and India came around looking for pickings. Especially after Pearl Harbor brought the United States into the war, theater commanders all over the area began calling on Dhahran for loans of trucks, pickups, welding equipment and drills, sapping the supplies which Floyd Ohliger had painstakingly built up in 1939 against the shortages of war. Now they could not use them and had to surrender them. Soon they were so pinched for transport that they organized a camel corps to transport material around the fields.

No one was more distressed by the curtailment of operations

* Abqaiq has since become one of the most prolific fields in the world.

than King ibn Saud, for it meant a drastic cut in his income and a blow to his hopes for its increase in the future. It could not have come at a worse time. Besides the oil royalties, his only revenues came from money brought in by pilgrims to Mecca. In a normal year more than a million devout Muslims came to Saudi Arabia for the *hajj,** but now nearly all the countries from which they came (the Middle East, India, the Far East and Africa) were involved in the war, and no ships were making the journey. To add to his distress, the winter rains had failed and there was a drought in the Al Hasa oasis, killing the crops; the wells were dry and the Bedouins were eating their camels before the vultures got them.

The king needed money badly, to pay off his old debts, of which there were always plenty, to relieve his subjects' distress and to pay for his own extravagances. Perhaps "extravagances" is not the right word, for though all the money the state received was paid directly to the king, what happened to it afterward was left to his trusted ministers—who, in fact, could not be trusted. When Ibn Saud asked for money—to pay for new wives or concubines, to subsidize tribes which might otherwise have become dissident, to satisfy the importunate demands of his myriad sons and their acquisitive appendages—he was always given it. But the purse-strings were in fact controlled by his finance minister, Abdullah Sulaiman, and if he kept a strict accounting of what was done with all the monies flowing into the king's coffers, no one else saw it. Ibn Saud was too bored to ask. He was not interested in accounts; all he wanted was money to pay his bills.

On January 18, 1941, Abdullah Sulaiman, writing on behalf of the king, sent a formal letter to Aramco stating the harsh reality of the financial situation in Saudi Arabia and requesting the immediate payment ("as an advancement on oil royalties") of $6 million. He asked the company to note that he was reserving the right, while making the present request, of continuing to call for further advance sums until such time as the emergency in Saudi

* Nowadays as many as two and a half million make the pilgrimage.

Arabia no longer existed. He also underlined the fact that the Americans had recently been granted an extension of rights and concessions, and that these had shown unlimited possibilities for successful exploitation.

Fred A. Davies, the president of Aramco, sent a telegram to the finance minister in Jidda saying that he had already arranged for \$3 million to be advanced to the Saudi government, but that at the same time he had to inform the king that "as long as the war continues and the market for Arabian oil continues curtailed we do not want to increase amount advanced nor to lead him to feel that this is precedent for further advances in future." Fearing this was too harsh, Davies and a fellow director went to Saudi Arabia from San Francisco. In a further message to the finance minister, the Aramco president attempted to mollify him by saying:

> In our many discussions with His Majesty and Your Excellency we have come to a full realization of the Government's need for extraordinary assistance during the present international emergency. Should this emergency extend beyond the year 1941, the Company will, of course, continue to assist as much as it can . . . Your Excellency need have no doubt that the Company realizes fully how closely its own interests are bound up with those of His Majesty's Government.*

Nonetheless, the truth had to be faced. The glorious future of Saudi Arabian oil had turned rancid, at the moment the company was \$34 million in the hole on its investment, and it was not prepared to pay out any more.

On the other hand, the U.S. government might be persuaded to do so. It was known that Washington was pleased at the inroads the Americans had made in Saudi Arabia, an area previously jealously guarded as a British preserve; perhaps the State Department could be persuaded that there was a danger of losing the advantage so handily gained.

It so happened that one of the directors of Aramco was James

* Quoted during a special inquiry before the U.S. Senate, 80th Cong., Petroleum Arrangements with Saudi Arabia (Washington, D. C., 1948).

A. Moffett, who had once been a member of the State Department and who counted President Roosevelt among his friends. At a moment when Abdullah Sulaiman was at his most insistent for the $6 million loan, Moffett went to the White House and had a discussion with the President in which he outlined the king's difficulties and the company's situation, at the same time stressing the strategic importance to the United States of the Saudi oil concession. He asked for a U.S. government loan to Ibn Saud to tide him over. Roosevelt replied that there was no way in which such a loan could possibly be made under existing American laws, but as a former Secretary of the Navy, he suggested that perhaps something might be arranged by selling oil products to the U.S. fleet. He told Moffett to write a memorandum to the government explaining the company's present plight and that of the king, and outlining a plan whereby a five-year oil sale might be made to the Navy which would provide the king with the $6 million he so sorely needed.

In the years to come, U.S. senators were to accuse U.S. oil interests in Arabia of being a government unto themselves, and in parts of his memorandum Moffett implicitly admitted the supranational nature of his company's commitment:

> It has now come to a point where it is impossible for the company to continue the growing burden and responsibility of financing an independent country, particularly under present abnormal conditions. However, the King is desperate. He has told us that unless necessary financial assistance is immediately forthcoming he has grave fears for the financial stability of his country.*

But now Moffett was giving the impression that his company was ready to abdicate its responsibilities, and that if the U.S. government refused to accept them, they would be passed on to the British. In fact, neither he nor his company was prepared to do any such thing, but they had 168,000 U.S. shareholders to think of, and could not continue to pay enormous subventions

* *Ibid.* Moffett quoted in memorandum submitted by W. S. S. Rodgers, chairman of the Texas Oil Company.

out of the oil company's funds at a time when its Saudi Arabian investment was stagnating. So what happened next was a shrewd ploy on the part of the company to relieve itself of the Saudi burden and to get the U.S. government involved in carrying it instead, without appearing to have done so.

In his memorandum Moffett duly outlined a plan for a U.S. naval purchase of Saudi oil, as President Roosevelt had suggested, but he had little hope that this was the way to find the money Ibn Saud was demanding. His suspicions were confirmed when he had an interview with Secretary of the Navy Frank Knox, who told him at once that the amount of oil that would be involved in a $6 million program was far beyond the needs of U.S. Navy ships in the Persian Gulf and the Indian Ocean. When Moffett expressed the view that the Navy could buy the oil as a future reserve, the Secretary replied that though it might be a sound idea, he had no funds for such a purpose.*

It was for this reason that the oilman included in his memorandum another way around the problem, one that would mean bringing in the British government. He was well aware of the fact that the British had strong political and strategic reasons for being concerned about the future of Saudi Arabia. Militarily they were in a bad way in the Middle East, with General Rommel and his Afrika Korps in the Western Desert knocking on the gates of Cairo and Alexandria. Their backs were to the wall—and the wall was the far from stable one of Saudi Arabia. Already there were strong indications that certain influences in Ibn Saud's court were in Hitler's pocket, while the king himself was reluctantly coming to the conclusion that Britain was losing the war. He did not particularly like the British, but he trusted them more than the Axis powers, especially Italy, whose activities in occupied Ethiopia, on the other side of the Red Sea, made him fearful of what might happen to his own country if it came within the Italian orbit. Therefore he was inclined to continue tingeing his neutrality with a pro-British hue as long as he possibly could, but in

* *Ibid.*

order to do so he must have enough financial resources to alleviate the distress in his own country and to prevent his people from being bribed or susceptible to Axis propagandists. To bolster him against enemy pressures, and to make up for the loss in oil revenues from the Americans, the British government had already begun lending the king an annual sum of £400,000 (approximately $1.6 million at the then rate of exchange). Moffett urged the State Department to persuade the British to increase their annual contribution to Ibn Saud's treasury, and to make loans of its own so that, together with what small revenues he did receive from oil and pilgrims, the total would provide him with the $10 million a year the oilmen estimated he needed to keep Saudi Arabia solvent and peaceful.

Moffett added one point to his memorandum which made it clear that his company, though dormant, was far from willing to see its proprietary rights in Saudi Arabia taken over by anyone else. He asked the State Department to make it clear to the British that in extending loans to the king, they were doing so for reasons of war strategy only, and their aid would in no way give them a future claim over the oil concession.

Fortuitously enough, it was a moment when the Federal Loan Administrator, Jesse Jones, was in the throes of working out a Lend-Lease loan to Britain of $400 million, and after a discussion with Harry Hopkins, the presidential adviser, and Navy Secretary Knox, President Roosevelt suggested that the best solution to Saudi Arabia's problem might well be for the British to apportion part of this sum to the king. On June 12, 1941, he sent a note to Jones saying:

> JESS: Will you tell the British I hope they can take care of the King of Saudi Arabia. This is a little far afield for us! F.D.R.*

Two days later Harry Hopkins wrote a letter approaching the matter from a slightly different angle, in which he speculated whether Saudi Arabia could itself qualify for the type of Lend-

* *Ibid.*, p. 24747.

Lease loan which Congress had agreed to supply to the United States' "democratic allies."

Dear Jesse:

The President is anxious to find a way to do something about this matter. I am enclosing confidential correspondence from the White House so you can see what goes on. Will you return it to me as soon as you have read it?

I am not sure what techniques there are to use. It occurred to me that some of it might be done in the shipment of food direct under the Lend-Lease Bill, although just how we could call this outfit a "democracy" I don't know. Perhaps instead of using his oil royalties as collateral we could use his royalties on the tips he will get in the future on the pilgrims to Mecca. . . .

Cordially yours,
Harry *

Eventually it was the British who agreed to take over the burden of keeping King ibn Saud afloat by using the U.S.'s Lend-Lease, and for a time Aramco officials were congratulating themselves on the way they had successfully plugged the drain on the company's resources. Their representatives in Jidda took pains to point out to Abdullah Sulaiman, and on two occasions to the king himself, that they must regard the British subventions as a substitute for their own aid, and not as additions to the loans and royalties the Saudis had been demanding from Aramco. In other words, the British were bailing them out of a most awkward situation, and saving them many millions of dollars which otherwise would have had to come out of the company's funds. How much that saving was is indicated by some figures which were quoted by Floyd Ohliger, vice-president of Aramco, at the Senate Inquiry into Petroleum Arrangements with Saudi Arabia at their 1948 hearings in Washington. Ohliger testified that the company paid the king, as loans against future royalties, the sums of $2,-980,988 in 1940, $2,433,222 in 1941, $2,307,023 in 1942, and $79,-651 in 1943. This downward trend must have brought joy to the

* *Ibid.*, p. 25415.

treasurer at the home office in San Francisco. On the other hand, over the same period the British government's loans to the king went upward even more steeply: $403,000 in 1940, $5,285,500 in 1941, $12,090,000 in 1942, and $16,618,280 in 1943.

Were the British motivated solely by their desire to keep Saudi Arabia in the Allied camp? That may have been true in 1941 and 1942, but by 1943 the danger that the Axis would sweep across the Middle East was over. Italy had been driven out of Ethiopia and Eritrea, the German army out of North Africa. Iraq, Syria and Palestine were securely in the hands of Britain or its allies. On the other hand, Washington's lack of interest in Saudi Arabia and the American oil company's abdication of its declared responsibilities had left a vacuum which London was only too pleased to fill. There is no evidence that a direct attempt ever was made by the British to take over the American concession, but what would have happened if King ibn Saud had offered it to them out of gratitude—angry as he was with the "defections" of the U.S. oil company? The British government held a 51 percent controlling interest in the Anglo-Iranian Oil Company, and by a lucky coincidence the director of military intelligence at British G.H.Q. in Cairo was none other than Archibald H. T. Chisholm, once an executive of Anglo-Iranian and negotiator of its Kuwait concession, now a brigadier, his monocle sparkling over his crimson staff officer's tabs. So far as intelligence operations were concerned, Saudi Arabia was well inside his territory.

Suddenly Saudi Arabia, which had once seemed "a little far afield" to Franklin D. Roosevelt, began to seem much closer as British influence began to make itself felt at King ibn Saud's court. It was the moment, Aramco decided, to sound the alarm and bring the U.S. government to the rescue.

"We didn't like it," testified W. S. S. Rodgers of the Texas Company later. (Texaco was a main shareholder in Aramco.) "We realized that he [King ibn Saud] had to have the money, but we didn't like the British Government making these advances. . . . And then at the beginning of 1943 it got so bad that we came down here to Washington and called on many people, trying to get the matter straightened out." *

Mr. Rodgers appears to have ignored the fact that it was his company which had suggested that Britain should increase her subventions to the Saudi Arabians in the first place; he now saw the situation in a very different light. In a memorandum submitted to the Senate investigating committee, he described it thus:

> The situation in Saudi Arabia had now [1943] reached the point that the British were backing the Saudi Arab Government so far as its finances were concerned. It would appear that they were looking for an opportunity to remain as the financial advisor and backer of the Saudi Arab Government without having to advance any actual gold or silver. This crystallized in March 1943 when they proposed a plan for a Saudi Arabian note issue. This plan provided for the creation of a Saudi Arab Currency Control Board in London, composed of the Saudi Arabian Minister, Government of Great Britain Representatives, and Bank of England Representatives.†

It was time to call a halt, Rodgers said in the series of conferences he had in Washington in February 1943 with members of President Roosevelt's Cabinet. The one on whom he seems to have made the deepest impression was Secretary of the Interior Harold L. Ickes; it was he who convinced the President that the British were threatening the U.S. company's oil concession in Saudi Arabia and that action must be taken before it was too late.

The maneuvering was well timed. There was much discussion going on in Congress and in the press about America's future oil reserves, and the Chiefs of Staff were making known their desire to have a reserve of petroleum in the Middle East for the use of the armed forces as the war moved toward the Indian Ocean and the Pacific. On February 18, 1943, President Roosevelt decided to twist the terms of the Lend-Lease Act even more than

* Testimony before the Senate Committee to Investigate Oil Resources. 79th Cong., 1st sess. Washington 1945.

† *Ibid.*

they had been manipulated so far. In a letter to Secretary of State Edward R. Stettinius he wrote:

My dear Mr. Stettinius:

For purposes of implementing the authority conferred upon you as Lease-Lend Administrator by Executive Order No. 8926, dated October 28, 1941,* and in order to enable you to arrange Lend-Lease aid to the Government of Saudi Arabia, I hereby find that the defense of Saudi Arabia is vital to the defense of the United States.

Sincerely yours,
Franklin D. Roosevelt †

This meant that King ibn Saud at once became the recipient of the bounty of the United States Treasury, along with America's fighting allies. Once more Ibn Saud's insatiable needs were appeased and Aramco's concession saved. First the British had come to the rescue, now Washington; and it had not cost the oil company a cent.

For such invaluable service the British naturally did not expect any gratitude, for their charitable gestures had not exactly been devoid of self-interest. On the other hand, the munificence of the U.S. government ‡—basically for no other reason than to relieve a commercial enterprise of a heavy financial obligation—might have been expected to produce some tangible evidence later of the company's gratitude. But when Washington asked the oil company for a gesture in return, it found that Aramco was, in Harold Ickes' words, "more disposed to thumb their noses at us."

DURING WARTIME, the U.S. government had some say in the control of almost every branch of American industry. Through boards established throughout the nation, raw materials were allocated, markets were controlled, prices were fixed and wasteful

* Which recognized Soviet Russia as a "democratic ally" for the purposes of the act.

† Senate Inquiry 1945, *op. cit.*

‡ Under Lend-Lease, it gave Saudi Arabia a total of $99 million.

competition was eliminated. The petroleum industry had its own committee through which production rates of crude oil and automobile and aviation gasoline were decided, and some oilmen—though by no means all—claimed that the industry worked better this way. The President and members of his Cabinet agreed, none more so than Harold Ickes. Indeed, he went one step further; he believed that the best arrangement for the U.S. government would be if it were to buy into the oil industry and secure a controlling interest in a major source of supply so that Washington would have a say in pricing, production and policy, and would cease to be at the mercy of commercial enterprises interested solely in increasing prices and profits. What the British government had done to Anglo-Iranian in 1914—that is, buy a share of the enterprise—and what the Arabian and Iranian governments were maneuvering to do to all the major oil companies in 1972, Ickes proposed that the U.S. government should do with Aramco in 1943.

In June 1943, Ickes wrote a note to the President proposing that a Petroleum Reserves Corporation be set up to "acquire and participate" in the exploitation of foreign oil reserves, and he suggested that the first task of the new corporation would be to acquire a "participation" in the Saudi Arabian concession. It would be a move which would once and for all prevent Britain (or "certain known activities of a foreign power," as Ickes put it) from taking over. To Roosevelt, who was not known for his love of oilmen (they were apt to vote Republican), it seemed a capital idea, and he forthwith set up the Petroleum Reserves Corporation with Ickes as president and chairman of the board. At once the Secretary summoned officials of the California-Arabian Standard Oil Company (which would soon change its name to Aramco) to Washington and announced to them that he proposed to begin negotiations for the purchase of an interest in their enterprise.

A year earlier, when the whole Middle East had seemed on the verge of being overrun by the Axis, and when the oilmen were mentally writing off Saudi Arabia, they might have welcomed this proposal as a form of government insurance of their property, and therefore worth sacrificing part of their control. By this time, however, the situation was altogether different. There was no question of a military debacle, the government had taken care of their financial obligations to King ibn Saud, and they could afford to think about postwar production and profits again. Hence, their attitude was hostile from the start. When they asked Ickes what he considered would be the extent of the government's participation in their company, the Secretary replied that it would solve a lot of problems for the Petroleum Reserves Corporation if the company were to sell its entire stock to the Corporation, for an amount to be agreed on, plus a royalty. Later Ickes said that this casual remark so frightened them that "they nearly fell off their chairs." The oilmen asked permission to retire to consider the Secretary's proposal; in fact, it was to get their breath back and to spread the alarm through the oil industry that the U.S. government was moving in on them.

Nothing is more calculated to unite a normally competitive and divisive industry than a threat to its future, and as word spread through the oil fields, tycoons from Texas to California and from Oklahoma to Venezuela began calling Washington and lobbying their friends in the Senate and House of Representatives. Aware that his casual remark had created panic before he was ready to pounce, Ickes gathered the oilmen together again and began reducing the amount of participation he thought the U.S. government should obtain. First it was "whittled down," as he phrased it, to 70 percent, then to 51 percent. As W. S. S. Rodgers, one of the directors involved in the discussions, later testified: "Naturally that did not appeal to us. Then they got down to a 33⅓ percent basis and we did not like it, or at least I did not personally, but we began to get a little closer. We did not get close enough, and one day Mr. Ickes said, I don't know why, 'The negotiations are off,' and I was very much relieved." *

* Senate Inquiry 1945, *op. cit.*

Mr. Rodgers may not have known why Ickes suddenly gave up, but most other people in the oil industry did. Too much pressure had been put on Washington by irate oil interests at a moment when the government needed all the co-operation it could get to ensure a steady supply of fuel for the armed forces in the crucial build-up for the last phases of the war. "They came up here to the Hill and built a fire under us on the theory that this was an attempt on the part of the Government to take over a private-business enterprise," Ickes testified later, "which, of course, was against the American tradition, as they put it, and perhaps it was. But this was more than a business enterprise; this involved the defense and safety of the country." *

Ickes had expected that at the very least the American oilmen operating in Saudi Arabia would be willing to grant some measure of U.S. government participation in their enterprise, if only out of gratitude for what the Administration had done for them during the dark days of 1943-1944. He was disappointed. "They felt in the meanwhile that since Rommel had been chased out of North Africa they were secure in their concession and more disposed to thumb their nose at us." †

SECRETARY ICKES made one last attempt to get the U.S. government involved in the Saudi Arabian oil bonanza. Calculating that if the front door had been slammed in his face there might still be a way in by a side entrance, he came up with the suggestion that the United States should finance and build a pipeline from the Persian Gulf to the Mediterranean, thus saving the long tanker haul by way of the Gulf, the Red Sea and the Suez Canal. In return for its outlay in building the pipeline (whose cost was estimated in 1944 at $120 million), the government would receive all the oil it needed at 75 percent of the market price.

* *Ibid.*
† *Ibid.*

This time the Secretary's proposal * was embraced with enthusiasm by Aramco's directors, who knew a bargain when one was waved under their noses. Not only would they get a pipeline for their products without spending a dime; they would also get it quickly, for the government would give priority to the project and the raw materials needed at a time when they were in desperately short supply. But once again the specter of government involvement in the oil business was too appalling a vision for the rest of the industry to endure, and such formidable opposition built up against it inside the United States (about 98 percent of the industry, according to W. S. S. Rodgers of the Texas Oil Company) that, as Ickes himself put it when writing about it later, the scheme was "done to death without benefit of clergy." †

Aramco decided to build a pipeline on its own—at vastly greater expense—and it went into operation on January 28, 1949, after agreements were signed with Saudi Arabia, Syria and Lebanon, the countries the pipeline traversed on its way to the Mediterranean. When one considers the way in which Tapline (Trans-Arabian Pipeline) has been used ever since as a blackmail weapon by the Syrian government and Palestine guerrillas,‡ it is fascinating to think of the mess in which Washington might have involved itself had the pipeline belonged to the government instead of a private company.

* The scheme was not new, but had simply been appropriated by Ickes. James F. Byrnes, director of the Office of War Mobilization, had recommended it to the President in January 1940. In 1943 Admiral Andrew Carter, petroleum administrator of the U.S. Navy, visited Saudi Arabia and conferred with local oil and government officials. He drew up a scheme which was approved by the Joint Chiefs of Staff, and was then taken over by Ickes. *Ibid.*

† Ickes, "Oil and Peace," *Collier's* (December 2, 1944).

‡ Its flow has been cut or part of the line blown up fourteen times in the last five years, usually as a protest against U.S. government policy toward the Arabs and Israel.

CHAPTER TWELVE

Breaking the Thick Red Line

ONCE THE U.S. GOVERNMENT HAD AGREED to the oilmen's plea to bail them out in Saudi Arabia, they came to the rescue in style. Two military missions flew in during 1943, the first led by General Ralph Royce and the second by Lieutenant Colonel Harold B. Hoskins. Supplies followed in their wake—desert trucks, bulldozers, road-making materials—and the job began of turning Saudi Arabia's rock-strewn desert tracks into some semblance of roads. Early in 1944 the first resident U.S. minister in Jidda was accredited to the kingdom in the person of Colonel William A. Eddy, a single-minded patriot, the son of a missionary, born and brought up in South Lebanon and fluent in Arabic. At the same time Dhahran, which had quadrupled in size thanks to the labors of thousands of Italian prisoners of war, had its first diplomat in the person of a U.S. consul general, who opened his offices on a small *jabal* overlooking Dammam Number Seven.

As if to recompense the oilmen for having failed to provide them with a government-built pipeline, President Roosevelt encouraged the Department of Commerce to heap other kinds of largesse on the company. The benefits did not come in the form of money but in something which, in those final days of war and early days of peace, was much more valuable—priorities for raw materials. All over the Western world everyone was fighting for materials to rebuild shattered factories and cities and for getting peacetime industry restarted. In the oil industry in the United States, and in British oil fields in Iraq, Iran and Kuwait, men were waiting for the machines and tools that would enable them to restart old wells and drill new ones, to be ready for the postwar boom in petroleum that everyone knew was coming. But all of them had to wait on Saudi Arabia, which was now the U.S. government's favorite son.

By 1944 Aramco had its old fields going full blast again and was drilling for new ones. Moreover, thanks to Washington's generous allocations, it had also started work on a vast new refinery at Ras Tanura, as well as an undersea pipeline from Dhahran to the refinery on Bahrain Island, and—the biggest undertaking of all—the Tapline project to the Mediterranean. For this, the Department of Commerce granted an export license for no less than 20,000 tons of steel. When there was an outcry from independent oil producers, who complained bitterly that someone in the Cabinet was unfairly favoring oil interests in Arabia, the Senate Small Business Committee called an inquiry on the subject. Several government departments spoke in favor of the plan, but the Defense Department refused to do so on the grounds that the export grant could not be justified. The Independent Petroleum Association produced a long and virulent attack on it, and was rewarded by a report from the committee declining to recommend the export license on the grounds that it was not in the public interest.* This delayed the matter for nine months, but at the end of that time the Department of Commerce quietly reactivated the license, and shortly afterward the 20,000 tons of steel were on their way.

While Aramco was enjoying these priceless fruits of government bounty, other oil companies in the Middle East could only

* Proceedings of the Senate Small Business Committee, March 11, 1946. *Congressional Record* (Senate), Part Two.

look on in frustration. Though oil began to flow in ever-increasing quantities from Saudi Arabia from 1944 on, Kuwait, for instance, though only two hundred and fifty miles north of Dhahran, did not get started until 1947, and in Iraq and Iran, expansion programs took even longer.

The summit of the U.S. government's love affair with Saudi Arabia was probably reached in 1945, when President Roosevelt and King ibn Saud met for a Middle East conference. According to U.S. spokesmen, the object of the meeting was the President's hope of enlisting the king's influential support in the Arab world for a peaceful solution to the Palestine problem, which was then, as now, bedeviling the Middle East. However, that it was more an exercise in the consolidation of U.S.–Saudi relations, to the undoubted benefit of American oil interests, is indicated by the secrecy with which it was arranged. It was not until the last day of the Yalta Conference in February 1945, when Roosevelt, Stalin and Churchill met in the Crimea to divide spheres of interest in liberated Europe, that the President told the British prime minister that he was going to see King ibn Saud on his way home. According to Colonel Eddy, the U.S. minister in Jidda, who had arranged the presidential meeting with the king, Churchill was "thoroughly nettled" and "burned up the wires of all his diplomats" with orders to arrange a similar meeting, but was too late to get there first.*

The meeting between Ibn Saud and FDR would take place aboard the U.S. cruiser *Quincy*, which was anchored in the Great Bitter Lake, in the Suez Canal area, and Ibn Saud was to be transported there from Jidda by the U.S. destroyer *Murphy*. To get to Jidda from his capital, Riyadh, the king had set off in mid-December 1944 with a cavalcade of two hundred cars, but almost at once, in the six-hundred-mile trek across the desert, the caravan ran into violent winter rains and flash floods which bogged it down for a week. The king had grown old in recent years, and had taken to dyeing his beard and hair, but as if to insist that his vi-

* *F.D.R. Meets Ibn Saud*, by W. A. Eddy.

rility was unimpaired, he now refused to travel without his harem, and eventually they all arrived, muddy but otherwise unharmed, in Jidda at the end of the month.

Eddy was horrified when he was informed by court officials that not only did the king expect to take two hundred of his retainers with him aboard the *Murphy,* but he also planned to have a selection of the beauties of his harem traveling with him. The U.S. minister had a message in his pocket from the captain of the *Murphy* informing him that the ship could not possibly accommodate more than the king, four advisers and eight servants. With visions of what would happen if a bunch of veiled Arabian houris were suddenly deposited amidst a complement of musclebound U.S. sailors, Eddy went to work to acquaint the Saudis with the realities of the situation. Ibn Saud finally compromised on a party of forty-eight—all males. Among others, they consisted of his second son, Prince Faisal,* Abdullah Sulaiman and two other ministers, his private physician, his chamberlain, his coffee servers, his cooks, and six enormous Nubian slaves armed with swords.

Since a U.S. destroyer has no accommodation for kings, the Saudi party covered the quarter-deck with an Arab tent, placed rich carpets on the deck and erected a throne for their monarch. There was a critical moment when a courtier tasting the ship's distilled drinking water pronounced it "dead," and sent ashore for supplies of Mecca well water (the only other well from which the king would drink was in Riyadh), but the worst crisis occurred just before the *Murphy* sailed out of Jidda, when an Arab dhow came alongside with a cargo of eighty-six sheep, alive and braying. The destroyer's captain, who by this time must have been wishing for less nerve-racking duty, like fighting the Japanese fleet, tartly informed the Saudis that he could not possibly take live animals aboard his ship and that there was plenty of meat for everybody in the *Murphy*'s amply stocked refrigerators. *Dead meat*, responded the Saudis; *rotting carcasses.* They insisted on their live sheep, to be slaughtered just before cooking. With the

* Now King Faisal, ruler of Saudi Arabia.

thought of his decks awash with the victims' blood, the captain finally consented to have no more than ten brought aboard.

The *Murphy* sailed, and luckily there were no more crises. The king saw a film of the naval war in the Pacific while his sons sneaked away to join the sailors who were watching a more worldly comedy with a heroine in her underwear. There were practice firings on the way up the Red Sea, and five times a day, after solemn consultations with the navigator, the king and his party knelt and prayed in the direction of Mecca. As the destroyer neared its rendezvous with the *Quincy*, the king presented each crewman with $40, and each officer with $60 plus an Arab costume and a gold watch. The captain received a gold dagger, and in turn presented the king with a pair of binoculars and two submachine guns.

Gifts were exchanged aboard the *Quincy* too, for King ibn Saud, who had grown lame himself, admired President Roosevelt's wheelchair and was immediately offered his "spare." * FDR also promised to send the king an airplane big enough to carry himself and selected members of his harem, and therefore convenient for traveling between Jidda, Mecca, Medina and Riyadh, which were now beginning to seem such long and weary journeys by car over bumpy desert tracks. In return the king gave the President jeweled swords, daggers and perfumes in richly decorated bottles. While the two heads of state talked of farming and of U.S.–Saudi co-operation in the extraction and marketing of Arabia's oil, they got on like old friends. On the solution of the Palestine problem they were rather less harmonious, and their conversation progressed to a misunderstanding that was to sully U.S.–Arab relations in the months to come. Roosevelt promised, first verbally and then confirmed by letter, that as President he would never do anything hostile to the Arabs, and that the United States government would make no change in its Palestine policy without consulting both Arabs and Jews beforehand. To King

* It turned out to be too small for the giant-sized monarch, but he brought it back to Jidda and had a larger model copied from it.

ibn Saud this was a binding promise which he expected would be kept as would one made by himself; he did not understand the American form of government and believed that when a President spoke, his word was law. He no more dreamed that it could be altered by his successor, or by Congress, than his own decisions could be reversed by the people of Saudi Arabia. Two months later Roosevelt died, and his pledge died with him.*

In the circumstances, Winston Churchill's meeting with the king was rather less cordial than that of the President. It took place at Fayum Oasis south of Cairo three days later, on February 17, and there was little that the British prime minister could offer Saudi Arabia and even less that the king could give in return.† The Americans were now paying Ibn Saud's bills and were getting Arabia's oil in return. As for the fraught question of Palestine, Churchill believed it was a problem on which Arabs and Jews would never agree; he could offer no solution and refused to make any promises about Britain's attitude. Moreover, if Ibn Saud was a proud and autocratic man, Winston Churchill was no less so, and as leader of a Britain which was soon to emerge as a victor in the war, he was not going to allow himself to be outranked by a mere desert king. It was explained to him that Ibn Saud was a strict Muslim and neither smoked nor drank; nor did he allow either vice to be pursued in his presence. President Roosevelt, it was pointed out, though a chain-smoker, had abstained in the king's presence so that the sinful odor would not reach the king's nostrils. Churchill would have none of this; to the interpreter he pointed out that "if it was the religion of His Majesty to deprive

* In breaking Roosevelt's promise, President Harry S. Truman used words which were henceforth to haunt every oilman or diplomat trying to do business with the Arabs: "I'm sorry, gentlemen, but I have to answer hundreds of thousands of people who are anxious for the success of Zionism. I do not have hundreds of thousands of Arabs among my constituents."

† Except the ritual presents, of course. Jeweled sword and dagger, genuine Gulf pearls and other precious stones from the king; a set of perfumes from the prime minister and a promise (to match Roosevelt's airplane) of "the finest car in the world"—a Rolls Royce which Ibn Saud never used because it had right-hand steering, and the king, who liked to sit in the front of a car, refused to sit on the left of the driver, a demeaning position for a superior Arab.

himself of smoking and alcohol, I must point out that my rule of life prescribed as an absolutely sacred rite smoking cigars and also the drinking of alcohol before, after, and if need be during all meals and in the intervals between them." The prime minister added, "The King graciously accepted the position," but the fact was that he was away from his own country and had no option.*

The king returned to his capital to be greeted by a great crowd of cheering Saudis, proud of what their ruler had brought back materially and politically from his meeting with Roosevelt. The material gains were more durable than the political promises, for they brought technical aid to Saudi Arabia on an even more generous scale than before, and an assurance that henceforth the oil from Arabia's wells would flow in ever-increasing volume.

In the first year after the end of the war, Aramco paid out $20 million in petroleum royalties to King ibn Saud. Soon the payments would grow to $4 million a week. The money was handed over directly to the king, and it came at a moment when he was beginning to lose his grip. At about this time Harry St. John Philby returned to Saudi Arabia from an enforced exile in wartime England to find the king a sorry shadow of himself. "He was a careworn man," he wrote, "already tiring under the strain of a vigorous life: with a crippled knee to which he was rapidly surrendering, and other signs of the inroads of the great enemy, which the most skilful dyes could not conceal. He was clearly following the path of least resistance through the dark forest of a world he had never seen and only knew by hearsay . . . the quick clear mind was blurred and hesitant as it struggled with the problems of a strange world." †

Ibn Saud now had in his hands more money than even he had ever dreamed of, and his nation was waiting to gain some benefit from it. But somehow it seemed to trickle through his fingers into

* Churchill's comments from *The Second World War,* Vol. VI, *Triumph and Tragedy.*

† Philby, *Arabian Jubilee.*

the pockets of the profligate hangers-on around his court. The wealth that should have been the making of Saudi Arabia was soon threatening to be its ruination.

NOW THAT Aramco rather than the United States government was undertaking to build the Trans-Arabian Pipeline, the company had to find the money to pay for it, as well as for the new refinery at Ras Tanura and the new undersea pipeline to Bahrain. This vast program was far beyond their financial resources, and at the beginning of 1944 the company had begun negotiations with two other U.S. petroleum corporations, offering them a 30 percent and a 10 percent participation in Aramco, respectively. These companies were Standard Oil of New Jersey and Socony-Vacuum Oil Company, and they were what Standard of New Jersey's official history calls "obvious prospects" to answer Aramco's needs. "They could command large capital and were willing to assume the risks. Both needed more oil for their markets —Jersey especially, for its outlets in Europe." * If the deal went through, Standard Oil of California, the original exploiter of the concession, and the Texas Oil Company, which had joined Standard of California in 1936, would each retain 30 percent participation in the new quadrumvirate, and Saudi's oil would be in the hands of what were now four of the biggest and most aggressive oil companies in the world.

But before the merger was consummated, there would be four years of hard arguing, legal threats, backroom meetings and thousands of miles of travel by the executives of the companies involved. For there was a snag—and it brought Calouste Gulbenkian, Mr. Five Per Cent, back into the story, angrily waving a map of the Persian Gulf area with that famous red line drawn around it. Because, of course, Standard of New Jersey and Socony-Vacuum were still signatories of the Red Line Agree-

* *History of Standard Oil Company (New Jersey)*, Vol. III, *New Horizons, 1927–1950*, by Henrietta M. Larson, et al.

ment, and it applied to Saudi Arabia, which had once been part of the Ottoman Empire. This meant that neither company could join Aramco without permission from their fellow members of the Iraq Petroleum Company, and though this might be granted by some of them, their number did not include Gulbenkian; he would want his inevitable 5 percent of any new deal that the Red Line signatories made. How to circumvent him?

Thanks to the Red Line Agreement, Calouste Gulbenkian had prospered and was now well on the way to becoming the richest man in the world. From IPC and other enterprises in which he was involved, he was receiving at least $20 million a year. In the 1930s he and his wife had moved to Paris, and there, except for keeping up appearances, went their separate ways. Theirs had been an Oriental marriage arranged by their families, and Nevarte Gulbenkian, plump and pretty, adored by her brothers, had never really liked the homely, humorless husband who had been forced upon her. Gulbenkian gave her a house on Avenue d'Iéna and visited her there at least once a day, usually for the midday meal. Once a week he carefully checked her household accounts of how she had spent her allowance, and she often had difficulty disguising the sums spent on presents for her succession of lovers, but there was never any question of a divorce between them. The supposition that they would remain married was so strong that Gulbenkian housed his formidable art collection in the Avenue d'Iéna house, and often brought collectors and curators from all over the world to see his paintings and sculptures there.*

However, Gulbenkian preferred to live in his permanent suite at the Ritz,† and it was there that he indulged a sexual appetite that was almost as voracious as that of King ibn Saud, though he did not require anything like as many women to satisfy it. As the years went by, however, his mistresses became younger and younger until, toward the end, nothing but a Lolita would stimu-

* They included some cigar-box nudes of the type oil millionaires seem to like, but also, thanks to his art adviser, Sir Kenneth Clark, Cézannes, Renoirs, Monets and several works by Degas.

† He was a director of the company which owned the hotel.

late him. "He used to say, and the late Lord Evans agreed in this," wrote his son, Nubar, "that while it is very unkind on a young girl to have sexual relations with an old man because she loses her youth, it does rejuvenate the sexual functions of an old man. This was always recognized in the harems of the East and even today at the court of King ibn Saud of Arabia, or that, until his death, of the Pasha of Marakesch, the harem contains one or two young girls just past the age of puberty who are kept, and regularly replaced, for precisely that purpose." *

Gulbenkian was miserly with his mistresses until such time as he was sick of them, when he pensioned them off generously to keep them quiet. As an example of his determination to keep hold of his possessions, his son tells the story of a rolling pin which Gulbenkian had obtained for his Turkish cook with which to roll out the wafer-thin Oriental pastry he was fond of. When the cook departed to go to work for the Turkish ambassador in Moscow, he took the rolling pin with him. It was worth only a few pennies, but Gulbenkian wanted it back and sent two detectives to interview the chef in Moscow and demand its return. He got it.

Though Nubar was his most trusted assistant and had acted as his emissary and negotiator in some complicated deals—including the Red Line Agreement—Gulbenkian continued to treat him as a wayward boy, and would stop his allowance or dock part of it if he was dissatisfied with his son's caprices. Gulbenkian was both mean and greedy, and ever watchful of his erstwhile partners in the oil business lest they try to do him in—as, admittedly, they often did. He had few friends, and none of them was in the oil business. "Oil friendships are greasy," he used to say. He believed only in one thing: the power of money. Money came to him through the Red Line Agreement, and he clung to it as if it were his lifeline. But then he fell into a trap.

When France was overrun by the German army in 1940 Calouste Gulbenkian did not flee the country, but moved instead with Marshal Pétain's government to Vichy. After signing an arm-

* Gulbenkian, *Pantaraxia.*

istice with Germany, the Vichy administration subsequently broke off relations with Britain and came under Nazi domination. At once the British Custodian of Enemy Property declared that he was taking over Gulbenkian's interest in IPC and that of the Compagnie Française des Pétroles, since both of them were now subject to enemy control. Their shares were declared forfeit. There was little that the French could do about it, but Gulbenkian was furious when he heard the news, for he held an Iranian as well as a British passport, and was free to move wherever he wished. In fact, in 1942 he moved to neutral Lisbon, and fought strenuously and successfully to regain his "rights" from the Custodian and be paid compensation for his loss of revenue.

But when, soon after the end of the war, Standard of New Jersey began to search for a way of breaking the provisions of the Red Line Agreement, which was holding up its participation in Aramco, the wartime confiscation suddenly pointed the way. Jersey consulted its legal expert, who advised that in his opinion "the fact that two of the owners of the Iraq Petroleum Company —Compagnie Française des Pétroles and an individual, Calouste Sarkis Gulbenkian—had come under enemy domination by the German occupation of France in 1940 had terminated the working agreement of 1928. As a result, the counsel held, the Red Line restrictions were no longer in force." * To obtain a broader judgment, the head of Jersey's law department, Edward F. Johnson, also consulted several eminent lawyers in London, including a famous barrister named D. N. Pritt. "He wanted their advice concerning the position that might be taken on this issue by the British courts, which under the bylaws of Iraq Petroleum had jurisdiction over matters in dispute between its owners. These British lawyers were unanimous in the opinion that the Red Line Agreement was no longer binding on the signatory companies." †

Looking for other allies to bolster its case, Jersey found one in the U.S. State Department. From officials there it secured an as-

* Larson, et al., *op. cit.*
† *Ibid.*

surance "that the Government would not support a new agreement incorporating the old restrictive features" and was against "the hobbling of American nationals by such devices as the Red Line Agreement." The Department stood for "what was essentially an opendoor policy in the Middle East, but with safeguards for the interests of the producing countries." *

Armed now with some heavy ammunition, Orville Harden, a vice-president of Standard of New Jersey, and Harold F. Sheets, an official of Socony-Vacuum, arrived in London with instructions to sink all opposition and wipe out the Red Line Agreement once and for all. They had only slight trouble with the two principal British signatories to the agreement, Royal Dutch–Shell and Anglo-Iranian, for they had also consulted counsel and agreed that the war had terminated the Red Line Agreement.

But Anglo-Iranian knew that once Standard and Socony-Vacuum were part of Aramco, the company would open all stops to get oil flowing, and the subsequent American flood might well affect Anglo-Iranian's own growing sale of crude oil on the world markets. Orville Harden had already been primed on how to deal with this, however, and what followed was a good example of the master strategy of which Standard of New Jersey was capable when its board got down to serious planning. First, Harden negotiated a deal with Anglo-Iranian whereby his company would purchase from the British a large amount of crude oil (it was eventually fixed at 110,000 barrels a day) for twenty years beginning in 1952. He also agreed to his company's participation in a new corporation to build a new pipeline from Iran and Kuwait to the Mediterranean, if the building of such a line was found feasible. These arrangements more than appeased Anglo-Iranian's anxieties about future competition and ended any doubts they might have had about the elimination of the Red Line.

From Standard of New Jersey's point of view, the arrangement was extremely satisfactory. It not only took care of British objections to its plans; it also enabled Standard to proceed with a

* Quoted in *ibid.*

cherished scheme of its own to establish itself solidly in the British petroleum market for private cars.

In this period of postwar austerity, Britain's gasoline supply was still under government control, and private motorists were stringently rationed. Whenever public opinion and the press demanded an end to the rationing system, the Labour government cited as its reasons for refusal the fact that the bulk of the gasoline would have to come from American sources and would have to be paid for in American dollars, of which Britain was desperately short.

These statements rankled Howard W. Page, the resident Jersey director in London. "It made the American companies look like the bad boys," he said, "and it was hurting us because the public thought we were the dirty dogs who were keeping their family cars off the roads." So while nursing a broken ankle from a skiing accident, Page worked out a scheme.

"It took me two months to work it out with New York," said Page, "but thanks to government help, I got it through the British Parliament in two weeks. It was a very simple scheme. With any increase in consumption of our oil, we announced, we would take payment in 100 percent sterling, and we would use that sterling for the purchase of British goods for use around the world. After the government's statement that ours was dollar oil and theirs was sterling oil, and they could only pay for sterling oil, this made ours sterling oil by the definition of the term. The Labour government was furious at first. They sent guys from the Treasury to try to explain that it wasn't as simple as that, but I told them I didn't understand what they were talking about, and they finally gave up. So did the government. They knew and I knew that 99.8 percent of the people were for the scheme and only .2 percent of politicians and black marketeers were against it, and it had to go through. The scheme was officially accepted, and petrol rationing ended in Britain. Of course both the Labour government and the Tories took the credit for it, but we didn't mind that." *

* In a conversation with the author.

Breaking the Thick Red Line

By making its agreement with Anglo-Iranian, Jersey now didn't even need to supply Britain with dollar oil because it had arranged for ample supplies from a sterling source. So Anglo-Iranian's agreement to waive the Red Line restrictions had brought them a double benefit.

Since Royal Dutch–Shell had also waived its rights under the agreement—in return for similar crude-oil contracts from Socony-Vacuum—there were now only the French and Calouste Sarkis Gulbenkian to deal with. Gulbenkian was permanently ensconced in a hotel in Lisbon, but his son was authorized to act on his behalf. Nubar was in Paris conferring with Victor de Metz, president of the Compagnie Française des Pétroles, the other main shareholder in IPC and signatory of the Red Line Agreement, when the two men were asked to come to London for an urgent meeting with Harden and Sheets. There they were told that the U.S. Department of Justice considered the Red Line Agreement of 1928 a restrictive agreement, contrary to American antitrust legislation, and that they therefore no longer considered themselves bound by it. Nubar at once pointed out that what the Justice Department thought about the agreement was hardly relevant, since all the signatories had agreed that any dispute over it would be settled under British law.

The Americans then changed their tack. From their briefcases they produced the opinions of various eminent British counsel that owing to the occupation of France, both Gulbenkian and the Compagnie Française des Pétroles had come under enemy domination, and agreements with them were no longer binding.

De Metz looked at Nubar Gulbenkian, who seemed quite unfazed by this thunderbolt. "*On va voir,*" said De Metz softly, and the meeting was adjourned. Some days later, through their legal advisers, the French and Gulbenkian formally demanded a share under the Red Line Agreement in Jersey's and Socony-Vacuum's projected interest in Aramco. When this was refused they announced in February 1947 that they were starting legal proceedings "to obtain confirmation of the validity of the 1928 agreement and a declaration prohibiting [Jersey and Socony-Vacuum] from obtaining an interest within the Red Line Agreement independent of the other owners of Iraq Petroleum." *

The news created something like panic not only in Jersey's board room in New York but also at Aramco, all of whose plans were now threatened. In March 1947 Jersey and Socony were forced by the impending action to work out "standstill" agreements with Aramco and its owners. "The purchase by the two American companies of participation in Aramco and Tapline was to be held in abeyance pending clarification of the legal issues." †

Aramco had the worst headache of them all. In anticipation of the sale of stock to the two newcomers it had gone ahead with its plans, and it simply did not have enough money to proceed on its own. However, after several emergency meetings, "The two prospective buyers agreed to guarantee a bank loan of $102,000,000, this being the total amount they were to pay directly for their shares in the company. They also worked out with Aramco and its two corporate owners the other terms of the purchase, which included their forgoing of specified amounts of dividends for a period of years. At the same time they arranged to purchase oil from Aramco in the interim and to guarantee their proportionate share of a loan for Tapline of $125,000,000." ‡

Still, the first round had gone to the French and the Gulbenkians.

From this point onward, the sheer number of backroom meetings, bribes, threats and blandishments makes it impossible to go into them in any detail here. The American and the British signatories wished at all costs to avoid legal action over the Red Line Agreement, for they knew that the Gulbenkians' lawyer, Sir Cyril (now Lord) Radcliffe, was a brilliant performer in this type of

* Larson, et al., *op. cit.*
† *Ibid.*
‡ *Ibid.*

case,* and they feared the dirty domestic linen that might be aired, thanks to the material which Nubar said his father could give him.

Therefore, the first objective was to divide the opposition. These were lean postwar years for the French and they needed to sell as much oil as possible to boost their economy. At this time their only oil came from the wells of the Iraq Petroleum Company. Suddenly the French heard reports that IPC proposed to cut its production, which meant that Aramco and Anglo-Iranian would benefit at the expense of IPC, thus drastically reducing France's income at a moment when she desperately needed it. But there was nothing the French could do about it, since their British and American partners outvoted them.

At the same time that this threat was waved over their heads, however, the French were offered a carrot. Since they had lost their shares in IPC to the British Custodian of Enemy Property during World War II, they had, naturally, lost the revenues which had since been earned from them. The astute Gulbenkian, claiming that he had been deprived under false pretenses, had got his back because there was a genuine legal doubt in his case. With the French, there was no doubt whatsoever that the seizure had been perfectly legal. On the other hand, the British and American shareholders indicated their willingness to advise the British government to make a gesture to the French—provided, that is, that the French were willing to co-operate in return . . .

At this stage the French government sent to London a new representative from the Compagnie Française des Pétroles with instructions to make a compromise with the British and Americans—with Gulbenkian's agreement, if possible, but without it if the Armenian remained adamant. It so happened, however, that this French emissary, Robert Cayrol, had enjoyed the friendship

* In a case which Nubar once brought, Radcliffe compelled the defendant (who happened to be Gulbenkian *père*) to produce 987,000 documents weighing a ton. Nubar was suing for an increased allowance from his parsimonious father. He got it.

of Calouste Gulbenkian for many years and got on extremely well with his son. It was Nubar who persuaded him not to hurry matters. "The time when others are pressing you for an urgent decision is the time to take it slowly," he said, quoting his father. It was the Americans who desperately needed the agreement, he pointed out, and every delay was costing them money—ideal circumstances in which to drive a hard bargain with them.

"We were both busy men so that most of these discussions [with Cayrol] took place at the end of the business day," wrote Nubar later. "Cayrol used to come round at about seven o'clock—to the Ritz when I was in London or to the Georges Cinq when I was in Paris—and we would have a good dinner together. There was no question of ice-cold negotiations but very friendly ones in which we sought a solution acceptable to all parties; true, we were careful to drink the same amount of wine to ensure parity. Session by session we worked gradually towards a basis for agreement until at last we reached the stage at which we felt it worth while to bring in the others." *

By this time the Gulbenkians had realized that there would have to be some relaxation in the restrictive powers of the Red Line Agreement, but both they and the French were determined to make it plain to the Americans that release from its provisions could only be obtained at a stiff price. So long as that price was not forthcoming, the threat of legal action would persist; in fact, the Gulbenkians instructed Sir Cyril Radcliffe to proceed with preparations for the case as if it were going through, and a date was actually set in the Law Courts in London for the hearing. In the meantime, Nubar and Cayrol worked on a formula that would produce for both of them *more* rather than less revenue when the Red Line was finally erased.

At last, in November 1948, negotiations with the British and Americans had reached a stage sufficiently advanced for all parties to agree to assemble in Lisbon, where, the Americans hoped, an agreement would finally be signed and they could join Aramco. The meeting place was the Aviz Hotel, a rococo palace with

* Gulbenkian, *op. cit.*

chandeliers hanging from high ceilings, statues of maidens crouching between marble pillars, walls tapestried with hunting scenes, guests (usually in exile) with noble titles, and servants who spoke only in whispers. Calouste Gulbenkian lived in suite number 42, and though the mountain had come to Mohammed for this meeting, Mohammed did not deign to put in an appearance but sent the oilmen messages through his emissary, Nubar. Cayrol was there for Compagnie Française des Pétroles, Orville Harden and Howard Page represented Standard of New Jersey, Sheets was there for Socony-Vacuum, and Morris Bridgeman for Anglo-Iranian. There was also a representative of Royal Dutch–Shell. All of them were anxious to have the new agreement signed as quickly as possible, because the Gulbenkians' action was due to come up in court at any moment, "and it was touch and go whether the case would open in the Law Courts in London before agreement was reached formally in Lisbon." *

The agreements were finally being typed out and were to be ready for formal signature at seven o'clock in the evening. It was Sunday, November 14, and since the Gulbenkians' English lawsuit was due to begin the next morning, arrangements had been made to telegraph London and announce a settlement as soon as the papers were signed. "It was at five minutes to seven," wrote Nubar, "that father found one more point which had not been covered by the Agreements. To say that there was consternation on the faces of all the men gathered there would be a piece of English understatement. But father was determined. Telegrams were sent to London, where the unfortunate Boards of the Groups involved had been kept waiting on tenterhooks for news that the Agreements had been signed. Now they must consider a new point and, in turn, send telegrams to Lisbon giving their acceptance or otherwise of the latest Gulbenkian demand." †

Nubar had ordered a dinner to celebrate the occasion, and had made sure that it was the best that the Aviz could provide.

* *Ibid.*

† *Ibid.* The author asked Nubar Gulbenkian what the last point was. He said he couldn't remember, but he thought it was a demand for an increase in IPC's output—the amount having been diminished to put pressure on the French.

A gay, dandyish, bearded leprechaun of a man, full of high spirits and a sense of fun, he was never one to take business setbacks too seriously, so he suggested to the others that they might as well eat while they waited for the telegrams from London.

"There were some twelve of us at that table," he wrote, "all men, with the sole exception of Cayrol's wife . . . No doubt she had looked forward as I had to a gay, convivial evening, but if ever there was a gloomy occasion, that was it. The meal was accompanied by long periods of silence, for no one was inclined to make the conventional efforts at conversation when all our minds were turned to what might then be going on in London. I realized hardly anyone was drinking, either, and my impression was confirmed the next day when I settled the bill for the dinner: just one bottle of champagne had been enough for twelve people." *

It was not until the early hours of the morning that word came through at last that the British and the Americans had conceded on Gulbenkian's last point. At two o'clock in the morning the agreements, newly retyped, were signed at last by everybody, including Calouste Gulbenkian.

Once more Nubar Gulbenkian called for the champagne, but it had been taken away and the Aviz kitchen staff had gone to bed, so instead the negotiators celebrated on sandwiches and cheap wine from an all-night café.

Nonetheless, there was just cause for celebration on everyone's part, even though it had cost the major shareholders in IPC a great deal of money. Henceforth all of them could pursue their development plans free of the restrictions of the Red Line, which now vanished from oil companies' maps. Anglo-Iranian and Royal Dutch–Shell had found new and lucrative markets for their crude oil. Gulbenkian and France had secured written assurances that under no circumstances would the Iraq field be restricted; on the contrary, its output would be considerably expanded. Gulbenkian was also given an extra allocation of free oil, to sell on the open

* *Ibid.*

market, as well as his 5 percent of the profits from Iraq and from the new fields in Qatar. These provisos would add at least $8 million to his annual income.

As for Standard of New Jersey and Socony-Vacuum, they were at last free to exercise their option in Aramco. Soon they and their two partners were masters of the richest oil quadrumvirate in the world. At long last oil was beginning to flow from the fruitful new field at Abqaiq by pipeline across the Arabian Desert to the Mediterranean, fourteen hundred miles away. A new refinery was in operation at Ras Tanura, and there were berths there for a great tanker fleet, which now began regular shipments to Africa and the Far East. By 1950, two years after the Red Line was erased, Standard of New Jersey's 30 percent share of Aramco Oil was 164,000 barrels daily, and its portion of the estimated reserves was no fewer than 2,800 billion barrels.

The future looked bright and everyone should have been happy, but two people were not. One of them was Nubar Gulbenkian. For three years he had used all the wit, charm and intelligence of which he was capable—and he had these qualities in plenty—to win his father's war with the giant oil combines. To be sure, in the moment of crisis, it was his father's nerve and tenacity which made the difference, but the son deserved some credit for his untiring work on Gulbenkian's behalf. Not only did he get no more than a perfunctory thanks from the old man, but when he left to return to London and asked for his bill, he discovered that he had been so free with his hospitality to the other delegates that the amount came to far more than the sum his father had allotted. He went to Gulbenkian's suite to complain, and finally persuaded his parsimonious father to allow him double his current expenses the next time he came to see him in Lisbon. Three weeks later Nubar returned to see his father. When he called for his bill at the end of this visit, he discovered that the price of his room had trebled. It was not until after his father's death that he learned that Gulbenkian owned the hotel.

King ibn Saud was also unhappy. He had watched from afar

the negotiations going on in London, Paris, New York and Lisbon, but no one had thought of consulting him in Riyadh. The future of his own country was being settled by foreigners and infidels, and he had played no part in it. He was old and in pain and beginning to lose his grip, but not so much as to blind him to the fact that he had been ignored. It was insulting, and he decided to react in the only way an infidel American would understand. He asked for more money, and invited another infidel to help exploit Saudi Arabia's oil and compete with Aramco. At which point an independent oilman named J. Paul Getty enters the Middle East story.

Part Four

ENTER THE INDEPENDENTS

PRECEDING PAGE: *J. Paul Getty at Khor al Mufatta in the Neutral Zone in 1955.*
PHOTO: JOHN WEBB/BROMPTON STUDIO

CHAPTER THIRTEEN

Neutral Zone

A FEW DAYS AFTER the big international oil corporations signed away the Red Line Agreement in November 1948, a small private airplane took off from a sand strip just outside the town of Kuwait and headed south toward the desert. It flew across the rolling sand sea for about fifteen minutes until it reached a range of small *jabals* that ran down to the Persian Gulf, and then it descended to an altitude of a few hundred feet. For the next two and a half hours, while the Bedouin tribes shook their fists and finally fired rifle shots at this noisy machine that was scaring their camels and women, it cruised around and around in circles. Finally, at a signal from the pilot to the man with the instruments beside him, the plane turned and flew back to Kuwait with a near-empty tank.

The name of the pilot has disappeared from the records, but his passenger was a clever geologist named Paul Walton. That night Dr. Walton went to the Kuwait post office and cabled his boss: STRUCTURES PROMISING.

It was on the strength of this message that Jean Paul Getty entered the Middle East oil business.

In 1948 J. Paul Getty had reached a turning point in his life.

A tall, gaunt Midwesterner from Minneapolis of Scotch and Irish descent, he had the long, dew-lapped face of a sad-visaged lion. His melancholy expression was due more to his private life than to any doubts about the oil business. His fifth marriage was going through a separation that would eventually lead to divorce, but he had reached the glum conclusion sometime earlier that he would never have a stable marriage,* and when this one ended he was already resolved to make different arrangements for female companionship. So far as the oil business was concerned, he had every reason to be pleased, even if he was not satisfied. At fifty-six, he was not yet a billionaire, but he had already shown the toughness and skill needed by an independent to play in the high-level poker game that was the U.S. oil business. The big major companies and the successful independents have always fought with every weapon at their command—secret boycotts, squeezes and market manipulations—to keep newcomers out of the competition, but Getty had beaten them all at their own game, and without giving away any of his overall control he had become master of two of the biggest independent operations in America, Getty Oil Company and Tidewater Oil Company. Even at that time the two of them were worth nearly $400 million.

There were those who thought of Getty as another Calouste Sarkis Gulbenkian, but there is no comparison between their respective roles in the oil business. It is true that both had fathers who had launched their sons into the oil business world with generous gifts.† Each collected expensive art and had a taste for pictures and sculptures of plump nudes. They also shared an intense interest in women ("A man's driving force is sex," Getty once said), a lifelong search for bargains of every kind, and a miserliness that compelled Gulbenkian to check his wife's house-

* "I blame my business interests for having been married five times," he declared later. "A woman resents a man being dedicated to his business. In fact, she resents anyone dedicated to anything but herself." Quoted in the *Sunday Express*, London, 1957.

† George Franklin Getty, a lawyer, accepted strips of Oklahoma Territory in lieu of fees and found himself landlord of an oil field. He launched his son with a gift of $1 million when he was twenty-one.

hold bills (Getty also put in pay telephones for his guests and rode around in four-year-old cars).

But Gulbenkian was a middleman, a fixer, the graduate of Oriental bazaars whose real interest was not oil, but the money it would bring him. Getty is an oilman who likes the smell and feel of it. Not only has he never grown rich on 5 percent of other people's earnings but he has always been in control of every enterprise in which he has been involved. He prefers to compare himself with John D. Rockefeller, a visionary and innovator as well as an entrepreneur who saw new ways of using oil before other men had thought of it. Like Rockefeller, he has kept a close check on every enterprise in which he has ever been concerned, and has explained his parsimony by citing the stern mother who nagged him all through his formative years to remember that "he who wastes not wants not!"

The end of World War II had found this humorless, lugubrious, lonely man walking around with what an acquaintance described as "a look in his eye as if something is missing from his life." It was not a woman, for he could always find plenty of willing companions, even though he would never waste a cent on them.* What he was missing was a challenge which would lead to the kind of encounter he most enjoyed: meeting a big business corporation head-on and winning the subsequent battle.

Getty had been looking for a stake in Middle East oil in Iran, Iraq and Egypt. In the first two countries the major oil companies had already got there before him, and in Egypt he did not think much of the prospects. Then one of his paid scouts quietly let him know that once Aramco had consolidated its setup with Standard of New Jersey and Socony-Vacuum, it was planning a major reshuffle of its Saudi Arabian concessions. Besides the areas where the company had leased oil rights at Dhahran, Abqaiq and Abu Hadriya, Aramco also held the concession in the Neutral Zone between Saudi Arabia and Kuwait (this area having been part of

* "Surprisingly, he is often like a little boy and brings out the mother instinct," said one of them.

the sale made to them by Major Frank Holmes back in the 1920s). Now they made a deal with King ibn Saud in which they relinquished all rights in the Neutral Zone in return for an offshore concession running out to sea from their Dhahran field. From Aramco's point of view it was much more convenient, economically and administratively, to keep their concessions centered around one refinery and tanker port, and the Neutral Zone had complications.

The main snag was the fact that these two thousand square miles of rocky and inhospitable desert belong not to one country, but to two. When the British high commissioner, Sir Percy Cox, visited Ibn Saud at Oqair in 1922, one of the agreements he secured between Arabia and the neighboring territory of Kuwait was the establishment of a neutral zone along their frontiers, where tribes from both would retain their grazing and watering rights, and no forts would be built. This meant that both King ibn Saud and Sheik Ahmad of Kuwait had a say in what happened inside the zone, and Aramco figured that there could be friction.

With the concession in the Neutral Zone relinquished, each ruler had the right to redispose of half of it for the highest price he could get. The ruler of Kuwait got into the act at once by announcing that his 50 percent would be auctioned off to the highest bidder, and that no "derisory" offers of the type which had won prewar concessions would be considered. An American syndicate called Aminoil came forward at once with an offer which was so generous that the major companies decided not to compete, and the bid was accepted. The head of the syndicate, Ralph K. Davies, was convinced that in the future, high prices would have to be paid for concessions and succeeded in persuading the fellow members of his syndicate to go along with him.

They were shocked at how high the price turned out to be. For a 50 percent share in the Neutral Zone's potential oil over a sixty-year term (and there was no positive proof yet that it was there in quantity) they promised \$7.5 million as an immediate down payment to Sheik Ahmad, plus an annual guarantee of at least $625,000 in royalties. This was far more expensive than the concession in Kuwait itself for the whole of which Kuwait Oil Company had paid out $170,000 in 1934, plus an annual royalty of $35,000. It was also different from Al Hasa in Saudi Arabia, for which Aramco had paid $200,000. Moreover, Sheik Ahmad exacted several conditions which looked onerous to Aminoil and outrageously exorbitant to their indignant major oil-company rivals, who immediately spread the word that the new syndicate was ruining the market. In addition to a royalty payment which was twice the Kuwait rate, Aminoil agreed to pay a 12½ percent royalty on all natural-gas sales, a tax of 7.5 U.S. cents per ton of crude oil in lieu of taxes, promised to construct a refinery and give the sheik a 15 percent interest in it, to run an educational program for Kuwaiti employees of the company, and finally to build a new hospital in Al Kuwait, the capital.

Moreover, since Aminoil's purchase was only 50 percent of the concession, no work could be started until the Saudi Arabian half was sold. Who would be prepared to pay the price that King ibn Saud would ask, which was bound to be at least as much as Aminoil had paid to the ruler of Kuwait? While the major oil companies were swallowing hard, and the main contender, Royal Dutch–Shell, was deciding whether it coud afford it, Paul Getty sent his lawyer, Barnabas Hadfield, to Jidda with instructions to buy the concession from the king no matter how much it cost. It cost plenty.

First, there was a down payment of $9.5 million. Second, there was an advance on royalties of $1 million, payable yearly, again for a sixty-year term, and not returnable if royalties failed to reach that sum. There was a royalty payment of 55 cents per barrel (Aminoil was paying 35 cents), a pledge to deliver crude oil or kerosene or aviation gas up to 100,000 gallons free of charge to the government, an agreed program for schools and education, and an appointment of the king's own delegate to the company's board.

At their headquarters in London, New York and San Francisco, the directors of the world's major oil companies read the terms with dismay. "This could change everything in the Middle East," said Howard Page, of Standard Oil of New Jersey.

Indeed it could. On February 20, 1949, King ibn Saud signed an agreement with J. Paul Getty's Pacific Western Oil Corporation (he had bought an 82 percent control of the company after his father's death) giving him his half-share in the Neutral Zone concession, and in return was handed a check by Barnabas Hadfield for $10.5 million,* a sum sufficiently high to make even a king thoughtful. If that amount of money could be handed over by a one-man company even before it had brought up a barrel of oil, why had Aramco paid him only $28 million in 1948, a year when oil was flowing out of Al Hasa as never before, and two of the richest companies in the world were now partners of Aramco? Why was Aramco paying him a royalty of only 21 cents a barrel when Getty was willing to pay 55 cents?

Since, as usual, the Saudi Arabian treasury was empty and debts were mounting rapidly, the king instructed his minister of finance, Abdullah Sulaiman, to summon Fred A. Davies, the Aramco president, to an emergency meeting. He wanted an answer to his question, and more money—much, much more money. A year later he had it: $50 million more. But thanks to the ingenuity of an American corporation lawyer, it was not Aramco that paid the bill, but the U.S. Treasury.

EVEN A SEMIBILLIONAIRE cannot just reach into his safe and pull out over $10.5 million, just like that. Since J. Paul Getty's assets were tied up in his oil and business interests in the United States, he had to go to the banks for a loan to pay for his Neutral Zone concession. Therefore he needed the money back as quickly as possible, and this meant getting wells drilled and oil flowing.

* Being the down payment and the $1 million advance royalty payment mentioned above.

But it wasn't as easy as that. There are not many more formidable spots in the world than the Arabian Desert in midsummer. It becomes so hot that a man who stays in the sun too long can boil to death, his blood vessels dehydrating in the intense heat. In the Neutral Zone there are only a few wells, and no underwater reservoirs which can be tapped, as there are at Al Hasa. Only flies, locusts, vipers and scorpions seem to survive the intense heat of mid-desert, and along the coast the humidity is so great that air conditioners spurt water and people bathe in sweat. When summer sandstorms come—and they are a feature of Kuwait—life is almost insupportable. Even winter, with its flash floods, spectacular hailstorms and sometimes bitter cold, can make existence a burden and a trial on the nerves.

Hence, the circumstances in which Getty found himself in 1949, and for the next few years, were not exactly propitious for the speedy return of his money. For a man of his parsimonious nature, the clause in the concession contract which nagged him most was the one that obliged him to pay $1 million annually to King ibn Saud, oil or no oil. Of course he had figured on delays and large financial outlays while water was piped in, communication lines constructed, and rigs, drills, crews and air-conditioned living quarters shipped in.

What Getty had not included in his calculations were incompetence, inexperience and stubbornness, and he soon discovered that his mandatory partners in the Neutral Zone concession were afflicted with all these vices. Aminoil, which held the other 50 percent of the concession, was a consortium comprising two individuals (Ralph K. Davies and J. S. Abercrombie) and eight American oil companies, the largest of which, Phillips Oil, held only 33 percent of the company's shares. Aminoil's president, Ralph K. Davies, owned only 8 percent, and had been nominated by the others as a front man because he was a lawyer, a former Standard of California vice-president, and a wartime controller of U.S. petroleum supplies. All these posts demanded a talent for successful negotiating in board rooms and government offices, and though

Ralph Davies was a charming and effective operator on his home ground, which was somewhere within walking distance of the Fairmont Hotel or the White House, he was less successful coping with the deviousness of Arabian oligarchs or making vital and costly decisions in the field.

Though Getty was complete master of his own half-share in the concession and could therefore speak more authoritatively than any member of his partners' consortium, he allowed Aminoil, through its president, Ralph Davies, to take control in the preliminary stages. It was a costly mistake. So was his decision not to visit the Neutral Zone himself but to send as his representative his eldest son, George Franklin Getty II. George Getty was twenty-five years old and had had a year in college after leaving the Army. He had neither the years nor the experience to make his presence felt, either with the Arabs or with Davies, who was apt to wave him away like a persistent desert fly. To backstop his son, Getty had dispatched to Kuwait his two geologists, Paul Walton and Emil Kluth. From surveys they had made, they were convinced that a commercial oil field would be found in the western part of the zone, where there was evidence of Eocene limestone formations. They recommended that operations begin there, but Davies listened to his own geologists, who were pursuing deep-hidden oil along the same seams as the rich Burgan field, just over the line in Kuwait proper. Unfortunately, their first wells were east of the Burgan line and came up dry or full of salt water. Since they were deep wells, each failure cost $250,000.

In an effort to persuade Davies that this strategy was wrong, Getty flew in an expert geophysicist from Anglo-Iranian, but his advice was ignored. Had Getty himself appeared, his driving personality would almost certainly have prevailed and they would have "drilled where I wanted" and "found the field three years earlier." * But because of his tangled domestic affairs and stock-market manipulations he had to stay close to a telephone and teletype. He could neither call from the Neutral Zone nor send

* *My Life and Fortunes,* by J. Paul Getty.

cables except *en clair,* and so four years passed while Aminoil poured away their and Getty's money without tangible results.

By 1953 Getty had spent $40 million in the Neutral Zone without getting a cent in return, and there were persistent rumors in the oil world that he was ready to pull out. Toward the end of that year, however, a viable field was discovered inside the Neutral Zone, but about twenty miles from Burgan. It was where Aminoil's experts had always said it would be, and to that extent they were justified, but the wells that were subsequently sunk were deep, enormously costly to maintain, and nothing like the bonanza for which Getty had been hoping.

It was not until an unlucky incident brought George Getty home (he had run afoul of the Saudi liquor laws) that Getty senior was forced to visit the zone himself. He drove out from London by car with his close friend of the period, an Englishwoman named Penelope Kitson, and on arrival he galvanized everyone. He was scathing about the laxness of Aminoil's operations and administration; he stiffened the backbones of his own demoralized staff, and both impressed and terrified everyone who met him. "Even the jellyfish seemed to get out of his way when he took a swim in the Gulf," one oilman said. "He was the only man I ever knew who bathed every day and never got stung once. The jellyfish wouldn't have dared—Getty would have stung them right back."

As a result of this visit, the drilling tactics changed, rigs were set in new locations, and within a year the bonanza for which Getty had been hoping bubbled to the surface of the Neutral Zone. At depths of only 600 to 1,500 feet, in seams of Eocene limestone, oil was found in abundance. It was soon being pumped out at the rate of 16 million barrels a year, and the drills were biting into a vast oil reserve of approximately 13 billion barrels.*

It had cost J. Paul Getty $40 million to tap this new field, but it was to make him a billionaire at last.

* From a report of De Golyer and MacNaughton, Inc., Petroleum Geologists and Consultants, Dallas, Texas, March 20, 1958.

CHAPTER FOURTEEN

Fifty-Fifty

FROM POLES AND BASKETS IN THE BAZAARS OF DAMMAM, Riyadh, Al Hufuf and Jidda, the severed hands and feet of petty thieves still hung, flyblown and festering, in the blazing sunshine. They were not the only things that were rotten in the Saudi Arabian state. It was 1950 for the rest of the world, but the desert kingdom was a medieval despotism where sheikly intolerance and religious fanaticism still held sway. Aramco wives were startled when they joined a crowd in Dammam watching two black slaves beating a sack, and heard anguished female screams issuing from it; their surprise turned to horror when blood began to leak from the sack and they realized that they were watching a wretched woman being ritually beaten to death. She was inside the sack so that the male watchers would not see her face. There were still strokes of the lash for persistent nonattenders at the mosque, and for stealing a camel or someone's wife, a man would lose his head to the executioner's sword.

In the early days of Ibn Saud's reign the system had been a cruel one, but at least it had been egalitarian, and alongside the savage punishment for wrongdoers, there was also an honor system based on ancient desert chivalry. However, that was now beginning to decay and there was a smell of corruption in the air. The king was going downhill mentally and physically, and the princes and ministers, who sensed his imminent demise, were beginning to make "other arrangements." In this last period of the king's life, a Dutchman named Dirk van der Meulen, an Arabist and admirer of the mighty warrior-monarch in his heyday, came to visit him in Riyadh, and was received in audience at the palace.

"When I entered the hall I saw a change," he wrote afterward. "The King at the far end no longer rose from his seat in welcome. Only the guards in Western uniform saluted. The beduin Shaikhs near the entrance moved silently backwards to make room but gave no greeting. No one rose from his seat because the King could no longer do so. He sat in what seemed to be an invalid carriage. The spare, curled beard and the few locks of hair that peeped from his headcloth were black—but dyed . . . It was the voice that disappointed most. The voice was still kindly but the music had gone out of it . . . When I left the audience chamber of Ibn Saud I felt, for the first time, unsatisfied. The unfailing spring had failed." *

What had set the king on the road to indifference and death, it was said around the courts, was the failure of that strength of which he was as proud as his skill and bravery in battle—his prowess as a sexual contender. The wick had burned down. He had been stimulated for a time by a visit to Egypt, and had commented to his Anglo-Irish friend, Harry St. John Philby, "There are some nice girls in this country. I wouldn't mind picking a bunch of them to take to Arabia, say a hundred thousand pounds' worth." † He had in fact brought three or four back with him, and enlivened by new faces and new bodies, he had enjoyed an Indian summer of sex in which he fathered two more sons to add to his formidable brood of children. But they were the last. Sud-

* *The Wells of Ibn Saud,* by Dirk van der Meulen.
† Philby, *Arabian Jubilee.*

denly he could father no more, and this produced a decline, it was said in Riyadh, that nothing could halt.

In his feeble state Ibn Saud did nothing to check the corruption, waste and wanton profligacy with which his country's enormous revenues were now dispensed by his sons and ministers. All the ministers now had Lebanese and Syrian secretaries, whose only job was to siphon money out of their budget allocations and invest it for them in Beirut real estate or Swiss banks. So much Joy perfume (supposedly the most expensive in the world) was being imported into the country that one of the company's salesmen said, "They must be taking baths in it." They were.

New buildings were springing up all over Riyadh and Jidda, but they were jerry-built, rushed up by foreign contractors for huge profits. Several of Ibn Saud's sons had gone to San Francisco for the inauguration of the United Nations at the end of World War II, and had excited many California women with their doe-brown eyes and romantic Arab robes. They brought some of the ladies back, and picked up others in London and Paris, and most of them turned out to be high-priced courtesans who quickly got their fingers into the princely incomes. According to Harry Philby, one of them who married a prince, shortly afterward spent a million dollars on a European shopping spree.

Moreover, the princes themselves returned from the inaugural with none of the high principles which the United Nations declared as its ideal in those days. When asked what had impressed him most in America, one of the king's sons cited the miniature mermaid floating in a tank in the lobby of a New York night club. They loaded ships with Cadillacs and spares, with gallons of perfume, and cases of liquor. They imported hundreds of thousands of colored electric lights with which to festoon their houses and gardens. They hired French chefs, but found the food insipid and threw it away.

"As Riyadh increased in prosperity, more and more waste food was thrown into the streets," wrote Van der Meulen. "The army of mongrels grew in numbers and as a nuisance. Their nocturnal fights made life in town unbearable so it was decided to get rid of these pests. Since no Muslim will simply kill an animal because it barks, the dogs were not shot or poisoned, but in the desert at some distance from the town high, square mud enclosures were built. At the entrance of each enclosure was posted a town official who paid three Saudi rials for every dog delivered to him. When we visited these pounds the guards told us that between four and six thousand dogs had passed in. I looked through the crack between the gate and saw a host of fly-pestered dogs. For long I could not take my eyes off the suffering animals . . . When I asked one of the government officials about the dogs he told me that they were given water and old dates, and would doubtless die when their time came." *

But as Ibn Saud was discovering for himself, death does not always come quickly or easily, and he had to sit by, enfeebled and helpless, and watch the country he had single-handedly constructed go to pieces around him. Scandals involving the princes became the talk of the kingdom. At a party a British consul was shot dead by a drunken prince; two European women and a number of Saudi men died of poisoned alcohol at another princely celebration. The king ordered death for the culprits, but in each case the sentence was never carried out. He blamed the Americans of Aramco for bringing sinful alcohol into the country and ordered a rigid ban on its use, even in the privacy of the oil company's quarters, but what he did not know, and what no one dared to tell him, was that the princes themselves were the importers of liquor and made money on the side by selling it.

And for all Saudi Arabia's vast income from oil, money was scarce and the bulk of the people existed at starvation level. There were schools, but they had been built by Aramco. There were hospitals and good medical services, but they were provided and maintained by Aramco. The oil company had even acceded to one of the king's last whims and built him a railroad across the

* Van der Meulen, *op. cit.*

desert from Dammam to Riyadh. But where the oil company had no wells, offices or living quarters, there was poverty and disease. The money that might have gone into public welfare was squandered by the court under Ibn Saud's single lackluster eye.

The U.S. government has been blamed for having allowed Aramco to pour so much money into Ibn Saud's coffers while doing so little to see that it was wisely spent.

"It was also wrong that the money given to Ibn Saud was given with hardly a word of reliable advice on the spending of it," writes David Howarth in his biography of the king. "Aramco could not be blamed for that; it was only an oil company and its only reason for being in Arabia, or existing, was frankly to make a profit . . . [It] could hardly have been expected to act like a mandatory power, and that was what was needed. Advice was a job for a government." *

And Van der Meulen writes: "Are the Americans to be blamed for the scale on which the millions poured into one of the poorest countries in the world were swept out again into private hands? Could they not have done something about it?" †

The answer U.S. Arabists give to that is that advice was offered from time to time, and that those who gave it were rebuffed. Aramco made it known that they would have preferred to pay Saudi Arabia's share of the oil revenues into a government fund rather than the king's private purse, and were crisply informed by Finance Minister Abdullah Sulaiman that the king would be so offended by any attempt to do so that he might cancel the concession. These were the postwar years when Britain, France and Holland were giving up their colonial empires. It is easy to imagine what their comments would have been had the U.S. government stepped into an independent Arab state and announced that it would henceforth be managing its budget.

Aramco itself was in a difficult position. Most of its directors were oilmen who had worked their way up from the rigs. They

* *The Desert King: Ibn Saud and His Arabia,* by David A. Howarth.
† Van der Meulen, *op. cit.*

had developed a love of the desert and a certain paternal affection for the Saudis with whom they worked, and they were proud of their achievements in training, building and improving. To an overwhelming extent, any evidence of enlightened development along the Saudi shores of the Persian Gulf was their creation. They had not just drilled wells; they had built roads, a railroad and whole towns. They had dug irrigation ditches and made the desert sprout vegetables and fruit. They had brought education and medicine to a backward people, and wherever they worked the standards of literacy and health among the Saudis was higher than where the king's writ alone ruled.

But they were a company with dual loyalties. On the one hand they had to satisfy the demands of their four parent corporations —Standard of California, Standard of New Jersey, Texaco and Socony-Vacuum (now called Mobil)—for highly profitable returns. On the other, they had to keep the king supplied with more and more money. There was little they could do about the fact that it was promptly wasted in foolish extravagances or poured into the coffers of the swindlers around the court. What Ibn Saud's ministers wanted was not advice but more revenues, and the more they were paid, the more they spent and the deeper the kingdom slid into debt.

By 1950, after J. Paul Getty had paid $10.5 million for the concession in the Neutral Zone, the king's ministers became exigent.

"The Saudi Government had just entered into the contract [with Getty]," said Fred A. Davies, of Aramco, "and the terms were much better than ours. Our concession had greatly increased in value. We had developed a big reserve out there. They asked us as early as 1948. 'Isn't there some way in which we can get a greater take?' " *

Now they began demanding so much money that Aramco's lawyers in the United States believed that in order to satisfy them

* Testimony before Senate Committee, Joint Hearings on the Emergency Oil Lift Program and Related Oil Problems, 85th Cong., 1957.

the whole nature of the company's concession might have to be altered in order to provide it, and this was something they wished to avoid at all costs. Moreover, if made out of Aramco's revenues, the payments would have wiped out a large proportion of the company's profits, and the parent companies at home were in no mood to accept that. As Davies pointed out, they had shareholders to satisfy.

But how could a way be found to satisfy the greedy demands of the king's courtiers without dealing a drastic blow to Aramco's profitability? The legal staff of Aramco knew that in 1949 the company had paid $38 million to the Saudi government. In the same period the company had given $43 million to the U.S. government in income taxes. The fact that the Saudis were receiving less from their own mineral wealth than a government thousands of miles away was something which Aramco officials had kept discreetly quiet until now, but suddenly, in 1950, on the advice of the company's lawyers, the figures were leaked to the Saudis. The result was, in the words of Fred Davies, that "they weren't a darn bit happy about it." * Soon they were asking the question that anyone with Aramco's profitability at heart would have wanted them to ask: "Isn't there some way in which the income tax you pay to the United States can be diverted to us in whole or in part?" †

At this point the company suggested that the Saudi government consult the U.S. Treasury. Aramco had already discussed its problems with George C. McGhee, of the Treasury Department, who, according to Davies, "appreciated our difficulties." The result had been the dispatch to Jidda of a Treasury Department official, George A. Eddy, who had conferred with Saudi officials about their money problems. When asked a direct question by a Saudi official about how more money might be raised from "foreign firms," Eddy had first consulted the U.S. ambassador in Jidda as to whether he might answer the question, and

* *Ibid.*

† Quoted by Fred A. Davies, *ibid.*

then, given permission, had pointed out that several methods were available. In the case of an oil firm like Aramco, one of them was to demand an increase in royalties on oil produced; the other was to institute an income-tax system and get more money from the company by direct taxation. Eddy added: "I did explain to him [the Saudi official] the difference of the effect on the company of a royalty and an income tax." *

By this he meant that if the Saudi government simply increased the amount of royalty it was receiving from Aramco per barrel of oil, it would have a direct and damaging effect on the profits of the company. If, however, the Saudi government were to start an income-tax system, any such money paid to them by Aramco could, under U.S. law, be deducted from the amount of tax the company was liable for in the United States.

Eddy returned to the United States and was succeeded in Saudi Arabia by a Washington lawyer, John F. Greaney, who subsequently drafted an Income Tax Law for the Kingdom of Saudi Arabia which was instituted by royal decree on December 26, 1950. It came a day late, but it was still a munificent Christmas present for both the Saudi government and Aramco. Under the last year (1950) of the old system of payment by royalty only, Aramco paid $56 million to the Saudi government. In the first year of the new system of royalty, lease and income tax combined, the government got $110 million.

From Aramco's point of view, the nicest aspect of the new system was that it didn't cost them a penny. They simply wrote off their Saudi Arabian taxes against their liabilities for U.S. tax.

> SENATOR MCHUGH: The net result is that the United States Government now gets nothing in the way of income tax. Is that correct?
>
> F. A. DAVIES: They haven't received any taxes from us for two or three years, I guess.†

Soon governments with oil concessions all over the Middle East were adopting this method of payment. It was called the

* *Ibid.*
† *Ibid.*

50–50 system, since it gave the government of the producing country approximately half of the oil-operating company's earnings, and in the beginning everyone was more than satisfied with it. Except, of course, the government of the United States, which now began to lose something like $200 million in tax revenues every year from American oil companies operating in the Middle East.

ON NOVEMBER 9, 1953, King Abdul Aziz ibn Saud, Lord of Arabia, died in his sleep in his palace at Taif, a hill station forty miles east of Mecca. As Harry St. John Philby was later to remark, there were few people around, besides himself and one or two other European Arabists, who mourned his death. The king had been one of the heroic figures of Middle East history, and had carved the country of Saudi Arabia out of the desert with his own sword. But the man who had led the tribes to victory in battle had allowed them to relapse into squalor thereafter, and there were few ordinary Saudis who could be said to have benefited from his reign. They had more to thank Aramco than the king for any progress they may have made.

As for his sons and his venal ministers, they had long since turned away from him to more profitable sources of influence. "A hush fell on his palaces long before his death," wrote his biographer. "The throngs of visitors and servants and supplicants diminished, dwindled, lost their urgent air. An air of stealthy secrecy replaced it because there were so many things that had to be hidden from him." *

The king's body was taken by night to Riyadh and buried quietly, with no ceremony, according to the custom of Wahhabi Muslims.

Ibn Saud was succeeded on the throne of Saudi Arabia by his son Saud ibn Abdul Aziz. King Saud made a speech to his people on the day of his elevation in which he said: "My father's

* Howarth, *op. cit.*

reign may be famous for all its conquests and its cohesion of the country. My reign will be remembered for what I do for my people in the way of their welfare, their education and their health."

"What he [the new king] has been doing in the past few years bears that statement out very much," said Fred Davies. "He's building schools, he's building homes, he's building mobile clinics, he's building clinics that aren't mobile all over the country. He's building roads, water-wells, developing agriculture, he's developing the mosques. Too much is said here in the magazines about the gold-plated Cadillacs. I've not seen any gold-plated Cadillacs. I've seen Cadillacs but I've seen lots of Cadillacs in Texas . . . I want here to subscribe very heartily to the program that His Majesty is putting over in his country to scotch ideas that there is nothing but waste there. There are many good things being done there, and each semester, each year you can see great increases." *

At the time the Aramco board chairman was speaking—and he must have known about it—the new king had surpassed anything his father had achieved by running Saudi Arabia into debts totaling just under $500 million. He had a harem that in numbers far outstripped that of Ibn Saud, though he may not have serviced it so efficiently. He had a vast army of hangers-on at court who were allowed to indulge in the wildest extravagances. In short, the new king, to whom Aramco's chairman had paid such eloquent tribute, was profligate, decadent and an unmitigated disaster for his country.

* Senate Inquiry 1957, *op. cit.*

CHAPTER FIFTEEN

The Man in the Pajama Suit

EARLY IN MAY 1950 the remains of Iran's shah-in-shah, Reza Khan, who had been toppled from his throne by the British and Russians during World War II, arrived back in his homeland from Johannesburg, South Africa, where he had died in exile. Though he had done much during his reign to regenerate Iran, the old shah had never really been popular with his people; still, his reburial was made the occasion for elaborate ceremonials. Detachments of Pakistani, Iraqi and Turkish troops marched with their bands past the decorated bier in the wake of a great procession of men, guns and tanks of the Iranian army. The streets of Teheran were packed with spectators staring in superstitious awe at the imperial coffin as it was trundled on a caisson drawn by four black horses to its final resting place on the outskirts of the city. For the most part the crowd watched in silence, but when the British representative was sighted in the delegation of diplomats marching behind the coffin, voices rose and fists were shaken, and there were cries of "Our shah has come back! Now give us back our oil!"

Suddenly, among the corps of marching members of the Majlis, the Iranian parliament, there shuffled by a tall, thin, bent old man with a haggard yellow face and a dripping nose, which he did not bother to wipe. Great cheers rose from the throng. This time the crowd began crying a name: "Mossadegh! Mossadegh! Mossadegh!" They were hailing the leader of a newly formed coalition party, the National Front. Dr. Mohammed Mossadegh was popular with the crowd because the main plank in his political platform was to wrest control of Iran's petroleum resources from the British.

Despite the fact that those resources were now among the richest in the world, in 1950 Iran was still a backward country racked by disease and corruption, manipulated by squalid politicians and landowners, and ruled by the young Shah Mohammed Reza Pahlevi, who was weak and vacillating—far different from the arrogant and self-confident monarch he was to become. Only in the southwest, where Anglo-Iranian exploited the nation's oil, could it be said that rewarding work and a decent standard of living had been given the populace. In Abadan, in Ahwaz, in Masjid-i-Sulaiman and all the other communities which had grown up around the wellheads, refineries and tanker ports, the British company had built houses, hospitals, schools, drains for sanitation, roads for communication, clubs, movies and playing fields. As a commission from the International Labour Office reported after a visit in 1950 to Anglo-Iranian's operations: "The observer cannot fail to be impressed by the vast number of modern houses and amenities which the Company has been able to provide in a comparatively short time, in spite of exceptionally unfavourable circumstances." *

But the British had discovered, not for the first time in their colonial history, that these bounties did not make them loved by the people to whom they were given. Nationalists know no gratitude, and the fact that it was British know-how, guts, persistence and money which had found the oil in the first place left them unimpressed. Why not? The British had already earned an enormous return from their outlay of muscle, brains and finance.

* Quoted in Longrigg, *Oil in the Middle East.*

To the pioneers all honor and a just reward—but why go on pouring out millions to men who had come in only after the spadework was done, and to shareholders for whom Iran was only a name in an annual report? The nationalist politicians lusted after the vast fortune Anglo-Iranian was making. The workers, especially the trained fieldmen and office staff, yearned for promotions to the higher echelons which they believed were only denied them because the British held all the important jobs.

To the Iranian people it seemed that by taking over Anglo-Iranian all their problems would be solved; there would be jobs galore, money would be plentiful, and their sense of humiliation would be wiped out because their resources would be back in their own hands. Hence, they had been listening avidly to any politician who, as Mossadegh did, advocated nationalization of Anglo-Iranian as the universal panacea, and who did not hesitate to call the British crooks for the way in which they were milking the company's earnings at Iran's expense.

Mossadegh had a point. If Anglo-Iranian was not exactly cheating its host country, it was certainly paying bargain prices for its enormous harvests of oil. The concession agreement which it had made with the old shah was based on a royalty payment for every barrel of oil extracted; the amount paid to the Iranian government for 1950 was £16 million. In addition to this, the company sold petroleum to the Iranian government at exceptionally low prices; the government in turn sold it to the people at high prices, earning it another £7 million. Further, through Anglo-Iranian's intercession with the British government, the Iranians were allowed to manipulate the exchange rate between sterling and Iranian riyals, and thus earn themselves another £5 million. In this and other minor ways, the Iranians received from the British about £32 million, which constituted practically half of the Iranian budget.

But—and this was what the Iranians and their friends resented most—Anglo-Iranian was earning nearly five times as much from the Iranian oil fields, and paying over £40 million in taxes to the British government—£8 million more than Iran was getting from its own oil. This was a glaring inequity, and the British were later to accuse the U.S. ambassador to Iran, Dr. Henry Francis Grady, of "gross disloyalty" toward them by pointing it out both to the young shah and to Mossadegh. They charged that the American diplomat was trying to "sabotage" them in order to get them expelled from Iran so that U.S. oil experts could move in. In fact, unlike some U.S. advisers who came later, Dr. Grady wasn't a friend of any oil company, either British or American; rather, he was keenly interested in helping the Iranians find more money. With the assistance of U.S. consultants, the Iranians had since 1947 been contemplating a Seven-Year Development Plan and had been promised U.S. loans with which to implement it. The more money they could provide themselves, the less burden there would be on the American taxpayer. In his messages to Washington, Grady pointed out that though Mossadegh might seem a fiery and eccentric fanatic—he had a habit of giving heated press conferences lolling on couches, dressed in pajamas—he expressed the sentiments of the people.

Aware of the way things were running, Anglo-Iranian had decided that the moment had come to make a gesture. A proposal was made and provisionally accepted in August 1949 to increase the royalty payments to such an extent that in 1950 they would bring Iran's income from oil up to £40 million. But General Ali Razmara, the premier, who presented this proposal to the Majlis, read it out in a statement that was so patently translated directly from English—and had almost certainly been written in collaboration by the British embassy and Anglo-Iranian officials—that Mossadegh's supporters accused the premier of being a British puppet and shouted him down. The proposal was finally rejected in January 1951, and Anglo-Iranian, by this time beginning to be alarmed, suggested that Aramco's example in Saudi Arabia be followed and a 50–50 deal be consummated between the company and the government.

But the British had seen the writing on the wall too late, and by the time they got around to making this offer the moment had passed. Almost everyone in Iran was hellbent on nationalization, but not quite everyone. For the young shah, who sat insecurely on his Peacock Throne, was in a dither. His assertive sister, Princess Ashraf, was bullying him to take a stand, but he could not make up his mind what to do about nationalization. His instinct was to support it, but his advisers kept warning him that it would bankrupt Iran and lose him the throne. Finally he sent instructions to his ambassador in Paris, Ali Soheily, to go to Lisbon and ask Calouste Gulbenkian whether he should oppose it or espouse it.* What about Anglo-Iranian's latest offer? Should he accept it?

"Mr Five Per Cent told Soheily to tell the Shah to accept Anglo-Persian's offer," said Nubar later. "Father pointed out that it was the best ever proposed by an oil company in the Middle East and ended by declaring bluntly, 'Our country is not yet qualified to take over the oil industry.' Soheily found this opinion very embarrassing. The British offered to put a special plane at the disposal of Father and me (and our suite of five persons) so as we could fly to Teheran to give the Shah our views personally. But Mr Five Per Cent refused to move. Soheily himself had to report back the unpleasant news." †

On February 19, 1951, Premier Razmara announced in the Majlis that according to the oil experts whom he had consulted, nationalization was completely impracticable. Once more his tone and manner indicated that the statement had been translated from English. Razmara was hooted down, and shouts of "Traitor!" and "Bootlicker!" were hurled at him. Shortly afterward, as he knelt to pray in the mosque, he was shot dead by an assassin.

On April 29, 1951, Dr. Mohammed Mossadegh swept into power as head of the National Front government, and immedi-

* Both Gulbenkian and his son, Nubar, held Iranian diplomatic passports as "honorary attachés" of the Iranian embassies in the countries in which they resided.
† Quoted in Hewins, *Mr Five Per Cent.*

ately presented a bill before the Majlis for the nationalization of the Anglo-Iranian Oil Company. It was acclaimed unanimously. The following day the bill was signed by the shah and passed into law on April 31. All Iranian oil fields were now Iranian. It was the end of Britain's control, after a monopolistic reign which had lasted nearly fifty years and fueled the British Empire in war and peace. Now the monopoly was no more.

Not that the British believed it at first. "They can't do this to us," said the British ambassador, Mr. Francis Shepherd, in a statement to reporters in Teheran.

MOSSIE GRABS BRITAIN'S OIL—
BUT NAVY TO THE RESCUE

headlined the *Daily Express*. As if the year were 1905 instead of 1951, the British were planning to send a gunboat to get back what the Iranians had taken away.

LOOKED BACK UPON from the more realistic standpoint of the 1970s, the events of the next few weeks have a true-blue tinge to them that turned what was, after all, a quarrel between a landlord and his tenant over the terms of his lease into the last act of a farce about imperialism. Though the Labour government, which had nationalized several industries of its own in recent years, was in power at the time, British reaction to the takeover of Anglo-Iranian was everything that a stout-hearted Tory could have asked for. Even today an echo of its imperial tones are to be found in some British accounts of the crisis. The British oil historian Stephen Hemsley Longrigg, for instance, who was once a member of Anglo-Iranian's staff, writes:

"The ratification of this naïve and totally inadequate [nationalization] law—henceforward to be the inviolable basis of all discussion and negotiation—was followed by an enthusiastic popular and political campaign of abuse and misrepresentation, directed against the Company . . . Forces of dangerous, indeed of bloodthirsty, fanaticism were allied to and largely dictated the moves of the nationalizers, who became prisoners of their own law and of their broadcast promises of wealth and happiness for all; and the contributions of certain American diplomats and consultants, who professed to see the situation as one of Anglo-Iranian "colonization" or mere reaction or as a legacy of unspecified Anglo-Iranian misdoings in the past, encouraged even responsible Persians to expect American approval." *

Soon both sides were striking attitudes which one might think had gone out of fashion with the death of Queen Victoria. For the British, the wogs were on the rampage. For the Iranians, a war of liberation had begun against the colonialists. What started out as reasonable discussions soon deteriorated into vituperation, threats and violence. Britain submitted the nationalization law to the International Court of Justice at The Hague, but when the Court (which included a British member) finally pronounced, the result was a deadlock. The Court refused to adjudge what it considered to be a "domestic" affair, and told the contenders to go back and talk it over. But it was no longer a question of a quarrel between a private business and a foreign government. The British authorities had taken over, and now all future moves were decided from 10 Downing Street. To begin with, Anglo-Iranian was told to halt all royalty payments. The arrangement by which the Teheran government could change riyals into sterling at bargain rates was ended forthwith, producing an immediate famine in foreign currency for Premier Mossadegh.

Meanwhile, back in the oil fields, the 3,000-odd British technicians who worked for Anglo-Iranian had been offered new contracts by Teheran to go on working for the new national company. Unanimously they refused. Tanker captains arriving at the oil ports to take on crude oil were told by Iranian officials to sign receipts acknowledging Iranian ownership; they, too, refused and sailed away with empty ships. Soon one field after another began to close down, as the pumping stations and pipelines ceased to function.

* *Oil in the Middle East.*

This was a signal for riots to break out. The home of Anglo-Iranian's chief in Teheran was sacked by mobs; there were mass demonstrations in Abadan and other centers of the oil fields. Lying torpid on his bed in his pajama suit, or dashing suddenly to New York to address the United Nations, Mossadegh called the riots signs of "the natural reaction of the people to generations of looting by the British." The British response was to evacuate all of their women and children, and to send ministerial messages of encouragement to the men left behind in the oil fields.

HOLD FIRM, BRITAIN TELLS OILMEN
WE'LL STAND BY YOU, MINISTER PROMISES

was a headline in the *Daily Mail.*

At which point the situation escalated again when the British government dispatched a parachute regiment to Cyprus and ordered it to stand by for further instructions. British army units were alerted in Iraq, just across the border from the Iranian oil fields, and suddenly the Royal Navy cruiser H.M.S. *Mauritius* hove to off Abadan and anchored in mid-river. It looked as if a landing and military occupation of the oil fields was imminent. Luckily for Britain, at the last moment this disastrous project was abandoned. Otherwise the defeat she was to suffer at Suez five years later might have happened that much sooner, for the Iranian army was ready and itching to fight, and had the will of the people behind it. Britain had made no landing plan, preferring an ad hoc operation. Moreover, her plans of communication would have been seriously compromised throughout the Middle East, for without exception, every Arab nation approved of Mossadegh's actions. More important than all these drawbacks, however, were the two biggest threats of all to the success of a military operation. One was threateningly negative: the United States let it quietly be known in Downing Street that under no circumstances would its government give aid or support to punitive action by the British. The other was threateningly positive: the Russians began making dispositions on their borders with Iran, and there were indications that in the event of a British landing in the south, the Red Army would march into the country and establish a puppet regime in Teheran.

Faced with these horrid facts of postwar life, the Labour government wisely decided to back down and stop acting like a gunboat administration. Instead of reaching for its guns, it started to use its brains, and woke up to the fact that time, the strongest weapon of all, was on its side.

In October 1951 he last members of the British staff in the Iranian oil fields were evacuated aboard H.M.S. *Mauritius*. They went out like true-blue Britons.

"On the morning of October 4, 1951," writes the official historian of Anglo-Iranian, "the party assembled before the Gymkhana Club, the centre of so many of the lighter moments of their life in Persia, to embark for Basra in the British cruiser *Mauritius*. Some had their dogs, though most had had to be destroyed; others carried tennis rackets and golf clubs; the hospital nurses and the indomitable Mrs Flavell who ran the guest house and three days previously had intimidated a Persian tank commander with her parasol for driving over her lawn, were among the party, and the Rev. Tyrie had come sadly from locking up in the little church the records of those who had been born, baptised, or had died in Abadan . . . The ship's band, "correct" to the end, struck up the Persian national anthem and the launches began their shuttle service . . . The cruiser *Mauritius* steamed slowly away up river with the band playing, the assembled company lining the rails and roaring in unison the less printable version of 'Colonel Bogey.' Next day Ross and Mason [the two senior officials] drove away. The greatest single overseas enterprise in British commerce had ground to a standstill." *

Behind them they left pipelines through which the pumping stations were no longer pumping oil, closed-down refineries, tank farms spilling over with crude oil that no tankers were moving, and 70,000 Iranian employees still on the payroll for a salary of £1.6 million a month. The responsibility for paying them was now

* Longhurst, *Adventure in Oil.*

that of Mossadegh's regime, but he certainly did not have the money, and his government had other obligations, no less pressing, which he hadn't a hope of fulfilling.

But for Britain the situation also threatened to deal an almost fatal blow to its economy. Anglo-Iranian had always been one of the United Kingdom's greatest earners of foreign currency, and the markets it had served were now lost. The company had also supplied the bulk of Britain's petroleum needs, and henceforth these would have to be replaced from other sources, almost all "hard" currency sources which would cost the British treasury $40 million a month at a time when dollar reserves were dangerously low.

It was a desperate situation, and it explains why even a Labour government had contemplated a punitive expedition to occupy the Iranian oil fields by force. The chaotic economic situation created by Iran's nationalization did nothing to help the Labour party during the general election which took place in Britain at the height of the crisis, and the result was that Winston Churchill and the Tory party were voted back into power. Churchill's first action was to reach for the telephone and ask the White House's aid in resolving the situation.

ON JULY 15, 1951, Averell Harriman had arrived in Teheran on a special mission for President Truman. He came with preconceived ideas about the Iranian situation, for Ambassador Grady had made his views quite clear in his reports, stressing the fact that Mossadegh enjoyed the support of most Iranians and castigating the British for the arrogance of their local diplomats and the stubbornness of their oilmen. As a result, Harriman arrived as the envoy of an Administration which, while taking no sides in the dispute, sympathized with Iranian aspirations, and plainly expected to be welcomed as a friend as well as a go-between. Instead what he got was a flea in his ear.

For Ambassador Grady's reports were now out of date; what

had started out as a genuine popular movement of protest against foreign exploitation (real or imagined) had been taken over by the ideologues. The Iranian Communist party, Tudeh, which was controlled and financed from Moscow, had now thrown its weight and money behind Mossadegh, and provided him with a street army of well-trained thugs. The Tudeh party was not in favor of oil nationalization, for it supported Russian claims to oil concessions in northern Iran and now hoped to engineer Russian control of the fields from which the British had been driven; hence it directed its campaign to attacks on "colonialism" and "dollar imperialism," which made the United States as much a target as Britain. Harriman, all prepared to smile and wave in acknowledgment to the cheering crowds he expected to meet on his way in from the airport, instead ran into an ugly demonstration from a column of Tudeh demonstrators armed with sticks and screaming anti-American slogans. He was also somewhat taken aback at his first interview with Mossadegh when the Iranian leader, in the midst of an outburst against the British, suddenly dissolved in tears. This unexpected emotionalism must have affected Harriman's sense of judgment, for he emerged from the meeting convinced that if only Britain would acknowledge Iran's right to nationalize its oil resources, all would be well and reasonable discussions could begin. On the strength of this, the British government rushed an envoy out to Teheran, only to be confronted by a demand from Mossadegh that they pay $140 million into his treasury immediately—and without the money there would be no talks. Harriman announced that he was surprised and disappointed and returned to Washington a man whose ideas had changed.

By the time Truman's Administration had given place to that of President Dwight Eisenhower, there was no longer any question of America standing aside and letting the British get out of their own mess. Churchill and Eisenhower were old friends and spoke the same language. The Republican Administration had always been responsive to the oil lobby, and though far from being discomfited by their British rival's difficulties, the oilmen saw the peril ahead if a Middle East nation were allowed to get away with nationalization. What Iran could do today, Saudi Arabia might do tomorrow.

SO FAR AS THE OIL SUPPLY WAS CONCERNED, time was on Britain's side; so far as Mossadegh was concerned, it was running out. The trouble was that the premier had never visited the Iranian oil fields and knew nothing about the situation there. He himself had never bothered to ask expert advice on whether there was a sufficiently trained cadre to run the fields once the British left. There was not. The British had schooled several thousand Iranians to handle the lower technical rungs of the petroleum ladder, but there was no one capable of climbing to the top and taking over.

Nor had Mossadegh realized that selling oil is even more important than getting it to the surface and refining it. Since no Iranians in the Anglo-Iranian Company had held important positions in the sales department, there was no one available to explain the facts of life in the international oil sales market. At this time, 12 million barrels of refined petroleum were lying in the brimming tanks at Abadan, and when the Iranian leader demanded that it be sold at once to earn desperately needed foreign currency to meet his budget, he was told that no one would buy it.

Then sell it at half price, even quarter price, Mossadegh ordered.

But still no one would buy. Immediately after withdrawing from Abadan, Anglo-Iranian had asserted its proprietorial rights over any oil coming from their former fields, and threatened action against anyone who tried to sell or buy Iranian oil. The first to try it were the Italians. Italy had no oil fields of its own and had always relied entirely on Anglo-Iranian for its fuel supplies. When nationalization took place the Italians indicated—Anglo-

Iranian's threats notwithstanding—that they would go on buying from Abadan. But then the ebullient head of the Italian national petroleum company, AGIP (Azienda Generale Italiana Petroli),* a one-man dynamo named Enrico Mattei, flew to London for conferences with Lord Strathalmond, chairman of Anglo-Iranian. Mattei emerged from the meetings declaring that he had not previously taken the British threats of legal action seriously, but that now he did, and AGIP would not be buying Iranian nationalized oil. It was rumored that Signor Mattei had been offered powerful inducements and substantial future rewards for having changed his mind.

A small Italian company, SUPOR, did take the risk and loaded two tankers at Abadan. One had a writ tacked to its funnel when it arrived in Aden, and the other when it reached Naples. The company that owned a Japanese tanker which took a load to Osaka was also sued by Anglo-Iranian in Japan. Nonetheless, these law cases would take months to come to trial, and if a sufficient number of companies had been willing to take the plunge, no legal threats from the British company could have stopped the boycott from being broken. But the international oil market did not want it broken. Once the flow of Iranian oil stopped, oilmen discovered that they were doing even better without it. Saudi Arabia, Kuwait and Qatar oil fields stepped up production and turned 1952 and 1953 into bonanza years for the big American companies and for the Anglo-American-Dutch-French combine, IPC. By the end of 1953, world oil production had risen to 637 million tons, compared with 535 million in 1950. In comparison, Iran (which had exported 54 million tons of oil via Anglo-Iranian in 1950–1951) sold a total of only 132,000 tons to foreign buyers under Mossadegh in 1952–1953.

Belatedly made aware of this miscalculation, Mossadegh looked around for someone who could sell his oil. He tried Enrico Mattei, but by this time Mattei had seen Lord Strathalmond. Then he sent an emissary to London to see Nubar Gulbenkian.

* AGIP is the retailing subsidiary of the Italian national oil company, ENI.

The Gulbenkians had strongly advised the Iranians not to nationalize Anglo-Iranian, and for their pains had been deprived of their Iranian diplomatic status the moment Mossadegh came to power. Mossadegh offered Nubar plenipotentiary powers and "the gratitude of the Iranian people" if he would take over as sales director or adviser of the nationalized company. Nubar refused ("though it would have been nice to put the CD plate back on my car") and went to Lisbon to let his father know what was happening. "I told them so," the old Armenian said, happy that yet another of his predictions had come true.

By now diplomatic relations between Britain and Iran had been broken off, on Mossadegh's orders. The United Kingdom complained to the United Nations, and also used America's good offices to make a last offer to patch up the quarrel. In February 1953, new proposals were sent to Teheran, and the U.S. government indicated that they had its full approval. In return for a settlement, both Britain and the United States would recognize Iran's right to control her own oil industry and policies; expect compensation to be paid to Anglo-Iranian for this change in status; give Iran the opportunity to enter into arrangements for selling oil in substantial quantities at competitive prices on the world markets; and lend the country sufficient funds (to be repaid in oil) to enable the government to meet its immediate financial problems.

The proposals were rejected in a fiery speech by the Iranian premier in which he launched into his most virulent attack yet on Anglo-Iranian and angrily refused to contemplate paying compensation. "What the Anglo-Iranian Company did in southern Iran," he cried, "was sheer looting, not business!"

Mossadegh was determined not to have the British back, but he was also disastrously in debt. Oil workers had been put to labor on the roads, but the government was running out of money to buy supplies to keep them busy. A U.S. loan of $23 million had been enough to pay for shipments of food, but its arrival was said to have resulted from a personal appeal from the shah to the U.S.

ambassador, Loy Henderson (who had replaced Grady in 1951), and since Mossadegh was bitterly antiroyalist and now quietly working for the shah's overthrow, he was anxious that the monarch receive no credit for further aid.

Therefore, on May 23, 1953, when the Iranian premier asked the U.S. government for further loans, he made it clear that this time they would be on his terms. He sent a personal letter to President Eisenhower soliciting the financial aid he needed, and in the final paragraph added what amounted to an ultimatum. Calculating that there was nothing more likely to speed up action in Washington than a threat of Soviet intervention, he made it clear that if the United States did not provide him with the help he needed, he would be forced to look elsewhere for aid. His ambassador in Washington underlined what that meant. The Soviet government had already made it known in Teheran that it was ready to help, and was prepared to sign a defense pact, as well as an economic agreement, with the Mossadegh regime.

This threat was just as likely to frighten the Americans as the British, for the prospect of Russian penetration into the waters of the Persian Gulf was something neither nation liked to contemplate. Mossadegh figured that this tactic would produce precipitate action, and he was right.

The CIA entered the crisis.

CHAPTER SIXTEEN

The Flight of the Shah

IN 1953 Teheran was so full of American and Russian spies that a local wit once suggested that they ought to share apartments so as to save themselves time and money in keeping one another under surveillance. There was a residue left of the Polish and Balkan refugees who had been stranded in the capital during World War II, and most of them now worked as *filles de la maison,* musicians or waiters in the cabarets sprinkled around the city. Many of them were on the payroll of one of the two embassies (several of them on both) and found a ready market for the gossip they picked up. The Russians had also infiltrated agents into most Iranian government departments and business offices in the city, and the bazaars swarmed with characters who carried documents to prove that they came from Iranian Azerbaijan, when in reality they were citizens of Soviet Azerbaijan, just across the northern frontier. A number of these, sent down from the oil fields in Baku, were also operating in Abadan and Ahwaz, trying to make order out of the chaos of the nationalized petroleum operations.

The Americans had a military mission attached to the Iranian army which Mossadegh, not yet wanting to break completely with the United States, had reluctantly refrained from expelling, and its members were in close touch with their Iranian confreres. There were also some strange Americans named "Jake" and "Red" and "Uncle Ami" (their real names must remain secret) who put in regular appearances around the cabarets and bars; they spoke the purest Bronx, Brooklynese or Panhandle drawls but turned out to possess an even more fluent command of Kurdish or the dialects of Kermanshah, Khorramshahr and Azerbaijan, learned at the knees of their immigrant parents. They came in to report and quench their thirst for whiskey and female company, and then disappeared again for weeks on end.

But the area of the Iranian government into which American agents had infiltrated most successfully was the Iranian police force. From 1942 until 1948, the Imperial Iranian Gendarmerie had been under the command of an American named Brigadier General H. Norman Schwarzkopf who had taken over with orders to recruit and reorganize it. General Schwarzkopf, whose name will be familiar to students of the Lindbergh kidnapping case,* had a roving commission for the CIA, under Allen Dulles. Both of them had been wartime members of the OSS.

When he departed from Teheran, Schwarzkopf had left several old friends behind in the Iranian police who could be relied upon in an emergency. In fact, they had already proved their efficiency in the spring of 1953. Aware of the pro-American elements in the police ranks, Mossadegh had appointed a new police chief and given him orders to conduct a ruthless purge. General Ashfar-Tus did not take his instructions lightly, for he was a savage and ambitious man. His character can be judged by the fact that a fellow general, with a reputation for being one of the sternest disciplinarians in the Iranian army, had sacked him "because his behaviour with the cadets was so harsh." Afterward he had been given charge of the royal estates in the north and there had "treated the peasants with brutality and even cruelty." †

* He was in charge of New Jersey State Police in 1932 and headed the hunt for the kidnapper.

† *Under Five Shahs*, by General Hassan Arfa.

General Ashfar-Tus had one weakness: he boasted. On the night of his arrival at police headquarters on April 19, 1953, he announced to a number of his subordinates that he had a list of the names of every American spy in the force, and promised that they would all be dead or behind bars by the end of the week. But the next morning he did not turn up to make good his threat. He had been kidnapped and shot through the head, and his corpse was later discovered in the foothills of the Elburz Mountains.

Mohammed Mossadegh sent his letter asking for a loan to President Eisenhower on May 23, 1953, but a month passed by and no reply came from the White House. One can only assume that the President had been advised to keep the Iranian leader in suspense until certain preparations had been made. Certainly there was a sudden flurry of activity among CIA agents both inside and outside Iran during the hiatus. Allen Dulles, the CIA chief, flew to Switzerland, where, it was said, he would be joining his wife for a short holiday. A few days after his arrival, who should turn up to join him but Loy Henderson, the U.S. ambassador to Iran, and a week after that, Princess Ashraf, the shah's twin sister and one of the most forthright people in the kingdom, also turned up in Switzerland and had several discreet meetings with the two Americans.

It was not until June 29 that President Eisenhower got around to answering Mossadegh's letter. The Iranian leader was jolted by the unequivocal nature of the reply, which also contained an ultimatum: either Mossadegh must agree to an immediate opening of talks on the future of the oil fields, or there would be no loan. The old man gave a bedside press conference during which, alternately quivering with rage and shaking with sobs, he swore that he would never give in to the imperialists who were trying to humiliate the Iranian people. Strikes were called and there were Tudeh parades in the capital and the oil fields, hailing Mossadegh and the Soviet Union, and promising to blow the oil wells sky-high should British or American troops set foot on Iranian soil. On American advice, the army and the police force stayed in their barracks and left the streets to the mobs. The nation was now completely divided into pro-Mossadegh and pro-shah factions.

In the midst of this turmoil General Schwarzkopf arrived in Teheran, armed with a diplomatic passport and a couple of large bags. As soon as this news was known, cries of fury and alarm rose up in the Tudeh- and Russian-controlled newspapers, and members of a Soviet economic mission which was in Teheran for discussions with the Iranians, formally protested the presence of "this notorious agent of American intelligence." Schwarzkopf, innocently surprised by all the fuss, pointed out that during his stay in Iran years before, he had made many good friends, and now, in the course of a world-girdling vacation trip, he was there to see one or two of them. They included General Hassan Arfa, a staunchly patriotic Iranian but a bitter opponent of Mossadegh who was suffering a diplomatic "illness" on his estate at Larak outside the capital, where he had a well-trained private army sworn to protect him; General Fazlollah Zahedi, another foe of Mossadegh, also in hiding on his estate; several officers in the police force; and of course the shah, Mohammed Reza Pahlevi. The U.S. general did not seem to have any difficulty getting speedy access to all of them.

Sensing the danger in Schwarzkopf's presence, Mossadegh began to tighten his grip on the country. Cheered on by Tudeh mobs, he announced the abolition of the Majlis, referring to the Iranian parliament as nothing but "a nest of thieves." He then declared that he would hold a referendum to give the people the opportunity to approve of his action. He also asked for power to deal with "sinister elements in the army, the police, and with other brainless agents of international reaction." * He made it quite clear that the latter phrase was a reference to the shah.

There were fights at the polls, and the few who dared to vote against Mossadegh were badly mauled by the crowds. In truth, the old man in pajamas did not need to manipulate the referendum, for the masses were with him, even if the army, police and landowners were not. But the strong-arm methods of the

* *Ibid.*

government guards gave the shah the opportunity of intervening. "Being a constitutional sovereign," wrote General Arfa later, "the shah could not dumbly assist at such a tragic farce and he understood that this tragicomedy [of a referendum] was leading the country directly to chaos and anarchy, from which it could emerge only to fall into the hands of the Tudeh." *

The shah therefore used his constitutional right and dismissed Mossadegh and announced the appointment of General Fazlollah Zahedi as prime minister in his place. No one (particularly the shah and General Schwarzkopf) expected Mossadegh to obey the order. As soon as he had made his announcement, the shah took off from Teheran to join Queen Soraya at Ramsar, where he had a palace on the Caspian Sea. General Zahedi did not even bother to emerge from hiding or attempt to assume the post to which the shah had appointed him. He was quite right. The emissary who was sent by the shah to tell Mossadegh that he had been sacked was immediately flung into jail by the old man, who promptly announced to the nation over the radio that he had refused to obey the royal order. A flood of anti-shah propaganda burst into print in the Communist newspapers, and mobs swarmed into the streets execrating his name. At the suggestion of Ambassador Henderson (now back in the country after his trip to Switzerland), the shah called for his private airplane, and taking Queen Soraya with him, flew off into exile.

While the royal party was en route to Rome,† Mossadegh unleashed a full-scale campaign against the monarchy. All the shah's pictures in government offices, cinemas and shops were removed, and the mention of his name in morning and evening prayers in military units was forbidden. On August 18 the Tudeh party surged into the streets and began razing all statues of the shah and of his late father.

This was the moment for which the anti-Mossadegh forces

* *Ibid.*

† Where the Iranian ambassador refused to meet the shah—and lived to regret the lack of courtesy.

had been waiting. First, four hundred cadets at the military academy camp at Aqdassieh announced that they were going on a hunger strike to protest insults to the shah. Then crowds of peasants, armed with staffs and daggers, picks and bicycle chains, suddenly emerged at the southern entrance to the city, and the fact that they came from the direction of General Arfa's estate was not without significance. As they moved toward the center of the city they gradually began to form into a procession, and were quietly joined by what General Arfa called "many off-duty army and police NCOs." Suddenly thousands of anti-Mossadegh leaflets appeared. Army units and police squadrons, ordered by Mossadegh to disperse the insurgent crowds, joined them instead, and "at ten o'clock, Parliament Square was full of people, and improvised orators standing on the base where the statue of Reza Shah had been, addressed the delirious crowds who shouted continuously: 'Shah! Shah!' . . . At twelve o'clock most of the town was completely in the hands of the loyalists, the ministers having all disappeared, hidden nobody knew where." *

Only around Mossadegh's house did an army unit resist the promonarchist mob, firing tank guns and machine guns into the crowd as it surged forward. Two hundred were killed and more than five hundred wounded. As the old man peeped out through the shutters at the screaming mob below him, he asked an aide, "Where are all those who voted for me at the referendum?"

"*Bad bord* (Gone with the wind)," the aide replied.†

By four o'clock it was all over, and General Zahedi, who had come out of hiding to lead the revolt, cabled the shah to return from exile. Still in pajamas, Mossadegh was arrested in bed and taken to jail, to be tried for treason later. The Soviet economic mission hurriedly returned to Baku by the next plane, and a ruthless purge of the Tudeh party begain in all parts of the land, not least in the oil fields and Abadan, where thousands were imprisoned and hundreds subsequently shot.

* Arfa, *op. cit.*
† *Ibid.*

The Flight of the Shah

SITUATION RESTORED. AS YOU WERE IN IRAN, announced the headlines in the British newspapers. In London the directors and staff of Anglo-Iranian got ready to take over once more, convinced that it would soon be business as usual.

But it turned out to be not quite that simple. For one thing, not even the anti-Mossadegh elements in Iran could stomach the idea of having foreigners back in control of their oil. For another, there was the question of the CIA. Thanks to General Schwarzkopf, the coup had been successful, but it had been extremely costly and millions of dollars had been paid out in bribes. It hadn't all been done just for love of the Union Jack; how was the United States to be compensated?

IT WAS HERBERT HOOVER, JR., who first thought up the idea of a consortium. He was President Eisenhower's adviser on petroleum affairs, and the moment the shah was safely back on the Peacock Throne he flew to Teheran to look over the oil situation. For the moment he had the field to himself. The Russians had taken temporary fright and withdrawn their economic and petroleum experts, though they would be back shortly. The British did not even have diplomatic representation,* and the nearest Anglo-Iranian oilmen were waiting across the border in Iraq but were not allowed into the country for the moment.

Hoover was well aware of Anglo-Iranian's attitude toward the concession, now that Mossadegh was gone: it was status quo ante. As the company's historian put it later: "In London Lord Strathalmond and his colleagues from the day the company left Persia stuck their toes in with stubborn determination in defence of their rights, both legal and moral. They were determined that . . . the company's position should be acknowledged and respected." †

* After Mossadegh broke off relations with Britain in 1952, the United States represented them in Iran.

† Longhurst, *Adventure in Oil*.

But Iranian pride was also involved. When Hoover reached Teheran he discovered that though Mossadegh had departed from the scene, many of his ideas remained. The country was deeply in debt, but in his talks with the shah and the prime minister, General Zahedi, Hoover discovered that both of them were united on two matters: never would they allow Anglo-Iranian to return as the sole controller of Iranian oil, and never would any other foreigners be permitted to operate in the oil fields unless they first formally acknowledged that the land and oil belonged to the Iranian people.

From Hoover's point of view, it was an intriguing situation. What better method of bringing the two conflicting parties together than by using American petroleum interests as a bridge? The Americans would acknowledge Anglo-Iranian's prior claim to the concession, but would come in as the dominant partners and perhaps invite others to join the new line-up—the French, for instance, and the Dutch. An international consortium, in fact. In this way they could demonstrate to the shah that no one company or nation would be exploiting Iran's oil resources, and that in any case the new consortium was prepared to acknowledge Iran's prior sovereignty over its petroleum and do all its dealings through an Iranian state organization.

To Hoover's amazement, not only were the British against the idea—they simply thought that America was trying to muscle in on a British oil monopoly, as indeed it was—but the American oil companies were not keen on the idea either. What Hoover had not realized was how well the big U.S. oil combines had done because of the Iranian disaster. Aramco's oil production and sales from Saudi Arabia had increased by fabulous amounts; so had the figures in Kuwait and Qatar. The prospect of Iranian oil coming back on the market, even though they would make money out of it as part of a consortium, filled the rival American companies with more dismay than joy. In 1954's economic climate they were producing all the oil that the world could buy, and as the experts saw it then, the return of Iranian oil to the market would only mean that production would have to be cut in Saudi Arabia, Kuwait, Qatar and other Middle Eastern areas to accommodate it. What would the ruler of Kuwait, the emir of Qatar and the debt-ridden King Saud, all of them voraciously demanding more and more from their oil revenues, say when they were told that their incomes would have to be cut instead of increased?

In point of fact, it didn't work out that way. The world would soon be lapping up all the oil it could get as greedily as the sheiks were lapping up money, and not even the return of the Iranian fields could satisfy all the demands. But even if the major oil companies' show of reluctance was not a deliberate ploy,* it went a long way toward mollifying Anglo-Iranian over the coming loss of its monopoly. By the summer of 1954 the British had accepted the idea of a consortium and gave the go-ahead to the Americans to start negotiations with the Iranians. The line-up had been agreed. Anglo-Iranian would have 40 percent of the shares in the new consortium, and would be paid £34 million immediately by its partners as compensation, plus 10 cents a barrel on all oil exports until a total sum of $510 million (or approximately £200 million) had been reached.† The remaining 60 percent would be divided between the other members. They were not exactly newcomers to the Middle East oil game, or to one another. Royal Dutch–Shell would take 14 percent; the Compagnie Française des Pétroles would get 6 percent. The remaining 40 percent would be split, 8 percent each, among the U.S. majors: Standard of New Jersey, Socony-Mobil, Standard of California, the Texas Company and Gulf Oil Corporation.

"At least we won't be working with strangers," said an Anglo-Iranian director after he had read through the list. If the Justice Department hadn't decreed otherwise, it would have looked uncannily like an international cartel.

A delegation set out for Iran pledged to get an agreement

* They had to be assured by the Justice Department that they were not contravening antitrust laws before they went in.

† In 1953 the value of the company's Iranian concession—including the refinery at Abadan—was valued at £550 million.

signed as quickly as possible and put the oil fields back in operation. But by the time they arrived, the Iranians had recovered their nerve and the Russians were back in the game.

WITH ITS ECONOMY IN RUINS and its anti-Western elements either dead or in jail, it might have been thought that Iran would make the quickest deal possible with the new consortium and start paying off its debts with oil revenues. But things had changed in Teheran, and so had the young Reza Shah. He had scurried out of the country in order to allow other people to overturn Mossadegh, but when he returned he was not so sure that the old man hadn't been right after all. Too stubborn, obdurate and xenophobic, perhaps, but in principle a true Iranian who was only standing up for his people's rights. For this reason the shah counseled the new government not to press too hard for condign punishment of the defeated ex-premier, even though he had conspired to topple the shah from his throne.* To his intimates, the monarch confessed his shame and humiliation that a change in government in his own country had had to be manipulated by the CIA, the intelligence arm of a foreign power. His courtiers noted a new firmness and certitude in his manner. No longer was he vacillating or fearful; he had even come to grips with the facts of his royal marriage and had decided to give his people an heir even if it meant changing his queen for one more medically likely to ensure a successor to the throne.† As for the oil concession, he called in Premier Zahedi, and the finance minister, Dr. Ali Amini, and told them; "The Americans are coming in search of an easy bargain. Anyone who gives it to them will answer to me and to the Iranian people. I want it known as widely

* Whereas others were savagely treated, Mossadegh was given a sentence of only three years' imprisonment and received special treatment.

† Queen Soraya, who failed to bear him a child, was subsequently divorced by the shah and sent into exile. He then married a young Iranian student (Soraya was half German, half Iranian) who has borne a crown prince.

as possible by my people that the oil will stay Iranian, and that they will decide the punishment of any negotiator who fails to bear that in mind." *

The shah then stole a leaf from Mossadegh's book and instructed the prime minister to let it quietly be known to the Russians that they were welcome to send their petroleum experts back into Iran, and that they need not yet lose hope of securing a concession. He presumed, rightly, that the Americans would hear about this.

That royal audience took place in February 1954, and henceforth the stage was set for horse-trading of a kind that none of the Anglo-American delegation had bargained for. A month of preliminary negotiations, led by Ronald Hardman, a vice-president of Standard Oil of New Jersey, ended with the Iranian delegation handing a memorandum to the delegates just before they flew back to London to report. This, the so-called Airport Memorandum, contained demands so stern and uncompromising that, to quote a remark a Standard of New Jersey official made recently in an interview, "practically everybody in London said, 'It's finished. We'd better forget about Iranian oil. You'll never make an agreement with those guys.' "

But if the oilmen were ready to give up, the British government was not; nor, at its urging, was the United States. The British needed the oil revenues and the compensation from their newfound partners; the U.S. government wanted to make sure that Iranian oil was kept out of Russian hands.

"They put the heat on us," the same official said. "They insisted that we go back and have another try."

There was only one man among the Americans who believed that an agreement was possible—Howard W. Page, the London representative of Standard Oil of New Jersey. Son of an oilman, born and brought up in the California oil fields, trained as a pe-

* Statement recalled by friends of Dr. Amini, now living "in retirement" on his estate outside Teheran.

troleum chemist but the veteran of several tricky and complicated concession deals,* Page walked into a conference room with the friendly air of an honest man—which he was—prepared to make a fair deal with men of equal probity (which his opponents often found disarming). Page spent weeks studying the terms of the uncompromising Airport Memorandum, and at the end of that time he was convinced that an agreement was possible, no matter how unlikely it appeared.

"I thought there were just enough elements on both sides, just enough flexibility, to make it feasible," he said in a recent conversation. "I thought that we could lose some things we could afford without giving away what we couldn't afford. Also my past experiences in Iraq had taught me that if you want an agreement you can get it. I sensed that the Iranians wanted and needed an agreement, but that their representatives were asking for the sky because they were scared of what would happen to them at the hands of the Iranian mob if they didn't. Then I heard from Dennis Wright, who was the new British chargé d'affaires in Teheran, and an excellent diplomat, that he believed there was more flexibility in the Iranian position than appeared on paper. It was at this point, because I knew the Airport Memorandum like my own face, and because I was an American and the Iranians wouldn't deal with a Briton, that I was selected to head the delegation that would go back to Teheran and reopen the talks."

The delegation flew out in chartered planes in April 1954. Page took William Snow of BP † and John Loudon of Royal Dutch–Shell along with him. Both were senior oilmen, but as an American and as an experienced negotiator there was never any question of anyone but Page being in charge. The Iranians had selected Dr. Amini as the head of their team—a wise choice. "I would say that there were not more than ten men in the world

* He wrote the terms of the 1952 agreement between the Iraqi government and the Iraq Petroleum Company.

† Since it no longer had sole possession of the Iranian concession, Anglo-Iranian had changed its name to British Petroleum, by which it has been known ever since.

who knew more about oil than he did," said Page later. Amini was a brilliant scholar, he never lost his temper, and he had a great sense of humor, but his greatest asset was his courage. Unlike his subordinates, who were fearful of both royal wrath and mob violence, he was prepared to concede points as well as demand them. "He was the one man," Page recalled, "who would take responsibility. He would keep his eye all the time on the object of getting an agreement, and he was prepared to give a little here, provided we gave a little there. Without him we could never have done it. None of their experts had enough guts, and you can't blame them. There was always the fear of assassination. Feeling was running high, and both delegations had bodyguards."

A few hours after Page arrived in Teheran he was called to the U.S. embassy for a conference with Herbert Hoover, Jr., who informed him that the Russians were back in the game, that they were desperately anxious to know what the Anglo-American delegation was going to offer the Iranians and what tactics it would adopt once the talks got under way.

"Wherever you talk," Hoover said, "the Russians are going to have you bugged. I recommend that you have your conferences in the embassy. We've debugged this place, but we can't guarantee you privacy elsewhere."

Page declined. He didn't particularly want the U.S. government to know what the delegation was planning, either, and he knew his British and Dutch colleagues would be even more embarrassed than he by such a location. But during the course of his conversation with Hoover and others, an embassy electronic expert made one remark which Page remembered: " 'You know, the only thing that can stop these guys from hearing what you're saying is running water.' "

Page knew just the place. The negotiators were staying at the Darvan Hotel at Shimran, in the foothills overlooking Teheran, and in his strolls he had noticed a garden. "There was a house in the garden. It wasn't worth a damn, but the garden had a swimming pool and running water—and where the water gurgled out of the swimming pool and into another pool was just the place for our conferences."

The house was rented, a gardener's cottage was fixed up for the secretaries, and they were in business.

The negotiations were both complicated and protracted. The Iranians held a poor hand of cards but Amini was a master player. Also it turned out—though the delegates didn't discover this until later—that the Russians had slipped a joker into the pack. One of the Iranian team, who was also a member of the Majlis and one of the committee which, once an agreement was reached, would have to sign the documents, was under Russian control. Mr. Salim (not his real name) objected to any sign of flexibility on Amini's part, talked endlessly for hours on end about Iran's "sacred oil" and her "inalienable rights to her precious heritage," driving everyone to distraction.

The talks, which had begun in April, dragged on through June, July and August. The shah began to be impatient. He called Page and Amini to an audience at the palace, told them that he was summoning the Majlis to session a week hence, and that by then an agreement must be ready for the members to examine and approve. Either all details must be ironed out by that time, or he would cancel the negotiations and take "other steps." "You have seven days, gentlemen," he said.

In the next week Mr. Salim surpassed himself. By this time a draft agreement had painstakingly been drawn up in English, and an Iranian translation was being made. Since both would be valid in law, it was necessary that the translation be exact, but this was difficult because there was no equivalent in Persian for certain technical words. Furthermore, Mr. Salim took it upon himself to supervise the translation, and he managed to argue over every point at the rate of about one an hour.

Page knew that the shah meant what he said about the time limit. It would take three days to get the Persian-language version to London, The Hague and New York, where the heads of the companies concerned would sign it, and then return it to

Teheran. Page had a charter plane standing by with a pilot on alert, but thanks to Salim's procrastinations, the pilot ran out of rest time and the flight had to be scrubbed. Then it was learned that the regular KLM service was due to leave in forty-eight hours for Amsterdam, and that the Dutch ambassador would be aboard. Would he take the precious agreement with him to Amsterdam, where Royal Dutch–Shell representatives would rush it to the other parties? It was agreed.

Despite further delays from the indefatigable Mr. Salim, the agreement was finished at last. The Majlis was informed and the members began to assemble, waiting to hear the document read to them the moment it was signed. The committee was summoned to an anteroom in the parliament building, where everything was ready for signature, and they all arrived, including Mr. Salim, making no secret of his resentment.

"I had it all nicely wound up," Page said, "and we had a good hour and a half to spare before we took the contract down to the airport. One by one, the committee members sat down and began to sign, but then, just before his turn came, Mr. Salim excused himself and said he had to go to the men's room. And he didn't come back for an hour and fifteen minutes. We were desperate, and nobody could find him. We just hung around and waited, and all I could think about was that plane at the airport waiting to take off. If we missed it, we might as well throw the contract in the ashcan, and jump in with it."

Eventually Mr. Salim returned. He made no apology or explanation, but simply sat down and signed.

"By now we were so late," said Page, "that I didn't even stop to wrap the papers up. I just grabbed them and rushed out without saying good-bye to anyone. Luckily I had some paper and string in the car, and I told the driver to take off. He was one of the wildest drivers in Teheran, the roads were pretty rough, and I was bouncing around trying to tie up the documents. When we reached the airport, a KLM man who was waiting for us jumped into the car and said, 'Drive out onto the runway.' He signaled to the control tower and we sped away. The plane was on the runway, already warmed up and just ready to take off. My driver was so imbued with his role that he surpassed himself, and we went down that runway like a cat with a can tied to its tail. It was one of the large old piston planes, and I don't think the driver even saw the propellers going round. We were coming head-on to the plane and were about to go under the wings when I leaned over and wrenched at the steering wheel, just in time to swerve out of the way of a propeller. We screamed to a stop and I flung myself out, package in hand. The door of the plane opened, I tossed the clumsily tied bundle in, and as the plane passed us I got a thumbs-up sign from the ambassador at one of the windows. That's how we got the contract through."

The agreement between the consortium and the Iranian government was approved by the Majlis in August 1954 and formally ratified in October of the same year. It was to run until the end of 1979. In the circumstances, British Petroleum had no reason to complain about the settlement. By installments its partners were paying the company £200 million in compensation for its loss of the monopoly. Also in compensation, the Iranian government had agreed to hand over £2.5 million a year for ten years to BP. As for the partners, they were now operators of one of the biggest oil fields in the world, and they had got it at a bargain price, which they could easily pay out of income.

For the moment, the Iranians were not grumbling, either, though they would later. In return for their compensation to BP, they were given the ownership of all the company's physical installations in Iran for producing and refining oil, and would also become proprietors of any other installations which the consortium built in the future. Henceforth these would be the property of the National Iranian Oil Company, the nationalized body which had been created by Mossadegh after he took over Anglo-Iranian. Through two newly registered companies, the consortium would get the oil out of the ground and then sell it to NIOC, which would in turn sell it to the separate trading companies set up by each of the partners. And from the profits which these trading companies made in marketing the oil abroad, the Iranian government would take 50 percent in taxes.

It was a profitable arrangement for everyone—too profitable for the companies, the Iranians would say later. But for the moment everyone was pleased. The oil was owned by a national company, and there were Iranian members on the boards of all the subsidiary companies of the consortium. In the next three years, the Iranian government received $490 million in direct oil revenues alone.

ONE MAN WHO EXPECTED TO BENEFIT from the establishment of the Iranian consortium was Enrico Mattei of Italy. Unlike France, Italy was one of the major European countries with no oil production of its own, and as head of the nation's chief petroleum company, AGIP, Mattei had made all Italian oil purchases through Anglo-Iranian. As mentioned earlier, when Mossadegh nationalized the company and sought world outlets for the fuel, he had made a direct approach to the Italian, but as Mattei's biographer and former business associate, P. H. Frankel, wrote afterward:

"It is worth recalling that during the years of conflict of Anglo-Iranian with Mossadeq, Mattei proved to be entirely loyal to the former, AGIP's exclusive supplier; even when the Italian government acquired 'stolen oil' from Dr. Mossadeq through a somewhat obscure firm by the name of SUPOR, Mattei would have nothing to do with them. Thinking that such an attitude had earned him gratitude which would be adequately expressed on a suitable occasion . . . Mattei appears to have inquired whether he would be included in the consortium to be formed."

Mattei was kept waiting for an answer until the consortium was formed, and then given a flat rejection by its members. "It was apparently said that only companies which were already holders of concessions in the Middle East would be eligible, and that to open the door to any others would mean creating a precedent which would call forth such a spate of claims of would-be participants as to render any solution impracticable." *

Afterward Mattei was to maintain that he was cheated by the big oil companies, and that they had made promises to him which they did not keep. To this, Howard Page replies: "At no time did Mattei ever approach me, nor did I ever hear that he had approached others, to get into the consortium. I'm sure I would have known if there had been a confidential approach. So how he had the nerve to say what he did about us I don't know. There was no need to pay him off for anything." †

Mattei might have taken his rejection with better grace—putting it down to the determination of the "big boys" to keep Iran to themselves—if there had not been a new development a few months after the formation of the consortium.

Although the American companies had been assured by the U.S. Justice Department that their participation in the Iranian consortium would not constitute a breach of the antitrust laws, the Department changed its mind after the contracts were signed. Responding to pressure from the independent oil companies' lobby in Washington, it abruptly announced that legal approval of American participation would be withheld unless independent U.S. oil companies were allowed a share of the partnership.

The five majors involved in the consortium thereupon met and decided to avoid trouble by ceding 5 percent (1 percent each) of their share of the consortium. Nine separate independent U.S. oil firms, seven of which had never been involved in the Middle East before, found themselves participants in the Iranian bonanza, and with no entrance fee either.‡ As P. H. Frankel

* *Mattei: Oil and Power Politics,* by P. H. Frankel.

† In a conversation with the author. Mr. Page does not appear to have been aware of Enrico Mattei's meeting with Lord Strathalmond of Anglo-Iranian in London during the Mossadegh crisis. Of course, it happened before the American companies were involved.

‡ Of the nine companies, two were Getty operations: Getty Oil Company and Tidewater Oil Company. The others were Richfield Oil Corporation, Standard of Ohio, Signal Oil and Gas Company, Atlantic Refining, Hancock Oil Company, Dan Jacinto Petroleum Corporation, and American Independent.

remarks: "It was almost grotesque to see that for no reason other than the desire to pay lip-service to free competition, a few American investors were, by courtesy of Washington, given what amounted almost to a free ride, although the oil they were to get played no functional role in their activities, whereas a country like Italy and a refiner/distributor like AGIP were to remain outside." *

For the newcomers it was "a license to print money," as Mattei put it. He never forgave the consortium companies for the snub, and from then on, until his violent and mysterious death in 1962, he embarked on what Frankel describes as "a collision course" against the major petroleum combines.

Meanwhile, the nine American newcomers to the consortium, who were not expected to take any part in its operations, sat back to enjoy their unanticipated profits. It was just like Gulbenkian all over again—except that the lucky nine had done even less to earn their 5 percent.

* Frankel, *op. cit.*

Part Five

THE BREAKTHROUGH

PRECEDING PAGE: *The rulers of the Federation of Arabian Emirates. From left: Sheik Khalid bin Muhammad of Sharjah; Sheik Zaid of Abu Dhabi; Sheik Rashid bin Said al Makhtoum of Dubai; Sheik Rashid bin Humaid al Naimi of Ajman; Sheik Ahmed bin Rashid al Mullah of Umm al Qaiwain; son of the ruler of Fujairah.*

CHAPTER SEVENTEEN

Trouble at the Oasis

IN 1953 CAPTAIN JACQUES COUSTEAU, the French undersea explorer, sailed his research ship, the *Calypso,* into the Persian Gulf and anchored off a small island about eighty miles from the shores of Abu Dhabi. The island is called Das, and at that time it was one of the wonders of ornithology. Each year millions of terns swarmed toward it from all parts of Asia and Africa and laid their eggs in clutches along the foreshore. Cousteau and his crew smelled the reek of guano from fifteen miles away, and were thrilled by the aerobatics of these sea swallows as they soared into the brilliant blue sky and then dove at the shoals of fish boiling close to the surface of the Gulf. For the next few months Cousteau and his divers were overside, mapping and photographing the sea bed below them, and soon they were joined by a vessel of the Geomarine Service International and a Swedish submarine charting ship.

For Cousteau it was one of the most fascinating missions of his distinguished career, for he found the area a paradise of animal, bird and fish life. The shores of Abu Dhabi and the islands were rich in fox, hare, gazelle, and a spectacular giant lizard whose Day-Glo colors changed with chameleon swiftness. The seas were alive with shark, squid, prawn, turtle, sailfish, horse mackerel, great schools of barracuda and massive shoals of multicolored fish of all kinds. The air quivered with the sound of wings.

Yet Cousteau was downhearted because he knew that the success of the mission on which he was engaged would doom this tract of pulsating animal life to a slow death by oil. This was the decade of expansion by the oil companies into the shallow waters of the Persian Gulf, and the five hundred miles of sea from the Shatt-al-Arab River in the north to the Strait of Hormuz in the southeast had been divided up between the states whose beaches were lapped by the waters. In Abu Dhabi the main concession had been granted to a company called Abu Dhabi Marine Areas (ADMA), in which British Petroleum and the Compagnie Française des Pétroles were partners, and already an advance headquarters had been established on Das. The first clutches of precious eggs had been trodden underfoot, and clouds of furious terns beaten off by laborers. By the end of the year, a 4,000-ton drilling barge named the *Adma Enterprise* anchored off the island and lowered its four retractable legs eight fathoms down to the sea bed at a spot Cousteau had designated. Built in Germany, the *Adma Enterprise* was a remarkable vessel, capable of submarine drilling to a depth of 15,000 feet, but it did not need to go that deep to strike oil. Biting through the sand and rock, it hit a pool at 5,500 feet; oil began seeping to the surface of the Gulf, and the first dead fish floated in the water.

By the time Cousteau sailed away in the *Calypso*, the bulldozers were biting into the small *jabals* of Das Island and carving out space for hutments, tank farms, an airstrip and a small golf course. Crude oil was rolled into the ground to lay the dust, and the smell of petroleum took the place of the reek of guano. Tern eggs were picked up and eaten as delicacies by the newcomers, and the birds that obstinately remained on the island became targets for oil company marksmen.

Soon even the shape of the island would change as the hillocks were shaved down for building sites, and every trace of vegetation would disappear, smothered by oil and the paraphernalia of a petroleum-collecting center and loading point. Where swordfish once broke the surface, gas fires would flare across the Gulf, and one of the most spectacular bird islands the world had ever known would be no more, killed by the civilized world's greed for petroleum.

TODAY THERE IS A HILTON HOTEL in Buraimi Oasis. It can be reached by driving ninety miles east across the desert from the capital of Abu Dhabi, on the Persian Gulf, along an immaculate six-lane highway, and after the glare and heat of the journey across the great sand sea, arriving there is an experience of physical bliss. The oasis is a green island of date palms and market gardens, irrigated from a great underground lake almost twenty miles long and ten miles wide. Gerbils (desert mice) scamper across the roads, and yellow-vented bulbuls sing sweetly in the trees. The men bathing in the irrigation ditches have ringlets of hair hanging down on their shoulders, and the women, as if to make up for the heavy veils covering their faces, daub their arms and foreheads with vivid green dye and line their eyebrows and hairline with henna.

One would never guess from the hotel itself that Buraimi is the haunt of giant scorpions, of camel spiders the size of saucers, and that the irrigation ditches house parasites that nest and grow into worms several yards long in the bowels of all who absorb them. The Hilton's lawns are immaculately green, cropped and constantly sprayed, and lined with oleanders and mallows. The rooms are walled with marble, air-conditioned, and staffed by willing Bedouins and obsequious Lebanese. The swimming pool is treated, filtered and deliciously cool; the beds are comfortable, the food is edible, and the French wines excellently chosen. In sum, unless you are apt to be bored with your own company, it adds to the pleasure that you are usually the only guest in the three-hundred-room hotel, outnumbered at least 150–1 by the staff, your merest wish their willing command. But if and when Buraimi becomes the capital of the new Federation of Arabian Emirates (as everyone in Abu Dhabi hopes), that will change and this desert Hilton will have all the guests it wants.

Oil paid for the Buraimi Hilton, oil paid for the desert road that leads to it—and twenty years ago, oil almost caused a war to be fought over Buraimi's green acres. Saudi Arabia has always maintained that the oasis lies within its boundaries, on the grounds that the Saudi tribes of the Bani Yas have used its wells as watering places for two centuries. Since Buraimi is at least six hundred miles from the nearest Saudi settlement and only ninety miles from both the capital of Abu Dhabi and the Omani port of Sohar, these last-named states have strongly resisted Saudi claims, and the ten or eleven villages which are spread over the oasis have almost always been under joint Abu Dhabi and Oman control.

But on September 1, 1952, a Saudi Arabian official named Turki bin Abdullah bin Utaishin arrived unheralded in Buraimi with a company of troops from the Saudi army. The trucks and pickup cars which had brought them and their arms and supplies across several hundred miles of desert belonged, it was subsequently discovered, to the Arabian-American Oil Company, and the expedition carried with them maps which had been drawn and prepared in the library and archive section of the Aramco operation at Dhahran.* These maps showed not only the whole of Buraimi as belonging to Saudi Arabia but also vast stretches of Abu Dhabi territory reaching down to the shores of the Persian Gulf. It was later to be considered not exactly a coincidence that this territory contained oil concessions which had been granted by the Abu Dhabi government to the Iraq Petroleum Company, Aramco's rival in the Gulf; IPC would, of course, have lost its right to them if Saudi Arabia's claim to the territory was established.

Turki bin Abdullah settled himself and his troops in the vil-

* The chief archivist at the time was Dr. George Rentz, now a professor at Stanford University.

lage of Hamasah, and promptly announced to the locals that he had been appointed emir of Buraimi by the Saudi Arabian government. He added that since September 1 was the day of Dhu-al-Jijja (the Feast of the Sacrifice), he proposed to celebrate the occasion with a great banquet and invited the inhabitants of the oasis to join him at the feast. Eating is something no Bedouin can resist, and great crowds converged on Hamasah from the neighboring villages to consume the scores of sheep which had been slaughtered for the occasion. At the end of the meal the guests were asked to sign a visitors' book which Turki's aides had set up on a table under the palms. Since few of the Bedouin could read or write, they could hardly be expected to realize that when they made their marks they were signing a document acknowledging Saudi Arabian sovereignty over the oasis.

Word speedily reached Sultan Said bin Taimur in Oman and Sheik Shakhbut in Abu Dhabi of what was happening at the oasis, and both reacted by summoning their British advisers. Like several other states in the Persian Gulf, both Oman * and Abu Dhabi were under British protection,† and the two rulers, terrified that the predatory Saudis were about to wrest from them not only the oasis but the potentially oil-rich territories beyond, urged their advisers to call out the troops and expel the intruders at once. Instead, the advisers alerted London, where the government, already being prodded by an alarmed IPC, decided to proceed with caution. Intelligence reports had already informed them that the Saudi invaders had traveled in Aramco trucks, and the U.S. involvement which this indicated was causing a good deal of concern at the Foreign Office. It was finally decided to take diplomatic steps; the British minister in Jidda was instructed to protest to the Saudi Ministry of Foreign Affairs and request the government to remove its troops from Buraimi at once.

The Saudis chose to ignore the protest. In the oasis, Turki bin

* The area's official designation is the Sultanate of Muscat and Oman.

† They had been ever since British warships cleared Arab pirates out of the area in 1853.

Abdullah gave more banquets, distributed bribes freely among the local Bedouin chiefs, and garnered more signatures for his "visitors' book." By this time rumors were rife in London that Aramco geologists had arrived in Buraimi and were using it as headquarters for oil searches in Abu Dhabi and Oman territory. IPC and Anglo-Iranian, the two British oil companies concerned, began to complain loudly that "the Yanks are trying to steal our oil." So far as the British press was concerned, Aramco was rapidly replacing the Saudis as the villains, especially after the British resident in Hadramaut (which was part of Aden Protectorate) reported trapping an Aramco crew drilling for oil thirty miles inside his territory. He announced that he had expelled the American members of the crew, taken the Saudi guards with them prisoner, and forced them to leave $3 million of equipment behind in the desert.* This was the way to treat oil grabbers, the newspapers commented.

After three weeks had passed, and the Saudis had still not given any signs of withdrawing from Buraimi, Sultan Said bin Taimur of Oman decided to take a personal hand in the affair. He drove his open yellow Cadillac down the coast from Muscat, his capital, to Sohar, where he had assembled an army of seven thousand tribesmen mounted on camels. With a golden sword buckled around his waist and a rifle over his shoulder, he harangued the cheering Bedouins and told them he was leading them into battle against the hated Saudis in defense of their homeland. But before he could give the order to march there was an interruption. Along the foreshore a small truck raced up to the sultan's Cadillac, and out of it climbed a lanky figure in a gray suit and brown fedora, the drab Western clothes making an odd colorless dab in the multihued sea of shouting Bedouins. Major F. C. L. Chauncey, British consul in Muscat, was bringing an urgent message from the Foreign Office. The sultan must call off his attack, in return for which the British government pledged to secure a peaceful settlement of the dispute.

* See *The Wind of Morning*, by Sir Hugh Boustead.

The sultan protested; the consul insisted. It was not until after an hour of fierce argument that the ruler regretfully ordered his levies paid off and dispersed.

An American friend of the sultan's subsequently wrote: "It has been said certain Americans expended great energy to persuade the British authorities to intervene and use their good offices in requesting the Sultan not to march on Buraimi. It would have been a real jolt to [Saudi Arabian] prestige if the Sultan had returned at this time to restore law and order by forcibly ejecting the Saudis." *

Chauncey himself said later: "We couldn't let them do it. The Americans were so mixed up in it all. . . . Suppose the Sultan had gone in under our auspices and some idiotic American geologist had got himself killed. There would have been hell to pay." †

On October 12, 1952, a "standstill agreement" was proposed by the U.S. ambassador to Saudi Arabia. It was accepted by the British government on behalf of Oman and Abu Dhabi, and formally signed on October 26. But soon stories began to emerge from Buraimi that Turki bin Abdullah, the Saudi representative, was using the truce to spread bribes and pro-Saudi propaganda among the tribes. The British therefore decided to use force at last, and the Saudis were surrounded in their village of Hamasah by troops of the Trucial Oman Scouts, Bedouin levies led by British officers and NCOs. There were some violent skirmishes in which a British sergeant and a corporal were killed, and British newspapers began to wonder out loud why soldiers so far from home had to get themselves killed in such a godforsaken hole. There were urgent talks between the British ambassador in Washington and the State Department, and a verbal promise was given that the United States would advise the Saudis to withdraw completely from Buraimi. But before such instructions could be sent to the U.S. envoy in Jidda, State changed its mind and decided not to interfere. It was later charged that Aramco lobbyists were

* *Oman: A History,* by Wendell Phillips.

† Quoted in *Sultan in Oman,* by James Morris.

effective "in driving the State Department back to the safe position of neutrality from which it had begun to emerge." *

On July 20, 1954, an Arbitration Agreement was signed under which Britain withdrew the Trucial Oman Scouts (they were replaced by a joint police force) and all oil operations were suspended in the disputed area. An International Arbitration Tribunal was set up in Geneva to settle "the location of the common frontier between Saudi Arabia and Abu Dhabi and sovereignty in Buraimi Oasis," and it was through its statement of claim to this tribunal that the Saudi government revealed that it was asking not just for the oasis itself but for "the beaches, banks and islands [on the Persian Gulf] of which have been resorted to for two centuries by the Bani Yas tribes for their fishing and pearling." †

The tribunal was presided over by a Belgian, Dr. Charles de Visscher, and its four other members consisted of a Cuban and a Pakistani jurist, plus one member each from the contending countries: Sir Reader Bullard for Britain and Sheik Yusuf Yasin for Saudi Arabia. Sir Reader had been British minister to the Saudi kingdom before World War II, and had known and admired the late King ibn Saud. Yusuf Yasin was described by a European in Jidda as "an island of integrity in a sea of corruption." Their reputations seemed to ensure a fair decision, and the Saudi government, its brief directed by a team of high-priced international lawyers from New York, its case backed by some formidable arguments and maps prepared in the archives of Aramco, was apparently prepared for a straightforward argument of its case and reliance on the justice and rightness of its cause.

But soon Britain was complaining, and not without reason, of "breaches of the conditions of arbitration which were agreed upon by the parties in an exchange of Notes." ‡ They produced

* See *Contemporary Arab Politics,* by G. E. Kirk.

† Statement submitted by the Saudi Arabian government to the International Tribunal on the Question of the Buraimi Oasis, Geneva, 1954.

‡ Statement of the aims of the tribunal by the Foreign Office, London, quoted in the *Times,* London, September 5, 1955.

evidence to show that attempts had been made by the Saudi government among dissident members of Sheik Shakhbut's family to organize a coup d'état against him. When these failed, they had sent an emissary named Quraishi to talk to Shakhbut's younger brother, Sheik Zaid, who lived in a fortress in one of the Buraimi villages.

> Quraishi told Shaikh Zaid [the note went on] that he had heard from the Arbitration Tribunal that the territory would be returned to the Saudis, and promised that if he would throw in his lot with the Saudis he would be assured of his position in Buraimi, and would receive funds from the Saudis and 50 percent of the profits from any oil that might be discovered there. Quraishi also said that should the arbitration go against the Saudis they would take the area by force.*

Zaid, who drove around Buraimi in a British car of pre–World War II vintage, was also offered an air-conditioned Cadillac ("from our friends at Aramco") and an immediate gift of $9,000 as a token of Saudi Arabia's future esteem. By this time he had let the local British military officer, Captain Clayton, know about the Saudi approaches and was told to encourage them.

> At a further meeting between Zaid and Quraishi on August 4 [the British declaration went on] Quraishi said that King Saud would give Zaid 400,000,000 rupees ($90,000,000) if he would prevent the Iraq Petroleum Company from operating in the disputed territories and leave the field open to Aramco. King Saud would like to give a written guarantee to this effect, but was afraid to do so lest it should fall into British hands.†

Later the Saudis were to deny scornfully the suggestion that they had offered such a fantastic bribe to Sheik Zaid. "With that sort of money," one of them said, "we would not have needed to buy a desert sheik. We could have bought the British." Sheik Zaid has since agreed that it was an unrealistic sum,‡ but its sheer size reinforced his conviction that the Saudis were not only trying to grab Abu Dhabi territory, but that afterward they

* *Ibid.*
† *Ibid.*
‡ Though by no means as much as Zaid possesses today.

would renege on all promises they had made to those whose support they were seeking.

Instead of indicating his willingness to accept the bribe, Sheik Zaid flew to Geneva with Captain Clayton to give evidence against the Saudis at the arbitration tribunal, and no amount of cross-examination by Saudi Arabia's high-priced lawyers could shake him. The Saudi political agent, Quraishi, was then called to the stand, but his evidence rapidly lost its validity when it was noted that Sheik Yusuf Yasin, the Saudi member of the tribunal, was repeatedly sending notes down to the witness, who, after reading them, would immediately raise new points of complaint against Zaid. When the Belgian president of the tribunal, Dr. de Visscher, mildly suggested that his fellow arbitrator should cease interfering with the witness, Yasin turned the whole proceedings into a farce by declaring that he was in fact Quraishi's employer and accepted full responsibility for the agent's conduct, because he, Yasin, was the Saudi government official responsible for Buraimi Oasis affairs. At this, Sir Reader Bullard was heard wondering aloud how, in the circumstances, he could be expected to judge the case with impartiality.

"Impartiality?" cried Yasin. "What has that to do with this case? I am here to arbitrate, not to be impartial."

At which point the whole proceedings collapsed. The next day Sir Reader Bullard resigned from the tribunal on the grounds that it "has been hopelessly compromised by the conduct of Shaikh Yusuf Yasin and by other distasteful matters which have come to notice. I do not think the Tribunal is any longer in a position to reach a unanimous or judicial conclusion." * A few hours later Dr. de Visscher and the Cuban member also resigned, and everyone packed up and went home.

Afterward Harry St. John Philby maintained that the venality of the new king of Saudi Arabia and the corruption of his courtiers had ruined what otherwise might have been a good case for

* From Sir Reader Bullard's letter of resignation, quoted in the *Times*, London, September 17, 1955.

the establishment of Saudi hegemony over Buraimi and its environs—and the ousting of IPC from the Abu Dhabi concessions in favor of Aramco. "If it [the Saudi claim] had been presented properly on the basis of the admirable brief prepared by the American advisers of the Saud government," he wrote, "[it] would have had a very fair chance of acceptance on the basis of law and historical precedent . . . the fountain of Arab chivalry has been fouled with oil; and the mouths of the preachers and prophets have been stopped with gold." *

In point of fact, it is doubtful whether any impartial tribunal in the world would have been prepared to accept Saudi Arabia's claim to an area so far from its own centers of population and so close to the capital of Abu Dhabi and the ports of Oman, but since the Saudis had ruined their own case, no one will ever know.

King Saud went on maintaining his country's right to Buraimi long after the tribunal had been abandoned. In 1956, a year later, a statement was issued from Riyadh, the Saudi capital, reading:

> The Saudi Arab Government hereby gives notice to all concerned that it regards the Buraimi Oasis as part of Saudi Arab territory, and that it will not recognize any concession for the production of petroleum or any other mineral or minerals or any concession whatsoever granted by any Government or authority in the said area. Any company or firm or person that enters into such an agreement will be doing so at its own risk and the Saud Government will regard such agreement as null and void, and will neither be bound by it nor responsible for any consequence.†

On the other hand, Aramco has eliminated all mention of Buraimi from its files, and a glazed look comes into the eyes of its American officials when they are asked for an account of the part played by the company in the episode. As for IPC, what it finds particularly reprehensible about the affair is the fact that throughout the protracted incidents, charges and proceedings, Aramco never once conferred with it, despite the fact that two of IPC's

* Article in the *Sunday Times*, London, October 23, 1955.

† Statement published as an advertisement in *Oil Forum*, New York (January 1956).

important shareholders are Standard of New Jersey and Mobil, both of which are major shareholders of Aramco.

"What else could we have said to them," replies an anonymous Aramco spokesman, "except 'We wuz robbed!'?"

As indeed they were, if Saudi Arabia's claim had any validity, for the areas of Abu Dhabi which the Saudis hoped to annex have since proved to contain some of the richest oil fields in the Middle East.

CHAPTER EIGHTEEN

Three Sheiks and an American

IN THE 1950S THE TRUCIAL STATES OF ARABIA (Abu Dhabi, Dubai, Sharjah, Ras al Khaimah, Ajman, Umm al Qaiwain and Al Fujairah) were ruled by an oligarchy of rapacious sheiks whose power of life and death over their subjects was unique in the world. Rarely has there been a greedier group of men, nor a more short-sighted one.

Abu Dhabi was on the brink of one of the most fabulous breakthroughs in the history of oil exploitation, and already large revenues were beginning to flow into the sheikly coffers. Yet Shakhbut bin Sultan, the ruler, had no intention of using his growing wealth to improve the lot of his people. He preferred to keep it in gold bars under his bed and in stacks of notes in a back room off his living quarters.* He was a mean and miserable old man, not given to cleanliness (the virgin slave girls for whom he had a passion must have found him a loathsome burden), suspicious, jealous, arrogant and subject to black rages, when his voice rose and he shrilled like a child.

Anyone who tried to get Shakhbut bin Sultan's signature on

* When he was ousted in the 1960s it was discovered that several square yards of money, worth at least $2 million, had been eaten by rats.

a document learned just how cautious he could be. His British adviser recalls an occasion when one of the managers of a large American bank was attempting to gain an entry for his branch into Abu Dhabi.

"The bank manager was accompanied as interpreter by a Lebanese of distinction," the Briton wrote, "who was on his knees before the ruler praising him, his wisdom, generosity and charm. It would really have been difficult for the ruler not to accede, and it was only when the bank manager put his hand to his inside pocket and started to extract a paper which was obviously intended for the ruler to sign that Shakhbut drew back with the look of a trapped animal. The Lebanese was forward on his knees like a praying mantis, hands raised in supplication. The ruler sat terrified on the back of the sofa and I shook my head to the bank manager signalling him to put the paper away quickly. So keen was he on his object that he was a bit tense and tried to press on, resulting of course in the ruler's even greater apprehensions and more solid resolution not to sign the paper." *

Even in the 1950s Shakhbut was receiving something like £20 million a year in oil revenue, and with a tenth of that sum he could have brought some measure of medical aid and education to the 15,000 people of Abu Dhabi. But that would have meant parting with his precious notes and bars of gold, and the sheik could not bear it.

The Trucial States of the Persian Gulf were all under the paternal control of the British. Its influence had been dominant in this area for a century, ever since British warships had attacked the buccaneers who terrorized these waters in 1853 and forced the rulers of the states along the Gulf shore to sign treaties pledging themselves to cease their attacks on passing shipping. The name of the area changed from the Pirate Coast to the Trucial States, and henceforth there had been British advisers on hand. They directed the foreign policies of the tiny sheikdoms, and provided officers and NCOs to stiffen their small defense

* *The Wind of Morning*, by Sir Hugh Boustead.

forces. They pledged not to interfere in the domestic affairs of their desert brood, and by so doing condoned rulers who at their best were petty autocrats and at their worst bloodthirsty tyrants.

Oil and the revenue it produced were beginning to bring outsiders and twentieth-century ideas into these stuffy little kingdoms, but rare were the sheiks who welcomed them. British political agents longed to see the money from oil revenues used for worthwhile projects, and in Abu Dhabi the local British agent, Colonel Hugh Boustead, pressed Shakhbut to employ part of his wealth in changing Buraimi from a disease-ridden, one-crop oasis into a market-garden showplace of Arabia. In furtherance of his hopes, he produced an agricultural expert and a detailed scheme for the sheik's inspection. He might as well have been talking to a deaf man, for Shakhbut would hear nothing of his plans. "My people know more about agriculture in their own land than you can tell them," he said to the crop expert. "The budget [for the agricultural plan] is a pure waste of money. We do not want it and we will not have it. What can you tell us that we do not know?" *

It so happened that present at the interview was the ruler's brother, Sheik Zaid, whose home was in Buraimi, and Boustead caught the look of anger and frustration in the younger Arab's fierce black eyes. "Shaikh Zaid was very disappointed," he wrote later. "He took on a very able Pakistani agricultural officer, Mr. A. H. Khan, and paid him from his own funds, but there was no supporting budget, and as a result agricultural development has been held up completely . . . The most disheartening feature of all was [the ruler's] simple refusal to give his own people, a pastoral, agricultural and fishing society, the easy assistance which his growing wealth could provide." †

One day there was to be a confrontation between Shakhbut and Sheik Zaid, as there always has been between brother and brother in the bloody history of the ruling families of the Trucial

* *Ibid.*
† *Ibid.*

States. No one could guess this at the time, or who would win, but Boustead had no doubt which of the two sheiks the people of Abu Dhabi would prefer as their ruler—if they had to have a ruler, that is.

THE GREEDY SHEIKS dreamed of the wealth and power that the discovery of oil would bring them, and no one more so than the tyrannical ruler of the wild sultanate of Muscat and Oman, whose shores stretched all the way from the Gulf of Oman in the north, along the Arabian Sea down toward the Gulf of Aden. Sultan Said bin Taimur was the kind of Arabian sheik that the playwright must have been thinking about when he wrote *The Desert Song*. With his Arab clothes and flowing beard he looked somewhat like a Biblical prophet, but he also liked to lave himself with unguents, groom himself meticulously and dress in Western clothes cut by Savile Row tailors. He spoke flawless English, which he had learned at school and university in the United Kingdom, and he had a devastating charm which he used—often to a degree they lived to regret—on young European women. He was also a despot who imposed on his people a regime so rigidly theocratic that even Saudi Arabia in the heyday of the *ikhwans* (the fanatic Muslim sect) had known nothing as hypocritically sanctimonious. In his humid capital of Muscat no music or dancing was allowed. Women were permitted on the streets only after darkness, and then only in the company of bodyguards. There was a curfew, at which time the gates of the city were shut tight, and before it all who moved abroad must carry a lantern with them—and by lantern the sultan did not mean a flashlight. Wrongdoers were mercilessly lashed or bastinadoed or slowly tortured to death.

But the sultan's greatest crime against his people was his neglect of them. In no other country in the Middle East were curable diseases so rampant: syphilis, dysentery, tuberculosis, malaria, fearsome intestinal worms, all of which tribal quacks treated with hot irons, and then rubbed the burns with camel dung and exposed to the flies. There was but one hospital, run by a few brave American missionaries, infidels tolerated by the ruler because they cost him nothing. It was a land of prohibitions, ferocious prejudices and fanatical cruelty, but those who suffered the most, their miserable lot approved by the tyrant ruler, were the women. An American woman missionary said:

> Once again comes the agonizing cry: "Come quickly, Doctor, my wife is in labor and cannot deliver!" Is this her first baby? No, her second. She had no trouble the first time. Atresia: that is what the doctor suspected and found. Atresia, a condition which causes the death of many a mother and her unborn child . . . is caused by fear —fear of divorce. Atresia is a condition where the normal tissue of the vagina becomes inelastic scar tissue. When this condition exists the child cannot be born. The woman in labor suffers agony, for in her body the irresistible force of birth contractions propels the foetus against an immovable barrier of rigid scar tissue, which has caused the birth canal to tighten like a purse string. Suddenly everything rips open.
>
> How was this scar tissue formed? The answer is that it results from the packing of rock salt into the vagina after childbirth—after the birth of the first baby. Why salt? The answer—to contract the vagina lest the husband, not deriving the satisfaction from his wife he experienced before delivery, should divorce her. . . . Surgery alone can save the mother and child. Every year a score of such operations are performed in the American Mission Hospital in Muscat.*

Only education could bring this barbarism to an end, and only education could persuade the sultan's subjects to seek medical care instead of savage tribal remedies for the loathsome diseases to which they were heir. But there was only one primary school in Muscat; it had few Arab pupils and was largely used by the sons of Indian merchants. When the British suggested a program for the building of schools, Said bin Taimur would have none of it. "That is why you lost India," he said, "because you educated the people."

* Account given by Mrs. Beth Thoms, wife of Dr. Wells Thoms, U.S. missionary in Oman. Quoted in *Unknown Oman,* by Wendell Phillips.

But oil he wanted. He had traveled north to Kuwait and seen there the gaudy opulence that flowed in the wake of oil revenues. He was convinced that in his vast kingdom of desert and mountain there must be equally rewarding troves of petroleum. But who would find them for him? So far he'd had no luck. The Iraq Petroleum Company was still searching along the coastal plain of western Hajar, in Oman proper, and though they would make valuable strikes in the years to come, to date their labors had produced no viable fields of oil. Farther south, in the mountainous dependency of Dhofar, IPC had given up its concession at the end of 1950, defeated by the terrain, the absence of roads, the lack of water, and harassment from rebel tribes infiltrating from Saudi Arabia and Yemen.

"Not too encouraging, is it?" said the sultan one day in 1953. He was talking in his seaside palace in Salala, in Dhofar, to a young American named Wendell Phillips. A year previously Phillips had led an archaeological expedition into Yemen, which lies to the west, in search of the sites of ancient Sabean cities. Suddenly the party was attacked by rebel tribes and was forced to flee into Dhofar, leaving most of their equipment behind them. Called to an audience with the sultan, the American had recounted the story of his misfortunes and to his surprise found the ruler to be a sympathetic listener. In fact, the two men established an immediate rapport. Phillips is a tall, pale, somewhat theatrical figure who likes to sport Arab dress and carry a couple of six-shooters in his belt, and he talks with the fluency and fervor of a revivalist preacher. He described his explorations with such enthusiasm that Said bin Taimur was moved to unexpected generosity; he invited Phillips and his party to continue their researches in Dhofar, and advanced him money to pay the cost of new equipment and native labor at the digging sites.

Now Phillips had requested this audience to tell the sultan that the money was used up and that the expedition was on its way back to New York. Obviously taken aback by the imminent departure of a young man of whom he had become extremely fond, the ruler thought for a moment, and then said, "We need oil here in Dhofar, not yearly payments. And by the will of God we shall have oil, for I am granting *you* the oil concession for Dhofar."

Recalling that moment, Phillips said in a recent conversation, "I didn't know what to say at first. All I could think of to ask was how big the concession was, and the sultan waved his hand and said, 'All of Dhofar. It is as big as your state of Ohio.' It gradually dawned on me that Said bin Taimur was offering the rights of this huge piece of territory to me. I thanked him for his gracious gesture, but then, my wits coming back, I pointed out that I knew nothing about oil, or oil concessions, or Arab oil concessions."

"You'll learn," the sultan said. "Now, shall we begin work on the oil agreement, or would you prefer tomorrow morning?"

As Phillips wrote later: "When you are a penniless explorer and have no knowledge of the oil business, the possession of an oil concession presents certain problems. Where does one go from here?" *

Phillips flew to New York convinced that he had a sure-fire fortune in his briefcase, and that all he needed to do was knock on the door of the nearest oil company and offer it to them. But it wasn't quite that easy. "For one thing, IPC had given up the concession after fourteen years," he recalled. "True, they had never drilled a well in the whole of that time, merely made geophysical surveys, but their failure to do so was a strike against the terrain. Also, Dhofar sounded so remote. I had to admit that when summer came it got really hot—though not as sticky as Muscat, where the climate is hell on earth—and that you wallowed in floods of mud when the tail end of the monsoon hit. I soon learned that the cost of development would be enormous. Since there were no ports or docks, everything would have to be flown in and brought ashore by dhow. Time and again, oil company executives told me to hang the concession on the wall as an interesting decoration."

* *Ibid.*

Then Sam Pryor, a Republican leader and a vice-president of Pan American World Airways, joined forces with the young explorer and in New York formed a company, Philpryor Corporation, to exploit the concession. Pryor had friends and know-how. In short order, he assigned the concession to Cities Service, a holding company with many subsidiaries, and through it to a new subsidiary. Through the sale of units, Phillips suddenly found himself in possession of his first million dollars.

"I remember I had a small account at the Chase Bank," he said. "It couldn't have been more than thirty dollars. So I went in and handed over the check I had just received from the brokers for a million dollars. Next thing I knew two policemen had grabbed me by the arms because the teller had rung the bell. Well, I managed to persuade them that the check was genuine, and the next day the manager invited me to lunch and brought his daughter along. It was my first indication of what it was like to have a million—and of the potentialities of the oil business. And we hadn't even started looking for oil in Dhofar yet."

Two years later, in 1957, at Marmul, thirty-seven miles inland from the Dhofar coast, the Phillips concession struck oil at 320 feet. After months of heartbreaking failure, it looked as if the company had at last found a textbook structure, an oil reservoir at least eleven miles long by five miles wide.

When the news was broken to Sultan Said bin Taimur, he was so overjoyed that he ordered an immediate reduction in the import duty on coffee, and allowed the curfew in Muscat to be delayed one hour. To his American friend he said, "Wendell, now you will be rich. How does it feel?"

"I am a modest man," said Wendell Phillips. "I shall not need to learn how to be a modest billionaire."

But both of them would have to learn that in the oil business, bonanzas are not always what they seem.

AL KUWAIT WAS NOW THE DREAM CITY that every Arab envied. In 1952 its population had been 310,000, and its income from oil $60 million. Two years later it was $210 million, and by

1955 it had reached $300 million. Each year more and more money flowed into the tiny state's coffers, until they were soon literally choking on it.

"At first we used to keep the revenue in an enormous safe in my office," said Colin McGregor, who is now financial adviser to the Kuwaiti government, in a recent interview. "We stuffed in stacks of overprinted Indian rupees, gold bars and sacks of sovereigns. There were millions and millions of them. The ruler's relatives would come in and pick up a million rupees at a time, and would get very snooty indeed when I insisted on a receipt. I remember how once one of Sheik Abdullah's most spendthrift brothers sent a message over saying that he needed a huge sum of money in gold sovereigns and Maria Theresa dollars. I put them in two large sacks and sent them over to him, but with strict instructions to the guards not to come back without a receipt. They came back an hour later with an old piece of paper on which was written: 'Received: 2 sacks of money.'"

Sheik Abdullah al Salim al Sabah, the ruler of Kuwait who had succeeded his father in 1950, was a shrewd man. He decided that if such a flood of money was going to flow into his possession, a large portion of it should be channeled off into the pockets of his people so that none could accuse him—as they did in the case of King Saud of Arabia—of keeping it all to himself and his cronies. Colin McGregor and other financial advisers persuaded him to keep one third of the yearly income as capital for investment, but the rest was lavished on the natives of Kuwait in proportion to their status. The ruler decreed that the squalid old town and port from which the pearl divers had sailed, and the bazaars where their finds had been bought and sold, should be razed to the ground and a new city constructed to take their place. He sanctioned an immediate expropriation of land, but at a handsome compensation, and soon there were few Kuwaitis who did not benefit from the boom, which established the price of land at $200 a square foot.

Henceforth Kuwait was to be a welfare state in which no citizen would lack for space and recreational centers while he was well, or medicine and hospital care when he was sick. A large rest home and hospital center was constructed for old people—this in an Arabia where traditionally the aged have been left to wither away in pain and squalor. Schools, wide roads, handsome squares and planted traffic circles began to rise out of the sand; and soon cars flooded in to fill them with traffic. From then on, if you saw a man with a camel or an ass, he was not a Kuwaiti.

All this cost an extraordinary amount of money (about $280 million in the space of five years), and since few Kuwaitis knew anything about construction, the city became the inevitable victim of foreign jerry-builders. Roofs fell in and roads buckled and split; cheap materials collapsed under the onslaught of sandstorms or winter rains. Entrepreneurs upped their costs in the course of construction, rightly calculating that the Kuwaitis would pay, pay and pay again to get a new swimming pool or high school finished. By the end of 1956, the head of Kuwait's development board, who happened to be the ruler's brother, had to confess that he had gone $200 million beyond his budget. He was reprimanded and told not to spend so lavishly in the future, but it was not until the 1960s that financial experts and accountants were called in to introduce some reality into the extravaganza.

From Sheik Abdullah's point of view, however, this was all in the good cause of keeping his subjects happy. As for his close relatives, he was prepared to pay them incomes of anything up to $1 million a year, on the theory that a young man with money to spend—and the relatives were encouraged to spend—would not be likely to waste time on plotting assassinations (hitherto a family preoccupation) or coups d'état. To show his gratitude, one of these relatives organized a feast which began with a bonfire started by igniting a stack of hundreds of 5-dinar notes (worth $12 apiece), and which was climaxed by a presentation to the ruler of a model of an oil derrick fashioned in solid 24-carat gold. In between, guests consumed young roasted camels, each camel containing inside it a roasted sheep, and each sheep stuffed with roasted chickens, ducks and pigeons. Each of the birds contained a boiled egg.

Along the shores of the Gulf today can be seen the traces of what the ruler's relatives did with some of their enormous incomes. They built palaces with huge filtered pools, air conditioning and great marble chambers, filled them with expensive modern junk from European department stores, and then let them rot under the onslaught of sandstorms, humidity, ants, mice and rats. Private golf courses were laid out in the desert and planted with shrubs, imported at vast expense from Europe and America, and then allowed to wither and die for lack of water. The cabarets of Beirut, Cairo and Istanbul were plucked of their prettiest *poules,* who were brought to Kuwait to adorn the tawdry palaces, and for a while the streets of the city were alive with female voices chattering in German, Hungarian, French and Rumanian from behind gossamer veils; then a sudden rash of venereal disease started a wave of revulsion, and the foreign tarts were expelled. Back they went to the Levant, well content with the fruits of their labors in this hot land, laden with gold bangles and precious jewelry.

The next wave of foreign females were rather more expensive, principally because their backgrounds were more respectable. Suddenly there was competition among the spendthrift princes to import the youngest Western girls they could find. They had to be beautiful, modern in outlook, but at the same time they must be both virgins and from bourgeois families. How costly this could prove was demonstrated by the example of the plump minister of education, Sheik Jabir, whose eye lighted on the nineteen-year-old daughter of a North German hotelier. Jabir, sixty years old, had already been married thirty-one times, his last wife being a beautiful Lebanese. The German girl's mother insisted that before her daughter contemplated a union with the sheik, he must first rid himself of all his wives, and the ousting of the Lebanese alone cost him $3 million. Eventually the fair-haired Rhinemaiden arrived in Kuwait to find wedding presents awaiting her of a stable of Arab horses (which she could not ride), twenty-three Cadillacs (which she could not drive), a load of precious jewelry (which she sent back to Germany to be reset), a rose-marble palace on the Gulf (which she hated because it was too humid, and the Gulf smelled), and a husband she could not stand. Within a year she was back in Germany, divorced, but worth several million dollars more.

Sheik Abdullah was indulgent about these peccadilloes. He himself had paid dearly for a new young wife, a young girl from Latakia whom he had seen in a swimming pool during a sojourn in one of his palaces in Lebanon. He was now past sixty, and it was said that girls from Latakia were wonderful with old men. As for the reckless waste of money on palaces and jerry-built schools, hospitals and roads, they could all be repaired, and one day they would all contribute to the rich, modern and prosperous image of Kuwait for which he was striving. What did it matter if the image had cost more than it need have, that there had been corruption and waste? Hadn't thousands of Kuwaitis grown rich in the process? And what did money matter? Along the road into the desert, the gas flares lit the *jabals* like candlesticks on an altar, and out in the Gulf, tankers lined up for Kuwait's inexhaustible supplies of oil. Why worry about expense when there was so much oil to help pay for it? "It is all part of the process of learning how to live with oil," the sheik said.

But in the meantime, the citizens of Kuwait were beginning to grow accustomed to their heady wealth. Bedouins who had once been proud possessors of a brace of camels now transported their families in brand-new American cars and refused to send their sons to the ornate secondary and technical schools which the ruler had built. What would they learn there? To make furniture and mend cars and repair electrical circuits? But they were demeaning jobs, not respectable positions for the prosperous new Kuwaiti. The sons were sent abroad to learn more effete professions, and into Kuwait to do the repair work and the dirty jobs came Arabs from other countries: hungry Iraqis and Yemenites, and resentful Palestinians and Egyptians. They worked for the rich new welfare state, but except for their wages they drew no benefit from it. They were the poor Arabs at the rich Arabs' table, and one day in 1956 they gave the profligate princes of Kuwait an ominous demonstration of just how bitter they felt about not being participants in the feast.

CHAPTER NINETEEN

Blowup

IN JULY 1596, at the urging of Secretary of State John Foster Dulles, the American and British governments withdrew their offers to finance the building of the High Dam on the Nile in Egypt. The Aswan Dam was the dream of Egypt's new ruler, President Gamal Abdel Nasser, and this slap in the face was meant to discipline him and make him agree to certain onerous conditions which the Anglo-Americans were demanding in return for the loan. They misjudged their man. Riding securely on a wave of Arab nationalism, Nasser replied by nationalizing the British- and French-owned Suez Canal. At the end of October, the British and French, having made a secret agreement with the Israelis, moved a joint army into Egypt with the intention of recapturing the Canal by force.

Even in the Victorian era when Britain was the most powerful nation on earth, it would have been a foolhardy operation. In the twilight days of the Empire, it was suicidal. Though sworn allies of the British, the Americans backed off at once and condemned the use of force. Even the British ambassador in Cairo winced as he heard RAF planes overhead and saw British paratroopers floating down onto Egyptian soil. "It would not be difficult to defeat the Egyptian forces," he commented, "but the difficulties would start after that. The Egyptians would organize guerilla warfare and it would be difficult for us to disengage against guerillas organized by Nasser or, if he had fallen, by his proclaimed successor. No Government set up by the occupying forces would last. Only a Government untainted by collaboration with the British could hold its position. If the British and French set up an international Canal management, they would find it difficult to withdraw their troops from the Canal and leave a foreign company, imposed by force, without their protection." *

This unhappy, ill-thought-out, ill-prepared operation had a vast number of melancholy world-wide consequences in the months to come, and British pretensions to global power would never be valid again. But in the Middle East the repercussions of this political and military blunder were immediate, and it was in Kuwait that they exploded most dangerously, for there it was not just against the British that they were aimed but against the Americans and the Kuwaiti ruling classes as well.

As has been suggested, it could almost truly be said that every native-born citizen of Kuwait was a member of the Kuwaiti ruling classes. They were all privileged, though of course some were more privileged than others. Soldiers in the small national defense forces were paid $150 a week, but were reluctant to volunteer because there was so much more to be earned with less effort. Loans with which to buy houses or businesses that would sound huge to British and American workingmen were freely available for every certified citizen. A Kuwaiti had only to claim some obscure disease for one of his children, and the whole family would be flown to Europe at government expense for expert medical treatment. But while "ordinary" Kuwaitis took it easy on government subsidies, while Kuwaiti businessmen garnered millions in government contracts and while Kuwaiti princes poured out astronomical amounts in high living, the Arabs lacking citizenship were enraged by the inequalities of the situation. The Suez

* *The Middle East in Revolution,* by Sir Humphrey Trevelyan.

crisis and the brush fire of anti-British anger which swept through the Middle East as a result gave them the opportunity for which they had been waiting to hit back at their arrogant masters.

At first it seemed that the protest marches which began on the streets of Kuwait were the expected anti-British reactions for which the Kuwaitis had been prepared. The ruler and his advisers had anticipated that British business houses would be smashed up, the British flag burned, perhaps even a few British subjects beaten up or killed. But that was to be expected; the British had asked for trouble, and now they would get it. So for the first forty-eight hours after the news about the invasion arrived in Kuwait, and the fulminations of Radio Cairo blared from the coffeehouse loudspeakers, the local authorities shrugged their shoulders.

Sheik Abdullah al Salim al Sabah took the whole business with great calm, though Gawain Bell, the British political agent, who brought him the news, was filled with apprehension. ". . . I found the ruler perfectly composed," he said. "His strength is to know when to procrastinate—how to avoid becoming involved in the dissensions of Arab politics. His last words to me in this meeting were: 'My boy, go back to your files and don't worry!' He then sailed away twenty miles to Falaika Island and stayed there for three weeks, incommunicado with his own government or Her Majesty's Government!" *

By this time anti-British and pro-Nasser banners were flying all over Kuwait, and threatening crowds had gathered before the British Political Agency. Huge Kuwaiti cars drew up and whole families leaned out of the windows to watch the fun.

While they were doing so, one of Kuwait's notorious sandstorms came roaring in from the Arabian Desert, and under its cover a number of saboteurs in small groups moved into Ahmadi, the oil town forty miles down the coast from Al Kuwait and the headquarters of the Kuwait Oil Company. L. T. Jordan, the company's British general manager, was driving back through the

* Quoted in Hewins, *The Golden Dream.*

storm from an inspection tour of the oil-gathering installations when he saw a sudden flash of red through the brown murk ahead of him. Soon his car was skidding and sloshing through a morass of escaping oil, and black smoke from an exploded pipeline had turned the sandstorm into a choking, impenetrable fog. He managed to turn his car around and make his way back to a sentry post, where he gave the alarm.

But by that time, nineteen other pipelines had gone up and were burning fiercely, their flames whipped by the gale-force wind. The saboteurs were even equipped for diving operations; they slipped into the water near the submarine lines which carried oil to the tankers waiting out to sea, and soon one of the lines went up and began pouring thick crude into the Gulf. The order went out to close down all the wells and to seal off the tank farm. All over the desert, the wells stopped working and the fires by which the company disposed of surplus gas guttered out. But the sabotaged lines burned on—and kept burning for days.

By now these incidents were not merely an anti-British demonstration; they were a bid to shut down Kuwait's oil supply. When news of what was happening in the desert reached Al Kuwait, the authorities woke from their complacent apathy and proceeded to act with a speed and ruthlessness which made Anglo-American oilmen whistle with reluctant admiration.

The man in charge of Kuwait's internal security was a young relative of the ruler named Sheik Abdullah Mubarrak. He had a pale, petulant face—he had inherited his complexion from his Circassian mother—which he adorned with a twirly mustachio, he gave sybaritic parties in one or another of the dozen palaces he owned, and he was said to have a solid-gold field marshal's baton in his office safe. Mubarrak ran his security forces as his personal bodyguard and paid them well over the wages granted to them by the state, and he forgave them anything if they kept law and order. Now he ordered them to the site of the demonstrations that were still going on in front of the British Political Agency, and personally harangued the crowds, urging them to disperse, and telling them they were victims of Communist agitators. Those who did not go quickly enough he flailed with his stick, and those whose appearance gave them away as Palestinians, Egyptians, Iraqis and Iranians, he ordered to be rounded up and taken away for questioning.

In the meantime, all Communist clubs and the cafés where foreign teachers and other workers gathered were being raided, and hundreds of young men brought in. They were beaten first and questioned afterward. Mubarrak knew he did not have much time. Al Kuwait is a city which not only lives by oil, but runs on its derivatives. Gas from the fields is piped into the capital and provides Al Kuwait's heat, light and power. If the wells were shut down for more than four days, the gas reserves would be gone and life in Sheik Abdullah's symbol of Arab progress would grind to a halt.* Also, police informers were reporting that other bombs had been planted not only along the Kuwait Oil Company's pipelines but along those of the Neutral Zone operators, particularly along those leading from "Gettysburg," J. Paul Getty's petroleum complex.

Thanks to Mubarrak's brutal methods, the saboteurs talked. In the next twenty-four hours, no fewer than a hundred time bombs were found attached to the pipelines and were defused by KOC explosive experts. One bomb went off at a wellhead in the desert southwest of Burgan, but fortunately it was of such feeble power that it failed to do more than superficial damage.

Still, Kuwait had had a scare that it would not soon forget. As one of the sheiks put it in a recent interview: "The saboteurs kept claiming that they were attacking British imperialism, but we were the sufferers—we Kuwaitis. It was our oil which was being sabotaged, our revenues which were being hit, our city which was threatened by a shutdown. There were a small num-

* A series of prolonged storms in the Persian Gulf in the spring of 1972 kept giant tankers away from the Sea Island loading points off Kuwait and forced a shutdown of wellheads. As a result, Kuwait was within four hours of a complete power failure when the storms cleared.

ber of Kuwaitis among the saboteurs—lazy, good-for-nothing dupes, mostly. We sent them back to their fathers with the suggestion that a good beating might alter their foolish ideas, especially since their families would have been the first to suffer in the event of sabotage. But the ringleaders of the affair were more serious. Most of them were Palestinians and Egyptians, and this made it difficult. Both were martyrs of imperialism and the Zionist conspiracy, and therefore heroes even to the Kuwaitis they were trying to harm. We could not punish them; it would not have been popular. We sent them back to their jobs but kept them under surveillance until we could more conveniently put them out of harm's way. With the rank and file of the militants, it was easier; they were nearly all Iraqis and Iranians. With them, we did not have to worry."

As the Kuwait Oil Company began to limp back to normal production—though its output would be severely curtailed for months to come—a purge of foreign labor began in the territory in mid-November, 1956. Scores of Iraqis and Iranians were rounded up, bundled into trucks and taken north to the Iraqi border, where they were dumped in the desert to find their way to civilization as best they could. With the more militant agitators, Mubarrak decided on even more brutal measures. Most of them were severely beaten, but half a dozen were trussed and bastinadoed, and to rub home the lesson, the punishments were carried out in public, an action which infuriated the ruler when he heard about it at his retreat on Falaika Island. "It will give Westerners a poor impression of our ideas of justice," he said.

Despite his professed imperturbability, Sheik Abdullah had been frightened by the sabotage. The way in which the anti-British demonstrations had turned against the Kuwaitis themselves came as a surprise to him, but he was shrewd enough to realize that as Kuwait continued to prosper, the envy of the foreign workers would grow. He did not propose to appease them by making them fellow members of the club—that would be going too far—but he did decide that it was time to call a halt to the reckless extravagances of his subordinate sheiks. In the next few months a regime of so-called austerity was imposed. The ruler himself inaugurated it by canceling the order for a new yacht and a private airplane.* Several plans for new sheikly palaces were canceled, and the most ornate of all, ordered by the ruler's son, was suddenly proclaimed a state guesthouse. No Kuwaiti suffered from the shortage of cash in the days to come, but they became far more discreet in the ways they spent it. Moreover, to show that all Arabs were brothers, and that Kuwait was wholeheartedly on the side of the anti-imperialists, Abdullah made clear his detestation of the Anglo-French action at Suez, his hatred of Zionism, and his fervent admiration for President Nasser.

As one Palestinian remarked about the changed climate, "We were still traveling second class, but at least they were beginning to admit that we were traveling in the same boat. That was enough—for the time being."

NO EUPHEMISMS could conceal the fact that the Anglo-French operation against Egypt in 1956 was a disaster morally, militarily and politically. It ruined the career of Prime Minister Anthony Eden, who had ordered the invasion. It convinced the world—and the Middle East in particular—that the sap was running out of the British imperial oak. Those who had sheltered in its shade began to look for better sources of protection. Others began to carve their names on its trunk. As for the oil business, it was inevitable that any reprisals against British interests were bound to involve U.S. companies. With the exception of Saudi Arabia, where the concessions were wholly American-owned, every major British operation had some U.S. participation. Saudi Arabia continued business as usual with Aramco, at the same time breaking off diplomatic relations with Britain, but elsewhere

* He "borrowed" one instead from the newly formed Kuwait Airways, which in turn went to the Kuwait Oil Company for the money with which to buy a replacement.

the Arabs hit back at the Suez invaders in the only way they knew how: by blowing pipelines in Iraq and Syria, by going on strike in the oil fields, and by blocking the export of oil to Europe.* American oil exports from the Middle East were held up equally with those of the British, and much of the contumely fell on their shoulders too.

No less than in the Arab countries, anti-British feeling was strong in Iran, where the memories of past intervention by British troops was still bitterly remembered. In this case the resentment was directed against the new oil consortium, in which the Americans now split with the British the major proportion of shares. The consortium had just begun to assume its portion of the world oil markets, and the fields were busy, though for a time Egypt's blockage of the Suez Canal forced the combine to redirect its shipments.

But the consortium was not popular. From the shah on down, most Iranians considered that the U.S. negotiators had driven far too hard a bargain, and that only Iran's desperate need to get the oil flowing again had forced them to accept the terms. Now resentment welled up and mingled with the anger over Britain's high-handed actions in Egypt; and had not the ink barely dried on the agreement, there might have been a campaign to get it amended. But it was too soon; changes would have to come later. In the meantime, the Iranians set about making sure that they would never have to accept such onerous conditions again.

In the summer of 1957, the Majlis in Teheran passed the Petroleum Act into law. It was the innocuous name of revolutionary new legislation, and it brought Enrico Mattei of Italy back onto the international oil scene. Mattei had been busy in the two years

* The U.S. oil industry provided an emergency petroleum export program to Europe to take care of the shortages. The resultant sharp rise in the price of domestic fuel in the United States led to a Senate inquiry which produced some interesting details of the interlocking nature of the international oil combines. See Senate hearings, Emergency Oil Lift: Program and Related Oil Problems, 85th Cong., 1957.

since the big Anglo-American oil combines had snubbed, and in his view, cheated him out of a share of the Iranian consortium. Most of his time had been devoted to devising ways and means of solving Italy's chronic power problem—its almost total reliance on the international oil companies for petroleum supplies. As president of Ente Nazionale Idrocarburi (ENI), the Italian national oil company,* he was more than ever determined to find oil supplies independently of the cartel, and if he could do so in such circumstances that he made life difficult for the British and Americans, so much the better. A former disciple of Benito Mussolini, Mattei later broke away and worked for the anti-Fascist underground during World War II, but he shared the erstwhile dictator's flair for dramatics. The sense of insult he felt at the way in which he had been humiliated by the consortium was, however, genuine. He was a proud, even arrogant man; he would never forgive the Anglo-Americans and any plan which would do them harm would be all the more rewarding, as far as he was concerned.

The post-Suez climate of open anti-British and subdued anti-American feeling throughout the Middle East was just what Mattei needed for his purposes. It produced a revulsion of feeling against the big oil combines, a conviction that the governments and people of the area were being exploited by the foreigners in their midst, and a widespread determination, as Walter Lippmann wrote at about this time, "that concession agreements for oil production and exploration in new areas must be based on different principles from those established by the big international companies." †

During the Suez crisis and in the uneasy months that followed, Enrico Mattei spent a great deal of time in Cairo and Teheran. In the Iranian capital he had several talks with the shah and with the petroleum minister, Dr. Ali Amini, and he was present in the Majlis when the Petroleum Act was passed into law. As a weapon

* Of which AGIP is a subsidiary.

† New York *Herald Tribune,* November 4, 1957.

for hitting the oil combines where it hurt most it could hardly have been better devised, for it introduced a principle against which the Anglo-American companies had been fighting for most of their lives: a joint enterprise plan in which the oil company and the oil-producing country participated on equal terms. This was very different from the so-called 50–50 arrangement which Aramco had worked out with the government of Saudi Arabia, an arrangement which they often euphemistically referred to as a "partnership." As an oil historian and consultant has remarked: "The 50–50 profit-sharing agreement . . . is about as much a true partnership as is the relation between the United States government and the thousands of corporations that it subjects to corporate income taxes." * Aramco paid the Saudi government 50 percent of its earnings in income taxes, but ran its company as a separate entity on behalf of its four chief U.S. holding companies. The Saudi government had one director (later two) on the Aramco board, but he had no voting rights or direct influence on policy. This was the way it was and this was the way the Anglo-Americans wanted to keep it.

But the proposals which Mattei outlined to the Iranians, and which were subsequently inculcated into the Petroleum Act of 1957, envisioned a joint venture in which the oil-producing country was involved from the beginning, although at no financial risk. The first of these agreements was signed between the Italian subsidiary AGIP and the government-owned National Iranian Oil Company (NIOC) almost immediately after the passing of the Petroleum Act into law, and the contract was approved by the Majlis on August 24, 1957. It was described by the Iranians as "the most progressive legislation in any Middle Eastern country," and by an American oilman as "a kick in our corporate guts." Between them, AGIP and NIOC formed a company called SIRIP,† which would have exclusive exploration and production

* Stocking, *Middle East Oil.*
† Iranian-Italian Oil Company.

rights over 5,600 square kilometers of territory north of the Persian Gulf, projecting out along the continental shelf; 11,300 square kilometers on the eastern slopes of the Zagros Mountains; and 6,000 square kilometers along the Persian side of the Gulf of Oman.

The Anglo-Americans were astonished when they read of the provisions which Mattei had accepted. The chairman of the board of the new company would be Iranian, and the directors would be Iranian and Italian, in equal numbers. All the labor and as many of the technical jobs as could be filled by qualified personnel would be assigned to Iranians, and a training school for others would be started at once. The Italians agreed to spend $22 million looking for oil: $6 million within the first four years and $16 million within the next eight. The moment one commercial well was discovered, the Italians were pledged to go on looking for new ones until the whole $22 million was expended. Only when wells were ready to go into production would the Iranians begin to share the costs. Thereafter, both the Italians and the Iranians would each take 50 percent of the oil, and would share the profits between them. But since SIRIP would be paying a 50 percent tax on their share, and NIOC would be getting 50 percent of the remaining profits, the Iranian government would in fact be making 75 percent of the return on the deal.

When oil-industry spokesmen had recovered their breath, they were openly derisive. Only a man desperate for oil at any price and motivated by a vendetta against the oil combines could possibly have accepted such onerous terms, they implied. The more extreme among them sharply suggested that Mattei's only purpose was to sabotage the holdings of the big companies in the Middle East. To which he replied:

> It should not be necessary for me to affirm my country's right to insure its own independent oil supplies. The supremacy of the international oil cartel is not taboo and Italy does not intend to respect it. Besides, today the cartel is everywhere being broken down. Through the State organization of which I have the honour to be president, Italy can now

enter the field itself to obtain its oil supplies . . . I have no doubt that Italy's agreement with Iran is not only in our interest but in the general interest also.*

Nonsense, replied the big companies. On such terms, no one would be willing to make a deal with a producing country. The risks were too great and the price was too high.

But three months later, the Iranian government announced that still another area of the country was being offered as a concession, on a joint-venture basis, under the new Petroleum Act. No member of the Anglo-American consortium bothered to apply, and their spokesman skeptically suggested that no one else would, either. But he was wrong; fifty-seven companies from nine different countries applied for the questionnaires which they had to fill out if they were to be considered, and Iranian experts who subsequently pored over their applications deemed eleven of them 100 percent "competent," financially and otherwise, to embark on exploration under the terms of the act, and ten others between 10 and 75 percent competent. All the applicants were eager enough to have paid $2,700 each for the opportunity.

One was chosen from among them. To the dismay of the consortium members, this time the winner was not a maverick of the Mattei type, but a respectable American oil company, Standard Oil of Indiana, bidding in the name of a subsidiary, Pan American International.

A week or two later an oil company vice-president traveling home on leave stayed overnight in a Rome hotel and saw Enrico Mattei walking through the main lobby. "We aren't buried yet," he said sourly, "but there goes the man who's driven the first nail in our coffin."

The Middle East oil business would never be the same again. Later it was estimated that snubbing Mattei had cost the Anglo-American oil cartel $1 billion in revenue. But it had done even more than that, as the companies would soon discover.

* From an article by Enrico Mattei in *Politique Étrangère,* Paris (last quarter, 1957).

CHAPTER TWENTY

Enter the Japanese

ENRICO MATTEI HAD REALLY STARTED SOMETHING. All over the Middle East the sheiks were beginning to repeat the phrase "joint venture" as if it were holy writ, and the major oil companies realized to their dismay that the interloper from Italy had changed everything. If they had been taken aback by the liberal terms he had conceded to the Iranian government, they were alarmed by what followed. For the conditions Pan American International accepted for its joint venture with the Iranians were even tougher. The company made an immediate payment of $25 million just to acquire the concession, whereas Mattei had paid nothing, and it agreed to spend $82 million on exploring and developing it, compared with Mattei's agreed outlay of $22 million. Otherwise the conditions were similar.

One thing was certain: the days of the old type of agreement were over. True, most of the concessions held by the Anglo-American major companies had several decades to run,* but for how long would the Arab and Iranian governments, now operat-

* The earliest to mature would be the consortium in Iran, which was due to expire in 1979. But some—Kuwait, for instance—extended beyond the end of the century.

ing the new joint ventures so much more profitably with their foreign partners, continue to accept the monopolist companies operating beside them? There were difficult times ahead, and the name of Enrico Mattei was cursed in the board rooms of the major oil companies in London, The Hague and New York.

In Saudi Arabia, one man who had followed the Italian-Iranian negotiations with fascinated interest saw in the adoption by his country of Mattei's new-style partnerships an ideal weapon for chipping away the foundations of the all-powerful American monopoly, Aramco. His name was Abdullah Tariki, and he had no love for the United States, Aramco or any other big capitalistic organization.

In 1957 Tariki was thirty-two years old. A son of a government official, he had been sent to Cairo University at the end of World War II, where he took a degree in geology, at the same time making friendships and absorbing radical ideology from the young anti-British, antimonarchist Egyptians who a few years later would be the most fervent supporters of Nasser. From Cairo he went to the University of Texas, where he took a master's degree in petroleum geology in 1947, and afterward he spent a year in a training course with the Texas Oil Company, one of the chief shareholders of Aramco. Although he will not talk about it, something apparently happened to him during this period that soured him on Americans. Offered a job with Aramco when he returned to Saudi Arabia, he turned it down; instead he enrolled as a minor official in the Bureau of Mines, and found himself acting as a liaison officer with Aramco at their headquarters in Dhahran. There, too, Tariki seems to have suffered some sort of traumatic experience which gave him a strong antipathy toward the American company.

In its treatment of its own Saudi technicians and employees, Aramco's American management is impeccable. The company's policy is to help them in every way possible to educate themselves and their sons, to loan them money for houses and to train them for promotion. Once they begin to climb the technical ladder, they are accepted and befriended. But with minor Saudi officials employed by the government, Aramco officials can be both curt and rough. Courtesies are reserved for the sheiks and the ministers, and less time is taken with small fry.

It may have been a verbal tussle with some exuberant Texan, or as some say, because one Aramco manager once kept Tariki standing for two hours in the hot sun while he stayed cool in an air-conditioned car; whatever the reason, Aramco was not Tariki's favorite company. By 1957 he had become director general of petroleum and mineral resources to the Saudi government, but the rise in his station did not mellow him. Among his close associates he talked bitterly of "American economic imperialism" and argued persuasively—he is a brilliant speaker—for the need to introduce new elements into the exploitation of Saudi oil in order to subvert Aramco's monopoly.

There was one promising area of Saudi Arabia well outside the vast concessions which Aramco controlled, and this was on the continental shelf outside the Saudi-Kuwait Neutral Zone.* Some of the major companies and a large number of independents had already made tentative bids for the rights to it, among them a company called Japanese Petroleum Trading, composed of a number of diverse banking, electronic, insurance and manufacturing groups (Mitsui and Mitsubishi among them). Its chairman, Mr. Taro Yamashita, had already had talks with a number of ministers and officials in Jidda, and to Abdullah Tariki he was a blessing. Afterward he was to say: "Yamashita was our Mattei." Like the Italian, Yamashita represented a nation without its own oil resources, till now dependent on supplies from the major Western oil combines. Like Mattei, he would be in direct competition with the Anglo-American cartel. Best of all, from Tariki's point of view, he was not an American.

From several preliminary conversations with Mr. Yamashita and his advisers, Tariki worked out the draft of an agreement

* Where between them Jean Paul Getty and Aminoil had developed highly profitable production.

which he was convinced would give Saudi Arabia its first joint venture and would include even more advantages than Iran had secured. But he anticipated trouble when he went to Riyadh to present his plan to King Saud and to his younger brother, Crown Prince Faisal, who was also prime minister. He knew how eager other applicants were for possession of the new concession, and such was the climate of corruption at the Saudi court that he feared that the Americans might already have bought their way in—and done so on much less rigid terms than he had secured from the Japanese.

To Tariki's surprise, neither the king nor the prime minister raised any objections to the agreement or to the entry of the Japanese into the Middle East oil business. Of course there was no reason why they should have opposed the agreement, and every reason why Tariki should have been congratulated for having driven such a hard bargain with the newcomers, but their lack of comment or criticism puzzled him, for he knew the ways of the royal household only too well.

When he returned to Jidda and told Mr. Yamashita that all was well, the Japanese indicated that he had known all along that there would be no royal objections. From his briefcase he extracted a document and handed it to Tariki, who read it in growing stupefaction. It was an agreement, signed the previous June, which stated in part:

> Whereas the Japan Petroleum Trading Company Ltd desires to obtain a concession from the Saudi Arabian Government that gives her the right to search and prospect oil in the Saudi Arabian territory, Mr. Taro Yamashita, president of the Japan Petroleum Trading Company Ltd and His Excellency Kamal Adham have agreed the following:
>
> 1. That Mr. Hassan Khalifa agrees to cooperate with Mr. Taro Yamashita to promote the exploitation plan under the supervision of His Excellency Kamal Adham.
> 2. That the contract regarding the said concession is to be concluded between Mr. Taro Yamashita, as president of the Japan Petroleum Trading Company Ltd, and the Saudi Arabian Government. . . .

3. That in case the above mentioned concession is granted by the Saudi Arabian Government . . . Mr. Taro Yamashita . . . undertakes to pay His Excellency Kamal Adham two percent (2%) yearly of the share of the total net profit accruing from the said concession . . .
4. That Mr. Taro Yamashita . . . will pay Two Hundred and Fifty Thousand U.S. dollars (U.S. $250,000) to His Excellency Kamal Adham within one month after the conclusion of said contract . . . and Seven Hundred and Fifty Thousand U.S. dollars (U.S. $750,000) upon sufficient amount of petroleum be [sic] found as to permit sound business enterprise. . . .*

His Excellency Kamal Adham was Crown Prince Faisal's brother-in-law and an official at the court.

Shocked, Tariki asked the Japanese why, with such royal allies in his pocket, he had accepted such onerous terms for the concession. Mr. Yamashita smiled. One of these days, he said, the big American companies and their old-style concession agreements would be in trouble, and he did not wish the Japanese to be associated with them when that happened. Joint venture was the way things would henceforth be arranged, and Japan would be content with it. "It must also be remembered," he added, "that so far, we have only fifty percent of the concession. We now must obtain the other fifty percent from the Kuwait government. And there we do not have a good friend like His Excellency Kamal Adham—at least not yet. It is as well to demonstrate in advance to the Kuwaitis how far we are prepared to go in order to secure an agreement."

IN MAY 1958 the Japanese Petroleum Trading Company Ltd successfully outbid all other contenders—and the competition from the United States and other countries was fierce—for Kuwait's 50 percent share of the offshore Neutral Zone concession, giving it control of all rights to drill in the area. "The

* Photostat in the possession of Research and Translation Office, *Middle East Economic Survey,* Beirut, Lebanon. See also *Al-Sayyad,* Beirut, March 19, 1959.

terms agreed to by the Japanese were in both [Saudi and Kuwait] cases unprecedentedly rigorous," writes Stephen Longrigg. "In the concession given by Kuwait the full concession period, including exploration, was 44½ years only, and in the Saudi concession two years more: and in both the right of renewal was only partially assured, and not at all if a rival Kuwaiti or Saudi candidate were in the field. In each agreement, a heavy surface rental was payable from the date of signature, and was to increase with the passing years if oil was found." *

In the agreement Tariki had drawn up, the Saudi Arabian government received 56 percent of any profits made; the Kuwaitis got 57 percent. These profits were to be calculated not just on sales of oil from the field, but on each step in the sales pattern and in any country. Moreover, the contract stipulated that no oil could be sold to "enemies of the Arabs"; that the Saudi government would have a substantial shareholding in the Japanese company and two directors on the board, who would supervise operations and policy; and that in the event of difficulties, the Japanese would agree not to appeal to their own government for protection.

Even compared with the Mattei agreement, these were stiff terms, and once more the old-style companies averred that the price was too high and that the Japanese would run out of money before they found oil. But, "undeterred by these provisions, and driven forward by their country's known and serious need for oil," † the Japanese went to work almost at once. Under the name of the Arabian Oil Company they assembled a mixed Saudi-Kuwaiti labor force and a cadre of Japanese and other foreign petroleum experts at a base on the shores of the Neutral Zone, and brought in a mobile drilling platform from Texas. On the afternoon of April 3, 1958, disaster struck. The mobile platform, moored eight miles from shore, had been drilling in 100 feet of water. At 1,500 feet below the sea bed the drill tore into a large pocket of natural

* Longrigg, *Oil in the Middle East.*
† *Ibid.*

gas, which bubbled to the surface in a sudden eruption and ignited on contact with engines on the platform. Fortunately, no one was hurt, but the subsequent fire raged for eleven days before it could be smothered by experts flown in from Texas. In the meantime, the badly damaged drilling platform had been beached.

It was an expensive start to the new venture, and the big companies, watching from the sidelines, prophesied that now even the Japanese would see that the terms were "unworkably onerous" and would get out before their reserves were burned up along with their equipment. But the Japanese were not giving up that easily. They sent for a new drilling platform, and began again, a few miles inshore from the gas pocket.

On January 1, 1960, the Arabian Oil Company brought in oil with its first well, which produced 6,000 barrels a day at 5,000 feet. Two years later, the company had thirty-four wells operating, and by 1964 they were harvesting 240,000 barrels a day. They were in the oil business for fair, and Mr. Taro Yamashita was proudly reporting to his shareholders that Japanese oil ventures in the Middle East had "a high future full of big dreams and hopes." *

Nobody was happier about this than Abdullah Tariki. Not that he was satisfied—his campaign against the big oil companies had only begun.

"THE TROUBLE WITH A LOT OF GUYS in the big companies," said Standard of New Jersey's Howard W. Page recently, "was that they were always talking about the sanctity of contracts. What they couldn't see was that in the light of what was happening in the Middle East, a lot of these agreements no longer made sense. In other words, if we had started out afresh and worked out new contracts we would never have insisted on many of the terms that were in the old ones. But there were, and still are, a lot

* Stocking, *Middle East Oil.*

of people in our industry who insist that a contract is a contract, and that the signatories have to live up to it. Actually, in contracts between companies, when things get out of whack we usually agree to adjustments in the interests of good business. But the same people wouldn't think of it when it came to contracts between countries. So when we got into negotiations with Iraq, some companies didn't agree to any changes, on the basis that contracts were unbreakable."

The Iraqi negotiations to which Page was referring were those of 1951, when, thanks to his patience and skill, a 50–50 agreement of the type which Aramco had signed with Saudi Arabia was accepted by both parties. But by 1958 the Iraqis had soured on this type of artificial "partnership" and were demanding a new deal. Thanks to the closing of the pipeline connecting the Iraqi oil fields with the Mediterranean—it had been blown up by the Syrians during the Suez crisis—shipments of petroleum had gone down drastically, and Iraq's revenues with them. The country was badly in need of money. Since the Iraq Petroleum Company was mainly British-owned and staffed, it had become the target of much of the anti-British feeling which swept the country after the Anglo-French attack on Egypt. To the more militant Iraqis not even joint-venture agreements now had any appeal; they were shouting for outright nationalization.

It was in this climate of seething discontent that IPC's managing director, G. H. Herridge, arrived in Baghdad on July 4, 1958, to begin talks about the concession agreements. The smell of revolution was in the air, but the IPC delegates appear to have mistaken it for the reek of petroleum and blithely proceeded to negotiate as if Suez had never happened. Eight days after their arrival, the British negotiators left again for London to consult with the head office, confident smiles on their faces, breathing optimistic words about "friendly atmosphere" and "mutual good will." They indicated that they had agreed to surrender certain parts of their concession and to double their production of oil within the next two years. These were gestures hardly likely to appease even the pro-British clique which ruled Iraq at the time; Nuri Said, the Iraqi strong man, needed much more liberal gestures to quell the rising tide of discontent.

It hardly mattered, for forty-eight hours after the British left Baghdad, the pro-British clique no longer existed. On July 14 Iraqi troops burst into the grounds of the royal palace in Baghdad and demanded the surrender of the young King Faisal II and his cousin Prince Abdulillah, who was crown prince and also retained the title of regent, which he had held during the king's minority. When the two men, accompanied by Abdulillah's mother and the king's two sisters, advanced toward the troops across the palace lawn, they were shot. Abdulillah, a petulant, perverse and arrogant character, was cordially hated throughout Iraq, and his body was thrown to the crowd outside, where it was joyfully seized for mutilation and "dragging." (Iraqi mobs have two favorite spectator sports during times of crisis and revolution: one is public hangings in the main square of the capital, the other tying the body of a designated public enemy to the back of a truck and dragging it through the streets.)

Abdulillah was not the only one to be "dragged" in the next few hours. Prime Minister Nuri Said, friend of Britain and an inflexible die-hard, was the main target of the army revolutionaries, but he managed to escape across the Tigris, where he hid in the house of a friend. A servant, thought to be faithful and devoted, was paid a large sum of money to reveal his whereabouts, and he was trapped while trying to slip away from the friend's house disguised as a woman. He too was shot and handed over to the mob.

In the meanwhile, the crowd was rampaging all over the city. A number of foreigners, including three Americans, were seized in their hotel and taken around the capital in trucks until the impatient mob swarmed upon them, beat them to death and "dragged" them too. One of the murdered Americans was a Eugene Burns, of Sausalito, California, who the rebels afterward claimed was a member of the CIA masquerading as a fund dispenser for a children's charitable organization. If so, he must have been singularly inefficient, for it turned out later that both the CIA and the British secret service were taken by surprise when the rebellion broke out.

Next the mob turned its attention to the British embassy, where it shot one of the staff, robbed the rest of their rings and watches, and burned the building to the ground. By nightfall the country was in the hands of the rebels, a group headed by an army officer, Abdul Karim Kassem, who had always been considered by Abdulillah and Nuri Said one of their closest friends. Sir Humphrey Trevelyan, who took over as British ambassador when the smoke of burning buildings and stench of dead bodies had been cleared away, wrote:

> The Hashemite monarchy, established by the British nearly forty years before on the ruins of the Turkish Empire and bound by alliance to the British, was overthrown. The romance of early Anglo-Arabism, the special relationship born in archaeology and the coincidence of wartime interests, pictured in Allenby's capture of Jerusalem and the contrived entry of Faisal I into Damascus, incarnate in those more than life-size figures of the romantic period, Lawrence, Gertrude Bell, Percy Cox, was finally shattered in the ruins of the British Embassy in Bagdad. The Hashemite-British alliance had outlived its time, while Nuri Said, the prime minister who had seen it all from the beginning, grown old, rigid and careless of security, had failed to adapt himself to the political needs of the time. The crash came, and in one day the old links with Iraq were severed, never to be restored.*

It might have been thought that conditions were now too fraught and unfriendly to warrant any resumption of negotiations between the Iraqis and IPC over the oil concessions; after all, a victorious rebel leader, eager to conserve the enthusiasm of the mob, was hardly likely to be in the mood for reasonable discussions. But suddenly the U.S. government inadvertently came to the aid of the British by landing American troops in Lebanon, an action it had so strongly deprecated when the British had done the same thing in Egypt two years earlier. True, President Eisen-

* Trevelyan, *The Middle East in Revolution.*

hower announced that the Marines had disembarked "at the request" of the Lebanese government, which was disturbed about the "explosive" situation in the Middle East; and at the same time British troops moved into Jordan, which was also in turmoil, and were put on the alert in Aden, but there were not many Arabs who accepted the explanation. The Lebanese prime minister (a Christian) announced that he was "surprised" when he heard the President's broadcast announcing that he had asked for help, but that may have been a ploy to appease his Arab constituents.

But if the Arab world deeply resented the American intervention, it was in no position to do anything about it—at least for the moment. And it was certainly convenient for Anglo-American oil interests in Iraq. The State Department sent Undersecretary William Rountree to see Kassem in Baghdad, and though the American was hooted in the streets by demonstrators, the Iraqi leader received him cordially enough. "Kassem . . . recognizing the importance of oil revenues," writes George W. Stocking, "and perhaps influenced by the landing of United States marines in Lebanon and the alerting of British troops in Aden, endeavored to put to rest Western Europe's fears over the continuity of its oil supplies and IPC's over the security of its properties." *

On July 18 he announced over Baghdad radio:

> "In view of the importance of oil for the Iraqi national wealth and the world economy, the Iraqi Government announces its wish for continued production and export of oil to world markets. It also upholds its obligations to all parties concerned. The Government has taken the necessary measures to preserve the oil fields and oil installations. It hopes the parties concerned will respond to the attitude taken by it toward the development of this vital source of wealth."†

Spokesmen for the new regime were soon assuring everyone that there was no intention of nationalizing the IPC concessions. Said Colonel Abdul Faik at the Iraqi embassy in London: "It is not the intention of the Republic to think about this, because we

* Stocking, *op. cit.*

† Quoted in *Middle East Economic Survey,* Beirut (July 25, 1958).

believe that if the oil continues to flow to the usual markets, it will be for the benefit of both parties—you get your oil and we get our pounds." *

In January 1959 a new delegation arrived in Baghdad from IPC to start talks with the Iraqi government. This time it was led by IPC's new chairman, Lord Monckton, no oilman but an amiable lawyer famed for his skill as a negotiator.† But more amiability than skill emerged from the meetings. Upon his return to London, Lord Monckton declared, "I do not remember any discussions conducted in a more amicable spirit, and they were continually saying that they wanted to work with us to our mutual advantage." ‡

But work with them on what, and in what circumstances? The Iraqis announced that IPC would immediately gets its expansion program under way, and would double the export of crude oil from Iraq by 1962. The company would also begin a construction plan for the erection of oil refineries inside the country. The following day IPC called in the press and read a prepared statement saying that the expansion program would be subject to "world market conditions," which, the spokesman reminded his listeners, were not very promising at the moment. As to oil refineries, IPC had no intention of building one in Iraq.§

It was a moment when IPC would have been wise to make a generous gesture to meet the political realities in the new Iraq. Company spokesmen and industry historians have since maintained that it would not have done any good, that Kassem was to prove a vacillating trickster whose word could not be trusted, and who in any case ended up with an assassin's bullet in him. Some of the sager petroleum experts dispute this view, however. "For quite a number of years," says Howard Page, "I was looked on as

* The *Financial Times,* London, July 22, 1958.

† In 1936 he acted as go-between during the abdication crisis in the discussions between the then King Edward the VIII (later Duke of Windsor) and Prime Minister Stanley Baldwin. He was also Minister Delegate to the Middle East during World War II.

‡ The *Times,* London, April 24, 1959.

§ The *Financial Times,* April 29–30, 1959.

a dangerous liberal for agreeing to study situations, particularly where you could analyze them and say, 'Well, if you were starting all over again, this is what you would have to give.' So long as these guys [the producing countries] didn't have control of us, we weren't prepared to give up anything worth a damn. I was looked on as a dangerous character for suggesting that we should."

And that was the trouble. If Kassem was to prove a fanatic and a trickster, it is possible that IPC helped to make him so. The Iraqis were in a poor bargaining position in 1959. There was a world surplus of petroleum; too much fuel was competing for too few markets. Oil producers in the United States had increased their output during the Suez crisis and were happily selling their petroleum unworried by foreign competition, because Congress had put a quota on imports of foreign oil. Russia had begun flooding the markets in Europe and Asia with Soviet oil at cut prices. The new independents like J. Paul Getty and Continental, a Delaware-based oil outfit, were beginning to market their products, and so far were under no control by the big combines. IPC had plenty of other places in the world where it could buy oil, and so did the companies which made up its membership; there was all the fuel they wanted in Saudi Arabia, Qatar, Abu Dhabi, Kuwait and South America.

So each demand that Kassem made of IPC—and in the light of present-day developments, they were not that formidable—was met not by concessions but with a deliberate, if unpublicized, tightening of the screws. If Howard Page was against this type of pressure, he was one of the few, for IPC's confrontation with the Kassem government was being watched by all the major Anglo-American companies, and a hard line was supported by all of them. "They realized that Iraq could scarcely have been unmindful of Iran's unfortunate experience with nationalization, which had shut it off from the world's market and eventually triggered the coup d'état which overthrew the Mossadeq regime. The Iran crisis occurred when surplus capacity was far less and the oil industry had readily met it . . . Perhaps more important to the companies was the realization that any major concession they made to Iraq would be promptly demanded by other Middle East governments, constantly guided in their relations with their concessionaires by the 'most favoured nation' principle. The companies were not ready apparently to upset entirely the pattern of Middle East concessions, already greatly weakened by the advent of the newcomers." *

So pressure was the order of the day for the IPC negotiators, with combined Anglo-American approval. The most obvious way to make its strength felt was for IPC to deprive the Iraqis of the increase in revenue upon which Kassem had been banking after the British company had announced its expansion plans. Far from doubling its output, the company quietly ordered its wellhead managers to slow down the flow. In the next three years, while negotiations were proceeding, IPC's production of Iraq oil increased only by between 0.5 and 1 percent. In the same period Iran's production went up by 12 percent, Saudi Arabia by 9 percent, and Kuwait by 11.5 percent. In one particularly critical month of IPC-Iraqi confrontation, the company's production dropped by 30 percent.

To stress the international character and solidarity of the negotiators, IPC's managing director, G. H. Herridge, was joined in Baghdad by F. J. Stephens of Shell and H. W. Fisher, a vice-president of Standard of New Jersey. They had been instructed to keep the existing concession contracts in mind as they bargained, and anything beyond the agreements they had made in 1951—with minor adjustments—were anathema to them. From Fisher's point of view—to echo Howard Page—a contract was a contract. When Iraqi leaders asked them for a gesture that would give the Iraqi people a say in the exploitation of their rights, Fisher replied, "This is a commercial arrangement between the two parties. Much as we feel for the hardships of the Iraqi

* Stocking, *op. cit.*

country and people, we should . . . adhere to the commercial basis . . ." *

It was this attitude of boneheaded inflexibility which gradually drove the Iraqi leader to the extreme positions that eventually led to a breakdown. As one of his ministers put it: "The other day, and for the umpteenth time, I was trying to explain to the company representatives just how negative and absurd their attitude was. You are making strenuous efforts, I told them, to help the underdeveloped countries by granting them long-term loans and technical aid. You are doing this, it seems, to raise their living standards and to shelter them from subversive doctrines. We don't ask this much from you for Iraq . . . Just give us our due. Try at least to behave in good faith. You take a malicious pleasure in misleading us, and in depriving us of our most legitimate rights. In the minds of the people, all this may finally rebound against you and against the principles of what is still called 'The Free World.'" †

What were the demands which Kassem was asking of IPC? They would certainly have necessitated a complete reconstruction of the concession agreement, but seen from the viewpoint of the 1970s, when most of the demands have been conceded or are being discussed, they do not strike terror in the heart of even the most conservative oilman. They can be summarized as follows:

1. The government pointed out that although it had the right to appoint two directors to the board of IPC and its subsidiary companies operating in Iraq, in practice the board had only limited functions, and policy was controlled by executive directors operating in London and New York. The government demanded that one of the executive directors should henceforth be an Iraqi appointee.
2. Desiring to enjoy the profits of the integrated operation of

* *Ibid.*, quoted from minutes printed in Platt's Oilgram News Service, November 24, 1961.

† Interview with Muhammad Salman, Minister of Oil, in *Le Commerce du Levant*, Beirut, February 2, 1963.

IPC oil business, the government demanded that the company give priority to the shipment of crude oil in an Iraqi tanker fleet that the government was planning.

3. All over the oil fields, IPC was flaring away surplus gas. The government demanded that IPC either pay the government for wasting it in this way, or arrange to have it piped to government depots.
4. The government pointed out that under the terms of the concession contract, whenever an issue of IPC shares was offered to the general public, Iraqis in Iraq were to be given preference in being allowed to buy at least 20 percent of the issue. But it had since been discovered that this was a trick, because IPC was a private and not a public company; therefore shares were never put up for sale and neither Iraqi individuals nor the Iraqi government were able to purchase stock. The government now demanded the right to acquire a 20 percent participation in IPC.

Clauses 2 and 3 of the Iraqi demands were not vital to IPC's monopoly, and the company was prepared to make mollifying concessions in these regards. Yes, they would advise the Iraqis on how to establish a tanker fleet, and would use their ships for transporting Iraqi oil, once the fleet was established (which, they suspected, would take a long time). And yes, they would make arrangements for the bulk of gas now being flared to be used by the government for powering its cities. But clauses 1 and 4 were much more fundamental. To have given way on either would have been to admit Iraqis into the board room of IPC, with a say in policy, prices and output, and this they would never accept.

Soon both sides were accusing each other of inflexibility or trickery. The more IPC procrastinated and cut down its production, the more the emotional Iraqi leader was driven into an extreme position. The less IPC was willing to give, the more he had to ask, in order to show his people that he could not be crushed and humiliated by these foreigners.

When the British company announced that for the time being

it was closing down operations in the potentially rich oil field of Rumeila, in southern Iraq—part of the vast concession which IPC had held since 1925—Kassem's response was swift. On December 11, 1961, he announced that under a new act called Law 80, he was withdrawing all but 1,937 square kilometers of IPC's concession. This was 0.5 percent of the total area, and represented the territory in which most of the company's wells were in active operation. The remaining 385,700 square kilometers would revert to the government and be offered to new concessionaires.

IPC promptly referred the confiscation to international arbitration, and both the British and U.S. governments sent notes to Kassem urging him to arbitrate. He brushed them aside. "Why should IPC continue to hold such vast areas of our country?" he asked. "The company is only utilizing three percent of it, and keeping other companies from exploiting the rest to the benefit of our people."

The confiscated areas were offered to world oil companies for immediate exploitation. But IPC was not worried; it was calmly confident that the Anglo-American cartel would hold firm, and no one would bid. The company was quite right. The international oil front held, and the confiscated areas lay fallow for the next seven years. No one would touch them.*

But IPC would live to regret its intransigence. In 1963 Kassem was shot down by assassins, and his severed head displayed on Iraqi TV screens to cries of "The pro-British puppet is dead! Long live the revolution!" The men who succeeded him were even more fanatical and demanding, and by this time the world oil situation had changed drastically. Outside Iraq, another kind of revolution had taken place, and for the big oil companies it was much more threatening than the demands of General Kassem.

* They were almost certainly inhibited by IPC's public announcement that it would take legal action against anyone trying to exploit "its" concession.

CHAPTER TWENTY-ONE

Over a Barrel

IN THE SUMMER OF 1960, Standard Oil of New Jersey did something which united the oil countries of the Middle East as nothing ever had before, not even their mutual hostility toward the State of Israel. When Jersey cut the posted price of its Middle East oil by 14 cents a barrel, news of the company's unilateral action reverberated through the Arab world like the shots at Sarajevo. Soon the other major companies began cutting their prices, too. From Baghdad to Riyadh, from Teheran to Kuwait, the sheiks and sultans were loud in their cries of distress and anger. Within a few weeks, they decided to take joint action against the cuts to save the posted-price system upon which Middle East oil production was now based.

It must be explained what the oil industry means when it talks about posted prices. Except for the new independents, all the oil companies operating in the Arab countries and Iran were—and still are—subsidiaries of the Anglo-American majors. The Iranian consortium is owned by British Petroleum, a quintet of American oil giants,* Compagnie Française des Pétroles, and Royal Dutch–Shell. IPC of Iraq belongs to BP, Shell, CPF, Standard Oil of

* Standard of New Jersey, Mobil, Standard of California, Gulf and Texaco.

New Jersey and Mobil. Aramco of Saudi Arabia is divided between Standard of California, Standard of New Jersey, Texaco and Mobil. Kuwait Oil Company is a condominium of BP and Gulf. These subsidiaries are producers of oil and have no hand in marketing it outside the producing countries. They sell their oil to their parent companies, whose responsibility it is, in turn, to sell it to the world. In the old days the subsidiaries would sell crude oil to their parent companies at low prices, and pay a royalty based on that deliberately low price to the governments of the countries in which they operated. Their parent companies would then sell this bargain-rate crude at the prevailing world prices, which were based on more expensively produced U.S. oil. The result was low royalties for the producing countries and high profits for the Anglo-American majors.

When the system of payment to the Arabs and Iranians was changed from royalties to the so-called 50–50 deals and then to joint ventures, the producing countries finally became involved in the sale of their oil. Their revenues now depended, as never before, on how much their oil was sold for, and they wanted assurance about this price.

During one of the negotiations in which he was involved, Howard Page of Jersey Standard answered this point by agreeing to formally posting a price for his company's oil in the Persian Gulf, and to promising that anyone could buy its crude oil at that price. This meant that there wouldn't be any price rigging, and that the producing country would be ensured of getting a fair share of the profits.

Since henceforth the formally and publicly posted prices were high—although not as high as the more expensively produced U.S. oil—the solution was generally accepted by Middle East governments, and for the first time many countries were able to calculate how much their oil revenues were going to amount to in a given year and fix their budgets accordingly. Not everybody was satisfied that the posted price was high enough, and there were those who accused the parent companies of making hidden profits by a series of bookkeeping tricks and transfers, but on the whole the posted-price system was welcomed. Until August 1960, that is, when, without informing any of the governments concerned, Jersey Standard knocked 14 cents off the posted price of a barrel of oil—a drop of about 7½ percent a ton—as sold in the Persian Gulf or at the refinery terminal in Tripoli. There was a buyers' market abroad for oil. Russian petroleum was flooding Europe, the emergency situation caused by the Suez crisis was over, and Libyan and Algerian oil was beginning to come on the market. From a strictly commercial standpoint, Jersey Standard had a good argument for the cut. It would have had a better one if it had cut not only the posted prices at which its subsidiaries were prepared to sell Middle East crude, but also the price at which it sold the refined petroleum in its gasoline pumps in Europe. But this it did not do.

At the time, it was suggested by some Arabs that Standard's unilateral action had been taken not strictly as a commercial price reduction but also as part of the industry's pressure being exerted by the Anglo-Americans on General Kassem in Iraq to force him into line. If so, Jersey Standard does not appear to have informed its British partners of its decision, and BP, in fact, strongly protested against the reduction before reluctantly dropping its own prices along with other companies.

But the most pained reaction came not from Iraq but from Saudi Arabia. It was a moment when King Saud was making one of his periodic attempts to reform; he had promised to balance the budget, to end corruption, and to bring about widespread improvements in education, health and communications. To this end, a program based on the estimated revenues the nation would receive from posted-price sales had been drawn up for vast building projects and several costly but necessary projects. At one stroke, Jersey Standard's price reduction slashed Saudi Arabia's income by $30 million for the year 1960–1961. At once Sheik Abdullah Tariki, who was now the Saudi minister of petroleum resources, rushed to Baghdad to announce his solidarity with

Kassem in face of the pressure the Anglo-Americans were exerting. Other Middle East oil ministers followed him to the Iraqi capital, where they were joined by Dr. Juan Perez Alfonzo, minister of mines and hydrocarbons in the Venezuelan government.* He urged the Arabs and Iranians to band together, but it was Tariki who was the leading spirit in the drive to form an official association. From the emergency congress of Middle East oil ministers in Baghdad in September 1960, over which he presided, emerged a body called the Organization of Petroleum Exporting Countries. Its membership consisted of Iraq, Iran, Kuwait, Saudi Arabia and Venezuela.† The members pledged to demand stable prices from the oil companies and immediate restoration of the cuts. They insisted that no cuts should be made by the major oil companies in the future without prior consultation with the governments of the producing countries. They acknowledged certain obligations to the companies (such as security of contracts) and demanded others in return: a steady income for themselves, a steady supply of oil to the consuming countries, and a "fair return on their capital to those investing in the petroleum industry." They pledged to maintain solidarity and remain united in all circumstances, and to spurn any special advantage that one among them might be offered by the petroleum cartel to persuade that individual country to break ranks.

It was the birth of OPEC. Not even the Arabs were confident of the baby's chances of survival. Too many times the peoples of the Middle East had tried to band together in a united front, and been broken up by external bribes or internal rivalries and dissensions. How could this one expect to be different? As for the international oil companies, after a preliminary *frisson* of alarm, they refused to recognize OPEC's existence; they would continue

* As chief supplier of oil to the U.S. domestic market, Venezuela had a vested interest in persuading the Middle East to maintain high prices for its oil. Venezuela charged the U.S. companies operating inside her territory a 60 percent tax on their oil profits, and could not afford to be undercut by exports from Saudi Arabia, Kuwait, Iran or Iraq.

† Qatar joined shortly afterward, followed by Abu Dhabi, Dubai, and (much later) Libya, Algeria, Nigeria and Indonesia.

to negotiate with individual countries, and under no circumstances would they deal with a combined organization. "We don't recognize this so-called OPEC," said Bob Braun, at that time president of Aramco. "Our dealings are with Saudi Arabia, not with outsiders." *

Most of the oil majors decided that if they ignored OPEC it would disappear, like other Arab organizations before it. Others considered that the danger to be faced came not from OPEC itself, but from the man who had been the leading figure in creating it, Abdullah Tariki. The word was passed around that if only Tariki could be shown up to the other Arab oil ministers as the anti-Western fanatic that he was, the whole revolt would collapse. The word went out that at the very next opportunity he should be discredited.

The opportunity came in October 1960, and never did the Anglo-American combines make a greater mistake than in seizing it.

ON THURSDAY, OCTOBER 20, the Arab Petroleum Congress opened in Beirut to hear Sheik Abdullah Tariki read a paper on "The Pricing of Crude Oil and Refined Products." Oil ministers from all the principal Middle East countries, heads of the major oil companies, and a flock of experts had been invited to listen to the paper and debate the points it raised in subsequent discussions.†

Tariki had hoped that the meeting could be held under the auspices of the new association, OPEC, but eager though they were to come to grips with the Saudi minister, the Anglo-Americans had made it clear that in such circumstances they would refuse to appear, since their presence would imply a recognition

* Quoted by Ahmed Zaki Yamani, at present (1973) minister of petroleum of Saudi Arabia, in a conversation with the author.

† This account is based on the records of the Middle East Research and Publishing Center, dated October 28, November 4, 11 and 18, 1960, and contemporary newspaper accounts.

of the organization. So though OPEC's name was on everybody's lips in the corridors, it was never mentioned in the speeches.

On both sides, the delegates were in a fighting mood. The cut in the posted prices by the oil companies had not only dealt a blow to the economies of the Arabs and Iranians, but the way in which it had been done was an affront to their pride. "They didn't even inform us beforehand," said one Arab minister bitterly.* On the other hand, the oil-company delegates, convinced that the Arabs did not understand the facts of commercial life, and not much concerned with Arab pride, were out to demonstrate that in the sink-or-swim world of competition between nations the governments of the Middle East had better learn that they could price themselves out of the market.

Tariki's paper was just what they hoped it would be: long on rhetoric and short on facts. Copies of his polemical attack on the methods of the major oil companies had been distributed to the delegates beforehand, and the assembled British and American experts had prepared their replies with what they hoped was devastating thoroughness. However, Tariki took some of the wind out of their sails by departing from his script in a sudden and dramatic aside in which he charged the oil companies with cheating by bookkeeping. By arbitrary and discriminatory pricing of Middle East oil, he claimed, and by transferring profits from oil wells to tankers, the oil companies had made an extra profit over the past seven years of no less than $5.5 billion, half of which should have been paid to the producing countries. The oil-company experts were so flabbergasted by this outlandish attack—after all, not even Rockefeller in his heyday had ever been accused of misappropriating nearly $3 billion—that when the first session came to an end they insisted that an extra day should be given to the debate. After a night working on their documents and telephoning London and New York, they were sure that Tariki could be exposed as a wildly dangerous windbag.

The next morning the big guns of the industry were wheeled

* "If we had told them beforehand," said Howard Page of Jersey Standard, "they would have tried to stop us from doing it."

in and opened fire on the Saudi minister. B. H. Groves of Socony-Mobil read out a carefully prepared list of figures, which had been painstakingly gathered by telephone from New York during the course of the night, to prove that far from making a hidden profit of $5 billion through juggling its tanker figures, his company alone had incurred a $30 million loss.

To the delight of the largely partisan audience, Tariki was undismayed. "What does it matter at what stage the profit or loss is entered in your books?" he replied. "You make a loss on your tanker transport operations in order to transfer the profits to your refining and marketing affiliates. The money was made somewhere, and you rarely sell to anybody who has no affiliation with your company."

Groves of Mobil gave way to George Ballou of Standard of California, who smoothly rebuked Tariki for using formulas for calculating oil prices which were never intended for that purpose, and assuring him that the oil companies were working in the best interests of the producing countries.

"How can we possibly take that for granted?" answered Tariki. "See what happened in Saudi Arabia." He then proceeded to describe how, shortly after the conclusion of the 50–50 agreement between the Saudi government and Aramco, the posted price for Saudi crude oil was set at $1.75 a barrel. Then his department discovered that Aramco was putting down only $1.42 a barrel in their books.

"When we asked why," Tariki went on, "Aramco said that this was a discount which they gave to the parent companies to build marketing facilities." He paused and then added, "But that was never mentioned to us during the negotiations, and it wasn't in the agreement. The effect was that what we were getting was not 50–50 but 32–68."

Tariki was beginning to warm up now. "They tell us that they have to sell to the parent companies at $1.42 to build marketing facilities. But while they were doing this, they were telling the U.S. government that $1.75 was the minimum price at which they could make a reasonable profit. You see? They can tell two governments two different things. Therefore we cannot accept anything they say without real investigation. They do not tell us what is going on outside the production phase. They say that this is complicated and is the business of the parent companies. They treat us like children. Now, if they really want the confidence of the people of the Middle East and their governments, it's about time that the oil companies divorce themselves from the controling end and work for the governments and peoples of the countries in which they are operating."

Tariki sat down to thundering applause, and leaned back to listen to a spokesman for BP, R. Anderson, assuring him that so far as his company was concerned, they were all partners together. Then he bounced to his feet again. "You say we are in it together?" he asked. "But when you make money, you make it alone. When you lose, you come out and cut the posted prices. Somebody here mentioned that we don't need a formula. We do; we need a formula to protect ourselves, because if we don't have one, you can easily sell our oil at fifty cents a barrel and make your profit through your tankers, refiners and marketers. You use the formula when you like it; you don't use it when we like it."

The spokesman for Royal Dutch–Shell, W. Nuttall, rose and waited for the renewed applause to die down. "May I say," he began, "that as far as the Shell Company is concerned, we have no intention of hiding things from him or anyone else? We would be delighted if Mr. Tariki would come to our office in London, and we will be able to explain to him the points which he considers are being hidden."

"Thank you very much," replied Tariki. He waved his hand at the Aramco, Standard of New Jersey and Mobil representatives in the audience. "The only thing I ask you to do is convince your friends here not to raise or lower the posted price before they have first consulted us and convinced us that it is the thing to do."

All the Arabs in the audience were on their feet now, applauding and cheering. The oil ministers crowded forward to grasp Tariki's hands. Only the Western oil experts remained seated, shaking their heads in feigned astonishment at the enthusiasm all around them.

Not only had Tariki not been discredited; by shifting his ground with what one commentator afterward called "considerable skill," and never allowing himself "to get caught in an indefensible position," he had neatly turned the table on the Western oil experts. As one English correspondent remarked, he "had brought into the open issues which have been festering for the past eighteen months or more . . . saying, in effect: 1) If you don't agree with our [Arab] figures, why not produce the correct figures relating to the profit and loss on each separate phase of the integrated operation; 2) If you don't agree with our pricing formula, why don't you come and discuss the whole matter frankly with us, as the Japanese have done; and 3) No matter whether you make a profit or loss on later operations, we want a share in them and want to know what is really going on." *

The basic instability of the relationship between the host countries and the Anglo-American oil companies which held concessions in their territories had been laid bare. "The chips are down," wrote Ian Seymour, "and what happens next is anybody's guess." †

ABDULLAH TARIKI was now the hero of the hour throughout the Middle East, and the association which he had done the most to create, OPEC, flourished in his reflected glory. The governments of the oil-producing countries, which had reluctantly accepted the idea but were skeptical about its survival, suddenly began to share the enthusiasm of their oil ministers. It is an indication of their new confidence that the shah and the sheiks even consented to advance money to the new organization. As a result, OPEC was able to vote itself a budget of £150,000 and establish its headquarters and secretariat in Geneva, Switzerland.‡

* Ian Seymour, in *Middle East Economic Survey,* October 28, 1960.
† *Ibid.*
‡ When Switzerland hesitated to give blanket diplomatic privileges to the secretariat, OPEC moved to Austria (in June 1965), where the government proved to be more amenable.

It only remained to choose a secretary-general for the organization. The logical choice would have seemed to be Abdullah Tariki, but while the members were debating this point, the fiery Saudi was back home savoring his triumph, and emboldened by the plaudits he had received, prepared to launch a crusade against his favorite target, Aramco. Each day he received Arab journalists from Baghdad, Mecca, Cairo and Beirut, and to each of them he poured out his complaints about the American company.

The Saudi people were being cheated, Tariki protested. Consider the Trans-Arabian Pipeline (Tapline) from Dhahran to the Mediterranean, for instance. Tapline was owned by Aramco, but did the company share its profits with Saudi Arabia on the oil it took from Saudi fields? It did not; instead, it passed them to the parent companies in the United States. "This is unfair practice," he declared. "We demand that Tapline's profits be split 50–50 with us. The company already owes us 180 million dollars which is our share of the gains they have made in the past few years." *

Tariki asked that this money be paid over at once, and followed by demanding a complete revision of Aramco's concession agreement, the relinquishing of territory, the stepping up of payments from 50–50 to 75–25, or even higher, and a swifter "Arabization" of the company.

Thoroughly alarmed over all the publicity these demands were getting, Aramco issued a statement setting forth the enormous efforts it had made in recent years to "Arabize" its operations, and detailing its welfare work on behalf of Saudi Arabia generally. It appealed for calm discussions of any differences, "in a spirit of co-operation and good will," which might exist between the company and the government. "On this issue the Government and Aramco hold honest differences of opinion," the statement went on, "since Aramco believes it owes the Government nothing in this connection. In view of this, Aramco wishes to assure the gracious Saudi Arab people that it stands ready to discuss or arbi-

* Quoted from an interview in *Al Zaman,* Baghdad, April 17, 1961.

trate, if necessary, any point of difference with the Saudi Arab Government." *

Tariki replied to this by hotly denying that Aramco had any right to arbitrate on tax matters which were the sole concern of the Saudi government, and threatened to report Aramco's "iniquitous exploitation" to OPEC.

But suddenly there was anxiety in court circles in Riyadh and Mecca. Funds were short again. Once more King Saud had allowed his corrupt friends to gain the upper hand, and his budget was again overdrawn. He needed allies, as well as discipline in his government. On March 15, 1962, he came to terms with his younger brother, Crown Prince Faisal, and announced a new Administration with himself as prime minister and his brother as deputy prime minister. Abdullah Tariki was no favorite of Crown Prince Faisal's, and when the names of the new ministers were announced, Tariki's name was not among them. No comment about his dismissal was allowed in the Saudi press, and he quietly left Saudi Arabia shortly afterward to go into exile and set himself up in Beirut as an oil consultant. Aramco was not the only Western oil company to breathe a sigh of relief at his disappearance from the spotlight.†

In the meantime, OPEC had gone elsewhere to find its new secretary-general. The members chose Fouad Rouhani of Iran, and made him both secretary and chairman of the organization. He was a very different character from the explosive Tariki, a quiet, urbane, cultivated man with a taste for Western culture, wines, food and literature. It would be Rouhani's job to persuade the Anglo-Americans to recognize the new organization, instead of forcing them to sit down as supplicants at OPEC's table, as Tariki had hoped.

* Quoted in Stocking, *op. cit.*

† In the winter of 1971 Tariki, who now advises the Algerian, Libyan, Kuwaiti and Iraqi governments on oil matters, was forbidden to reside in Lebanon—at the request of the Saudi Arabian government, it is believed.

Part Six

THE CHALLENGE OF THE SIXTIES

PRECEDING PAGE: *King Faisal of Saudi Arabia.*
PHOTO: PAUL CONKLIN/CAMERA PRESS LTD.

CHAPTER TWENTY-TWO

Death of a Gadfly

IF ANYONE COULD BE BLAMED for precipitating the cuts in the posted prices of Middle East oil in 1960, and thus for the resulting Arab attacks on the major petroleum companies, it was Enrico Mattei of Italy. The Italian oil chief was still the implacable enemy of the big Anglo-American combines, and his determination to make life difficult for them had now become a crusade. He buzzed angrily over the operations of the big companies like a gadfly over a cud-chewing cow, swooping down to sting whenever the animal looked too self-satisfied. In truth, the viciousness of his attacks was a reflection not simply of how much he resented the combines but of his chagrin at the lack of success of his own oil ventures as well.

Though it was Mattei who had pioneered the joint venture with Iran which had changed the nature of Middle East petroleum exploitation, neither he nor Italy had benefited much from the partnership. To be sure, no money had been lost on the outlay, but strikes in SIRIP's concession area had been modest rather than bountiful, and Mattei had been forced to try new pastures in Egypt, Libya, the Sudan and Morocco, where he hoped to reap the rich harvest of oil which he needed to keep ENI, the Italian national oil company, of which he was still president, and Italy itself supplied with fuel. So far the bonanza had not been forthcoming, and it did not lessen Mattei's antipathy that despite difficulties, the major oil companies continued to thrive. He missed no opportunity to do them damage, and in the late 1950s he hit them with a very low blow indeed by entering the marketplaces where the big companies sold their oil and gasoline, and undercutting them. The chief target of his operations was Standard Oil of New Jersey, operating in Europe under the name of Esso International. It was a classic case of the biter being bit, for Jersey was a chip off the old American oil bloc—the original Standard oil—which had won its way to international power by using its vast resources to undersell and bankrupt its competitors.

Mattei did not have the vast reserves that John D. Rockefeller had once used to destroy his rivals, and Standard of New Jersey was never in any danger of going bankrupt as a result of his operations. Nevertheless, he was able to make life for Jersey extremely uncomfortable.

How was this possible, since ENI had so far failed to find a productive oil field of its own? The answer is Russia. It was a period when the Soviet Union was searching for a market outside its East European satellites for its surplus supplies of crude oil, of which it possessed enormous quantities. It needed foreign currency. When Enrico Mattei flew to Moscow for a series of talks with the Soviet leader, Nikita Khrushchev, there was a meeting of minds—or rather of mutual antipathies. From both their points of view, it would be a productive relationship. Russia had the oil, which it was prepared to sell cheap; Mattei had the outlets, and would be buying at bargain prices. Each would earn the extra dividend of damaging the Americans.

In 1959, 1960 and 1961 the crude oil began arriving in Italy from Russia in ever-increasing amounts. It was processed in ENI's refineries, and then, as a result of a series of deals made all over Europe by Mattei, flooded on the market at prices well below those maintained by the major oil combines. Particularly hard hit by the undercutting were Standard of Jersey and IPC, both of which transported vast quantities of crude oil to the Mediterranean from the Persian Gulf and Iraq especially for the European market.

In the days of John D. Rockefeller a savage price war would immediately have ensued in which—backed by unlimited supplies of oil—Standard Oil would have flooded the market with cheap gasoline until the enemy ran out of gas and money. But Mattei had access to all the crude oil he needed, and under the circumstances the Soviet Union was not worried about how cheaply it was sold. On the other hand, as a member of Aramco and other combines in the Middle East, Jersey Standard had to maintain the posted-price system and could not cut the price of its crude without reducing its own profits and the revenues of the Middle East oil-producing countries. To some degree this was a dilemma facing all the major companies, but for Standard of New Jersey it was by far the most economically damaging.

Finally, however, as the flood of cheap Italian-Soviet oil rose and reduced Esso's markets, the American company decided to take the risk and cut the posted price of Gulf-produced crude, first by 10 percent and then by another 7½ percent. As has been demonstrated earlier, it was the worst tactic to take, for the major companies were eager to retain the status quo of their concessions in the Middle East. It united the Arabs and the Iranians in fury against them as never before, and produced OPEC.

Moreover, it did no good in the battle against Mattei; with Russian connivance the Italian cut the price of his refined crude to meet Esso's challenge. By now it was obvious even to Standard of New Jersey that the company dare not slice its prices again—and again and again—until the Italian admitted defeat. The Middle East governments would not stand for any more cuts; they had made that plain at Beirut, and since. In any future change of price, they would first have to be consulted, and it was obvious that they would never give their permission.

Through the bleak commercial months of 1961–1962, Jersey

brooded about what to do about Mattei's "unfair competition," and hard were the words that some of the executives used about him. "Moscow's stooge" and "tool of international Communism" were two of the politest pejoratives. Finally, cursing the stupid blunder which had made an enemy of him in the first place, the executive committee of Standard of New Jersey decided that if they couldn't beat Mattei, they must persuade him to join them. But how? He was touchier than an Arab oil sheik. He never forgave snubs, and Jersey had snubbed him three times over the years, two of them personal ones. Before World War II, when Mattei was an oil official under Mussolini, Jersey officials had discovered that he was standing in the way of their obtaining a blanket concession for the oil resources of the Po Valley, in northern Italy, and some hard words had been said by an American executive about his antecedents. The son of a peasant, half proud, half ashamed of his humble origins, Mattei had taken the words seriously and had not forgotten or forgiven. Years later, as head of Italy's ENI, he had visited the headquarters of Jersey in New York, where he expected to be received as the important personage he had become. But to Jersey, ENI was just another small oil company, and he had been kept waiting. He did not forget that, either.

Under the circumstances, it was no moment to expect that all would be forgiven over a handshake; some more subtle approach would have to be made. But it would have to be done soon, before the flood of Mattei's cheap crude played havoc with the markets.

THE MAJOR OIL COMPANIES were not the only ones who feared Enrico Mattei. General Charles de Gaulle and the French had even more valid reasons for objecting to his methods. At that time Algeria was still French, but the fight for its independence was reaching a bloody climax. In the Algerian Sahara the French had begun intensive exploitation of the desert oil fields, and no matter what happened to the rest of the country, they were determined to hold on to them. To give an international flavor to the operations, an emissary of General de Gaulle had gone to Rome, where he offered Mattei and ENI a concession in the desert in partnership with the French.

Mattei turned down the proposition in no uncertain terms. He gave the French two reasons which infuriated them, because they were eventualities they feared but were determined would not happen: they would lose the Algerian war, and they would then lose the oil fields.* When news of his remarks leaked out in Paris and Algiers, the French government officials preserved a tight-lipped silence, but members of the army, particularly the "ultras" of the OAS, the right-wing military terrorist organization, were loud in their denunciations of him. Soon, when rumors spread that Mattei was in touch with the Algerian rebels, the OAS began to make threats against his life. These were not to be taken lightly, since the ultras had already shown their ruthlessness in several particularly nasty kidnappings and assassinations in France, Switzerland and Belgium.

Mattei was unperturbed. An interview he gave to reporter Gilles Martinet reads in part:

> INTERVIEWER: There is at least one point on which General de Gaulle's ministers and those we call the "ultras" find themselves speaking in identical terms—namely when they talk about you. In their eyes you are the man who plans to destroy France's interests in the Arab world and who, more specifically, has his eye on the immense resources of the Sahara.
>
> MATTEI: What I am really being blamed for is my refusal to commit myself to a certain policy; in other words, for having refused to set myself up in the Sahara along with the French, English and American companies. The offer was made to me several times, but I consistently declined to accept a concession. I don't want my technicians to find themselves faced with the necessity of working under the protection of machine guns. Italy lost her colonies together with the war. Some people considered this a misfortune; in fact it turned out to be an

* As it turned out, he was right in both cases.

immense advantage. It is because we no longer have any colonies that we are so warmly welcomed in Iran, the UAR, Tunisia, Morocco, Ghana and other countries. I don't see why we should endanger this position by joining in an operation which everyone knows cannot be continued indefinitely in its present form.

When the Algerian war comes to an end, I shall see what can be done.*

Shortly after this interview was published in the magazine, Mattei flew to Tunis, where ENI had oil interests and concessions, but where there were also several scores of thousands of Algerian rebels training in camps. It seems likely that Mattei met some of the Algerian leaders during this visit, and that he had further talks with rebel representatives in various parts of Europe during the next twelve months. He was reported to be on the OAS death list, but was not worried about it, though he now moved everywhere with a bodyguard.

AGAINST THIS BACKGROUND, as well as the more mundane one of cheap Russian oil, Standard of New Jersey set about wooing Enrico Mattei over to the American side in the summer of 1962. The depredations of Soviet crude under Mattei's control were now such that the oil lobby had been operating behind the scenes in Washington, and the aid of President John F. Kennedy had been secured. Remembering Mattei's secret shame over his humble beginnings, and the poverty which had prevented his attending a university, a bright Jersey executive had come up with a brilliant idea. A plan had been devised to invite him to the United States to receive an honorary degree from Stanford University, which has always been interested in petroleum affairs; afterward he would go on to Washington, where he would be received by the President. From there on, the oil-company executives would take over.

In furtherance of the plan, Kennedy sent his undersecretary at the State Department, George Ball, to Rome, where he dined

* *France Observateur,* Paris (August 10, 1961).

in secret with Mattei at the U.S. embassy.* A few days later the Italian consented to meet Standard of New Jersey officials. Apparently flattery had got them everywhere. Soon Mattei was confiding to his friends, "The Americans are now offering to sell me crude at a price which matches Russia's."

Of subsequent talks, a document in the archives of the *Middle East Economic Survey* dated November 2 reveals that an agreement with Esso was being discussed: "[It] was to cover such questions as long-term supplies of crude oil to ENI (probably from Libya), the purchase from the engineering company of ENI group Nuova Pignone of vast quantities of petroleum-industry equipment, and the supply to the ENI marketing company, AGIP, of refined products in areas in which ENI still had no refinery."

According to a statement made later by Standard of New Jersey, the agreement was to be signed at the beginning of November, when Mattei would go to the United States for his honorary degree. But on October 27, 1962, he was killed. He took off from an airfield in Sicily, where he had been visiting ENI installations, in a company plane. It crashed shortly afterward, killing all aboard. And no one has discovered since whether it was an act of God or sabotage.

Standard of New Jersey was quick to issue a statement on October 29 (through a spokesman for its Italian subsidiary, Standard Italiana) deeply regretting Mattei's untimely death and emphasizing that the state of belligerency between the company and the Italian had been on the brink of changing to the closest collaboration. But a cow does not genuinely mourn the demise of a gadfly which has been buzzing around her for years, and there were no long faces in the front offices of the big oil companies when the death was announced. Executives of other companies read Standard's statement with a certain skepticism, for they doubted that any such agreement could have kept Mattei in check for long. He had spent most of his life fighting the Anglo-American oil bloc and had never made any secret of his feelings toward

* See *Financial Times,* London, October 29, 1962.

it. "In concentrating in a few hands the control of oil production and marketing," he had once said, "in maintaining with the consumers the relations of supplier and customer in a closed and rigid market, in granting only financial returns to the countries that own the oil, and in barring all international agreements for rational organization of the market, the international companies have increased their own power, but they have also created the conditions for either the breakup or the transformation of the system under the pressure of new forces." *

Mattei had always considered himself one of the "new forces, and no agreement with a member of the international cartel would have been likely to relieve the pressure on them for a moment longer than he found it expedient for his purposes. His death was a tragedy, but it would be hypocritical to pretend that the major oil companies were anything but relieved by the news.

* In an article in *Politique Étrangère,* Paris (Spring 1957).

CHAPTER TWENTY-THREE

Overthrow

IN THEIR PUBLIC DECLARATIONS, the men who ran Aramco portrayed King Saud of Saudi Arabia as an enlightened monarch interested only in the welfare of his country and his people.* But everyone in the company knew that the king actually was a disastrous failure whose regime was riddled with corruption and racked with scandals that were the talk of the Middle East. Though the oil revenues from Aramco and the independents operating in the Neutral Zone were still, as in his father's day, paid to him personally, and were now amounting to $350 million a year, he had managed to put both himself and his country hopelessly in debt. By 1960 Saudi Arabia owed 1.250 billion riyals abroad and over 500 million riyals at home, and so much paper currency was being printed that the value of the riyal had fallen from 3.75 to 6.25 to the dollar.

Yet Saud did not consider curtailing his wild expenditures on cars, planes, wives and concubines. He now had scouts in Cairo, Beirut, Teheran and Karachi buying up girls for his harem, and the export of young females and small boys from Cairo to satisfy the demands of the king's five thousand spendthrift courtiers

* See the testimony of Aramco's president, F. A. Davies, before a Senate committee, p. 197, above.

had become so notorious that the Egyptian press was demanding its suppression.

Aramco was not overly worried about the king's morals,* but his greed for money and his willingness to agree to any scheme in order to acquire it caused them considerable concern. Company officials had had to work hard to sabotage a scheme in which Saud planned to give Aristotle Onassis a monopoly for the use of tankers to transport all oil out of Saudi Arabian territory. The scheme had, in fact, been announced in the last days of the reign of his father, Ibn Saud, but it was his son and his cronies at the Saudi court who had arranged it with Onassis' go-betweens. (These included the late Hjalmar Schacht, once the financial adviser of Adolf Hitler, and a Greek named Spyridon Catapodis, who later sued Onassis for promised fees amounting to $14.210 million.) And it was Saud who had taken most of the bribes, amounting to several million dollars, which are said to have been passed over to secure the tanker contract. Aramco's reaction to the grim possibility that Onassis might soon be in a position to decide in which ships and at what rates they could export their oil was to threaten to take King Saud and his government before an international court. At the same time Aramco's parent companies started to put the pressure on Onassis before he could do the same to them, by instituting a quiet but effective boycott of the Greek's tankers; it was this rather than the threat of legal action which finally sabotaged the deal.†

It was not only a costly episode for Aramco, but King Saud's lies and deception during the course of the negotiations filled the directors with fear for the future of their operation under the

* Though they suffered badly from his hypocrisy. While indulging in the most extravagant excesses himself, he imposed a puritanical regime on oil-company employees, particularly the Americans and other non-Saudis, forbidding wives to drive cars, Christians to have a church, and banning all alcoholic drinks, including beer.

† Afterward Onassis was to describe this as the most critical moment of his career. "Because I was in the doghouse with the oil companies and could get no new charters, I had a great number of idle ships on my hands," he told his biographer, Willi Frischauer.

aegis of such an untrustworthy monarch. He was likely to cause trouble not only for the oil company but for his country, because his overweening vanity and the way in which he flaunted his money were making him notorious in the Arab world. He was reputed to be jealous of the influence of General Nasser, feeling that he and not Nasser should be regarded as the leader of the Arab world, and he had allowed himself to be involved in an inept plot against the Egyptian's life. As a result, powerful organs of the Cairo press and radio were now mounting attacks against him.

What Aramco feared most was the unrest that this barrage of antiroyalist propaganda might cause in Saudi Arabia itself, for the company was now so closely linked with the throne that any revolt against the regime would almost certainly turn out to be against the oil company as well. "Every time the king sneezes, Aramco builds another hospital," one British critic commented.

Still, none of the Americans dared raise his voice against the royal excesses. The only non-Saudi who had had the courage to do so was Harry St. John Philby, but despite the fact that he had been the trusted adviser of the old king, in the spring of 1955 his son and heir served Philby with an order of expulsion and vilified him as a liar, a traitor and a cheat. The scandalized and angry old Anglo-Irishman went grumbling off to Lebanon, mourning the depths to which his adopted and beloved Saudi Arabia had fallen.*

Fortunately, there was one man in the land whom King Saud both envied and feared: his younger brother, Crown Prince Faisal. The two men shared only one family characteristic, a tendency to sickliness; otherwise they were the opposite in appearance, temperament, education and outlook. Beside his fleshy, flaccid brother, Prince Faisal had the lean look of a famished eagle, and indeed a desperate need to appease some ravenous worm inside him had always been one of his troubles. "A handful of rice goes into his

* When Philby proved to be an even more vocal critic in exile, King Saud invited him to return, on the theory that he was more easily controllable inside the country.

mouth like a cannonball," his father once said of him. "He is a hero of the spread. He eats—he eats!" *

Otherwise, Faisal was rarely given to excess. In 1960 he had eight sons, compared with Saud's forty or more, and he lived happily with one wife in contrast to Saud's ever-changing wives, concubines, slaves and cabaret girls. He was one of the few sons whom the old king had encouraged to travel abroad, and hence had a social and political horizon beyond the minarets of Mecca and the sand dunes of the Empty Quarter. He lived modestly in a small residence which seemed humble compared to the king's flashy palaces, and though he shared the Saudi family's love for gold, he rarely displayed his considerable wealth.

There was never any doubt in either brother's mind that Faisal would have made the better king, and this knowledge sometimes embittered relations between the two men. But each time King Saud faced a crisis, he called in Faisal to help him, and a regime of austerity would follow, with the crown prince established as prime minister or deputy minister, empowered to cut the allowances of the vast numbers of feckless princes, and even to limit the king's own privy purse to a maximum of $60 million a year. Then, when the crisis was passed, and everybody had begun to complain of the miserliness of the regime, Faisal would be banished into obscurity and the orgy of spending would begin again.

It was not that the crown prince was an enlightened or liberal element in Saudi Arabian life. He did not disapprove of his brother's profligacy because he wished to replace it with a more democratic regime, and he certainly did not plan to allow the oil revenues of the country to pass out of royal hands. He had listened to the polemics of the Saudi oil minister, Abdullah Tariki, with prim distaste, and disliked the minster not simply on personal grounds but for political reasons as well, for his talk of takeovers made him a danger to the regime as well as to Aramco. Faisal was convinced that Tariki was a secret spokesman for the

* Rihani, *Ibn Saoud of Arabia.*

foreign-trained Saudis, most of them educated abroad, who had started an underground National Liberation Front, and he had tolerated him only so long as he was forcing the oil companies to pay higher prices for their products. What it came down to was that Faisal was simply a tidy, efficient, money-minded conservative who abhorred the waste of his brother's regime, but who wanted no change in its basic structure.

As mentioned earlier, in the spring of 1962 Saud appointed himself prime minister, and Faisal accepted the post of his deputy —but only on condition that Tariki be banished from the country. Faisal knew that it would only be a matter of time before he was in charge, for King Saud was neither physically nor mentally equipped to cope with the tasks of the premiership. The king had recently returned from a visit to the United States, where he had undergone operations for trachoma, that endemic Arabic infection, at the Peter Brent Brigham Hospital in Boston, and it was all he could do to shuffle from the festive board to the bed of his latest favorite, without having to handle urgent affairs of state. They were urgent because, at Faisal's insistence, a program of school and hospital building had begun, and yet another austerity program would be necessary to pay for it. At the same time, Egypt was backing the republican rebels in the neighboring kingdom of Yemen, while Saudi Arabia was discreetly but expensively subsidizing the Yemeni royalists. Still smarting from Nasser's attacks upon him, Saud was all for sending his forces across the desert frontier to confront the Egyptian army and had indulged in petulant outbursts of temper, when Faisal succeeded in circumventing him.

By 1964 the tantrums had become too much even for the patient Faisal to bear. He sensed that one of those moments was approaching when the king would once more dismiss him and revert to his feckless ways. But this time, Faisal decided, the times were too fraught, and Saud too irresponsible. In the last days of October he sent members of his entourage around the salons and coffeehouses of Riyadh, Mecca and Jidda to discover who his enemies were and how much he could count on his friends. In the past months he had bound a number of the more influential princes to him by quietly restoring their allowances and promising them posts in any new regime which might be established should "our beloved brother fall victim to his increasingly dolorous afflictions." He had also talked with the Muslim priests, the *ulema,* assuring them that should he ever be ruler of Saudi Arabia, the state would more than ever be subject to the laws of the Koran and the strict rules of the Wahhabi sect.

On the night of November 2, 1964, a meeting of princes and *ulema* was held, and the conspirators unanimously voted to depose the king and acclaim Faisal monarch in his place. Shortly afterward King Saud's palace was surrounded by crack troops, and Crown Prince Faisal personally presented an ultimatum to his brother: either abdicate and consent to leave the country, or he would never leave his palace alive.

King Saud capitulated and signed the instrument of abdication at once. He did not reproach his brother, and made only one request—that he be allowed to take certain members of his harem (of which he would prepare a list) into exile with him. They and others would help him spend the $3 million a year which Faisal gave him. But not for long: Saud died in Greece in February 1969.

Spurred on by government propaganda, the Saudi people hailed their new monarch. In Dhahran, the American oilmen of Aramco also hailed him. They knew that his regime would be theocratic and puritanical, and that there was less likelihood than ever that they would be allowed to drink an occasional glass of beer. But at least the new regime would be a tidy one, and hopefully the days of waste and corruption were over. Happily they returned to the business of pumping oil, and gave the order to step up production. The company was still looking for new fields, but for the moment it had all the crude it could handle.

In this Aramco was luckier than some rival Americans working across the desert to the southeast, in the mountains of Dhofar.

Overthrow

IT IS ONE THING to be given an oil concession for a province the size of the state of Ohio, but quite another to turn it into a viable oil field. As Wendell Phillips and his friend Sultan Said bin Taimur of Oman discovered, oil is a peculiar substance. Developments in geophysics, the use of magnetic, gravity and seismic surveys, and all the appliances of modern science make it possible to plot from the air the structural traps in which oil may have accumulated. Thereafter, geologists on the ground can judge from rock samples and structural prospects whether these traps look promising. But only by drilling wildcat wells can it be discovered for certain that a deposit of oil is definitely there, and even if it is, the amount may be small and the quality poor. The more wells drilled in any area, the better the chance of finding rich accumulations, but their cost is enormously high in men, materials, time and effort.

This was what Phillips and his associates were discovering in the province of Dhofar. By 1961 the Dhofar–Cities Service Petroleum Corporation (the name of the group which had taken over the Phillips concession) had spent $30 million on wildcat wells without finding a viable field. They were beginning to run out of money, and the sultan out of patience.

At first all had seemed to go well, and the sultan had proved an urbane host in his efforts to make life easy for the influx of oilmen. The coastal strip of this part of the Arabian peninsula is a paradise compared with the bleached and baking desert of the Empty Quarter, which lies behind it to the northwest. The sultan's castle stood on the seashore outside the port of Salala, hidden behind high walls at the end of a palm-lined avenue, and here, in a crenelated building looking down on the Indian Ocean, cooled by the tail end of the monsoon, Phillips would forgather with his bearded friend to report progress. His camp had been established in a small rocky cove some ten miles up the coast at Risut, at the foot of a cliff from which jutted the remains of an old Portuguese fort. When Phillips had first started working the concession in the mid-fifties, the crews had built a road winding north through the Qara Mountains, a range 4,000 feet high, to the arid desert plain beyond, and there they had drilled into a promising structure at Marmul. From the first two wells came crude oil in abundance, and the sultan had been summoned to see the liquid flow. He had promptly sent greetings to all his fellow Arab rulers in Kuwait, Dubai, Abu Dhabi, Riyadh and Yemen, proudly announcing the discovery and informing them that henceforth he would be a member of their exclusive club.

Unfortunately for everybody, the oil which flowed so freely was not the right kind. Had it been discovered in the United States, or close to a center of civilization, it would have made its exploiters rich. It was low-quality crude, marvelous for building roads but of no value as fuel, and not worth building a pipeline to take it to a port from which to ship it to the outside world.

By late 1961, Cities Service had brought in Richfield to join the hunt for better-quality crude, and they had drilled twenty-three wells, six of them over 10,000 feet deep, without important results. To make matters more difficult, a local imam with rival claims to Dhofar had begun a rebellion against the sultan, and the drilling crews now worked surrounded by armed guards to protect them against tribal marksmen hidden in the mountains around them. Phillips—tall, pale and frail-looking—wandered around the camp with his two six-shooters on his hip, potting at stones or teaching the sultan's son, Qaboos, lessons in marksmanship. For one period, Phillips was joined by his sister and mother. The former was not only a beauty but a crack shot, an expert rider of racing camels, and a daredevil driver of trucks over sand dunes. His mother, who rejoiced in the name of Sunshine, was a veteran of the Sierra Nevada gold rush in the early years of the century, and though in her seventies, could outwalk and outclimb almost everyone around her. Their presence heartened the crews and made Phillips seem paler than ever, and as long as they were on the scene, Sultan Said bin Taimur kept his temper. But when they left, he disappeared into his cool palace and allowed his officials to plague the company with petty restrictions. The group had now spent $40 million, and Phillips was beginning to feel guilty about the $1 million in his account at Chase Manhattan.

In 1962, a larger-than-life Texan named John W. Mecom appeared on the scene, a man who had once scornfully remarked about J. Paul Getty that "he doesn't know where his next half billion is coming from." He described himself as "a modest billionaire," but in fact he rarely hid his light under a bushel. Indeed, he had no need to, for his life would make a classic script for a John Wayne film. He had started as a roustabout in the Texas oil fields and progressed to wildcatting in South Texas and southern Louisiana, where he drilled the world's deepest well, at 25,000 feet. Since he was a one-man corporation who, like Getty and Gulbenkian, carried his office around with him and risked his own money rather than someone else's, he did not have to worry about other people's opinions. It was Mecom's experience that of every twenty wells drilled, all but one would be dry, full of salt water or otherwise unproductive, but that the one successful well would yield so much money that it would more than make up for the expense of the others within six months of going into production. "Oil is a hell of an expensive quantity to find and exploit," he once said, "but once you've found it, you've gained possession of the philosopher's stone."

Mecom looked over the Dhofar concession and then proposed to Phillips and his associates that he drill a separate well (in association with the Pure Oil Company of Ohio, one of the "big five" independents which Mecom also owned), on a site of his own choosing. Some months later he took over interest in the Cities Service's operation. In the next three years he drilled five wells and found promising showings of crude in three of them, but no bonanza.

By 1965, even Mecom was discouraged, Wendell Phillips was hugging his $1 million for comfort, and Sultan Said bin Taimur was crushed. So far the concession had cost close to $50 million, and the fortunes of the unlucky subjects of Oman had not altered a whit. But only the departure of the sultan would do that.

CHAPTER TWENTY-FOUR

The Libyan Jackpot

ALL THE IMPORTANT ARAB oil countries, plus Iran and Venezuela, now belonged to the Organization of Petroleum Exporting Countries, and as the combines were discovering to their cost, their combined weight was becoming considerably greater than the sum of the parts. With every month, OPEC's experience and know-how grew, and it was learning, slowly and painfully, though not without backsliding by one country or another temporarily seduced by British and American blandishments, that in union there was strength. In the years since its formation in 1960, OPEC's progress had been impressive, but there was still no formal recognition of its existence by the major oil companies; in their negotiations with the various governments of the Middle East they refused to contemplate global agreements and insisted that any contracts were limited to the borders of the countries in which they were made.

To strengthen OPEC's position, the new oil minister of Saudi Arabia decided to resort to a stratagem. Ahmed Zaki Yamani was a Harvard-trained specialist in international law who had taken over after the dismissal of Abdullah Tariki in 1962 at the insistence of the man who was now King Faisal. One of the coolest and most persuasive Arabs in the Middle East, totally unpolemical, anxious to demonstrate the basic reasonableness of his attitude, Yamani is a handsome and soft-spoken negotiator of the utmost charm and guile. It was he who had helped OPEC surmount its first major obstacle in 1962, when the association passed a resolution designed to force the Anglo-Americans into a radical new system of disbursing their oil royalties. This would considerably increase the incomes of the producing countries, and OPEC wanted to negotiate the new system as a body, both to get the organization accepted and in order that smaller or weaker member countries would not be tempted to settle with the companies for less than the others.

"Aramco was chosen as the first company with whom we would reach an agreement," Yamani said in a recent interview, "and I informed the company that an OPEC committee would like to have a meeting. I think Aramco might have accepted if it had been left to them, but they informed their member companies, and it scared them out of their wits. So Bob Braun, the Aramco president, refused to have the name of OPEC even mentioned in the negotiations. He said, 'We don't recognize the so-called OPEC,' rather in the same stupid way that we Arabs nowadays say, 'We don't recognize the so-called Israel.' "

Yamani went ahead, anyway, and formed a three-man committee on behalf of Saudi Arabia and the other member countries, and then marched them into the conference room. "The Aramco people at once challenged the legality of our three-man committee. But I was the chairman and I said I could choose whom I wanted, and we were not going to begin talks until they had accepted us. I told everyone to sit tight. We forced it on them. I did it just to impose OPEC on them. And by the end we had de facto if not yet de jure recognition of the organization."

By 1962 OPEC members between them were the landlords of 90 percent of the world's exported oil. (The United States is not a major oil *exporter.*) However, one country which was beginning to emerge about this time as one of the most important oil exporting countries in the world was the North African state of Libya. Libya was the Achilles heel of OPEC, because it was not a member of the organization and had refused repeated invitations to join. In the meantime, it was selling crude oil well below the posted prices of the OPEC countries, thus threatening to sabotage every agreement that organization made.

The truth was that through a mixture of economic ignorance, naïveté, and the venality of its administrators, Libya's oil resources had been sold short. It would be too harsh to say that the big American companies which were now established in the deserts of Tripolitania and Cyrenaica, the two Libyan provinces, had cheated their hosts, but they had certainly driven some extremely sharp bargains.

Until it was liberated by the Allied armies in 1943, Libya had been an Italian colony in which the native Senussi tribesmen had been degraded, and all the more fertile areas of the country—particularly the Jabal Akhdar, or Green Mountains, in Cyrenaica—turned over to Italian émigré peasants for farming. On Mussolini's orders, Libyans had been denied education or any say in the administration of their land, and their spiritual leader, King Idris I, driven into exile. At the end of World War II, the king had ridden triumphantly back into his capital, Tripoli, on the same white horse on which the Italian dictator had once boasted he would lead a victory parade through Cairo.

Libya was now free and independent, but what can you do with freedom and independence when you have only a handful of administrators capable of running a country, and no schools or technicians to manage even the smallest industries? The country had no alternative but to call in European, principally British, experts to advise it. Italian craftsmen and mechanics were allowed to stay on, even though they had once been hated oppressors, to keep the lights burning and the trucks running. In return for their aid, the British were given the right to station troops in the places where they had fought some of the most spectacular battles of the Western Desert, at Tobruk and Benghazi, and here they also maintained a base for the Royal Air Force and a staging post for trans-African air services. In addition, as part of the so-called Eisenhower doctrine of Middle East peace keeping, the U.S. Air Force had a big base at Wheelus, near Tripoli. King Idris ruled, but in little more than name.

Then, in the mid-1950s, oil was discovered in Cyrenaica. As in the Biblical lands of Arabia, there had always been gas fires and bitumen seepages in the Western Desert, and it seemed certain that beneath them were sizable accumulations of oil. From some of the samples which had been tested, there was every reason to believe that when brought to the surface, the crude oil would be of high-grade quality, with a low sulfur content, and suitable for sale to the European market.

When news of these possibilities spread through the oil world, companies hovered like bees before a hive in search of concessions. But Libya was uncertain about how to grant them. The king's ministers were told by their foreign advisers (none of them, unfortunately, oil experts themselves) that they should at once pass a general petroleum law defining the terms under which concessions would be granted. By this time the Libyans were extremely anxious to see their potential petroleum wealth turned into hard cash, so they were grateful when American oil companies offered to speed their progress by lending legal experts from their own organizations and helping them to draft a law that would expedite operations.

As a result the Libyan Petroleum Law of 1955 looked like an equitable piece of legislation that would benefit both parties in any concession agreement, and at a quick glance its provisions did not seem much different from those of the old Middle East concessions (though not the new ones). Profits from the concessions would be subject to a tax of 50 percent, and a royalty of 12½ percent on every barrel of oil would be treated as a partial payment of the tax on profits. But the Libyans were to discover that they had also accepted some fine print whose significance they did not appreciate until it began to affect them. Written into the law by the American experts were two provisions which were especially favorable to their operations. The first was a depreciation allowance, which permitted the concessionaires to deduct a charge of 20 percent on all physical assets acquired before production; in addition they had a choice of a 20 percent amortization of all their expenditures before production began, or a depletion allowance of 25 percent of their gross income, whichever saved them most money.

The second provision tied Libya's oil royalties not to the posted price, as it did in the rest of the Middle East, but to the price paid for the oil on the market. Since, at Arab and Iranian insistence, the posted price stayed rigid even when the market price dipped,* this provision was to cost Libya many millions of dollars in the years to come.

The moment the new law was passed, the rush began. It made operations especially attractive to independent oil companies, for the lack of a posted-price provision meant that they could sell their product at cut rates and secure a hold in the markets still largely controlled by the majors. Within a few months, fourteen different applicants, including six of the major groups,† had been granted forty-seven different concessions. Esso International was the first to make a big strike in their concession, Area Number 20, followed shortly afterward by an independent group called Oasis, which was a combination of three U.S. companies, Marathon, Continental and Amerada. Soon, when it became evident that Libya was about to become one of the most prolific oil countries in the world, King Idris and his ministers realized belatedly that the oil companies had pulled a fast one and that they were being short-changed on their own law.

In an attempt to rectify matters, in 1961 the Libyans enacted

* As has been mentioned, the oil companies only lowered the posted price once; Arab anger was such that they never dared do it again. Since then, the posted price has always been higher than the market price, and it is on the posted price that OPEC members' income is calculated.

† Esso International (a subsidiary of Standard of New Jersey), Shell, Standard of California, BP, Mobil and Texaco.

an amendment to the petroleum act which pegged the oil companies' taxable income to the posted prices. But they also allowed a clause to be inserted stipulating that "marketing expenses" could be knocked off the new tax rate. Once more they found themselves victims of some fast legal footwork. The Libyans had meant to specify that the amount allowed for marketing expenses was to be no more than 2 percent, but somehow that figure was never inserted in the amendment. Soon many of the companies were including enormous amounts for the rebates they were offering in the market in order to undercut the majors. And when *this* loophole was plugged, another amendment was mysteriously sneaked in stipulating that no amendment would affect the existing contractual rights of the concession holders without their consent.

WHY DID LIBYA ALLOW ITSELF to be continually outsmarted by the exploiters of its resources? Ignorance and innocence were not the only explanations. King Idris was a simple-minded Senussi who had grown old waiting for the liberation of his people, and by now much of his fighting strength had left him. His principal pleasure was to take his young tribal wife to a small palace he had built at Tobruk, where they went swimming and relaxed. His only interest in oil was a negative one; he became enraged when crude oil, being loaded into tankers near his palace, spilled and drifted in to pollute the beaches.

However, Idris had allowed himself to be surrounded by ministers whose venality knew few bounds, and the habit of corruption had permeated most ranks of Libyan officialdom. There is little doubt that much of the loose legislation which cost Libya so much during its early years was due to the large sums of money which changed hands between certain companies and the oil ministry. The majors did their share of bribery—they maintained that they had to in order to get any business done at all—but on the whole behaved with much greater circumspection than the independents. Some would say that in view of their global agreements, they had to; nevertheless, the fact remains that Esso International, for instance, began to sell its Libyan oil at Middle East posted prices (about $2.16 a barrel) the moment the Libyan government pointed out the inadequacies of the petroleum law, and paid taxes accordingly amounting to about 90 cents a barrel. On the other hand, the Oasis independent group continued to sell its oil at cut-rate prices (about $1.55 a barrel) and paid a tax of less than 30 cents a barrel.

Since they did not recognize the organization themselves, the major companies could not say so officially, but they realized that membership in OPEC was the only hope that Libya had of solving its problems and ending the inequity in oil payments. With OPEC behind them, the Libyans could talk from strength. But each time OPEC approached the government, one of the independents opened a new private bank account in Switzerland for a Libyan oil official, who then used his influence to veto membership.

The independents had a case for opposing any legislation that would bring Libya's taxes into line with those of the Gulf states and Iran, and during a conference with the Libyan premier, Husain Maziq, they argued that OPEC membership would not only harm the interests of the independents but cut into Libya's global oil sales as well. "They pointed out a conflict of interests between the independents and the majors and between the major companies and Libya itself," writes George W. Stocking. "They observed that the majors' interest in producing relatively high-cost Libyan crude was slight compared with low-cost Arabian crude. The increase in their Libyan costs, which the new royalty and tax provisions would exert on their total average costs, would be slight, but it would have a disastrous effect on the independents. They suggested that this would not be displeasing to the majors, who would welcome an opportunity to buy them out at bargain prices. And they argued that once the independents disappeared from the Libyan scene, the majors would have a cost incentive to increase Persian Gulf output at Libyan expense." * In other words, the independents maintained that any legislation which altered the present pricing arrangements would result in "a short-run gain and a long-term loss."

But then the independents made a mistake in psychology. Convinced of the validity of their arguments, confident that they had bought the support of the ministers concerned, they let it be known that in any case they would not accept any new legislation. It was the old story: a contract was a contract.

This was too much even for the corrupt members of King Idris' Administration, and news of the independents' arrogant attitude was leaked to the Egyptian and Arabian press, which began to talk about "colonialist exploiters" who "use the USAAF base at Wheelus to bolster their outrageous demands." † Shortly afterward, Libya announced that she was joining OPEC at last. Prime Minister Maziq issued a statement in which he said: "If they [the independents] continue this attitude, we shall be compelled to issue new legislation which will oblige them to abide by the new system [of taxation]." ‡ Such legislation, he made clear, would be drastic, and if necessary it would be enforced by Libyan arms. Any independent company which refused to pay the new taxes would have its oil exports prohibited and its installations taken over until payment was made. To bolster Maziq's threat, OPEC hurried to the aid of its new member. In a resolution passed at an emergency meeting in Vienna, the member countries pledged to refuse to give any concessions "to any company or the subsidiary of any company refusing to comply with Libya's new oil policies." §

With some prodding from Washington (where the government made it clear that the companies concerned were on their own in the event of trouble), the independents came into line and accepted the new tax rates. Libya was now a member in good stand-

* Stocking, *Middle East Oil.*
† *Al Ahram,* December 2, 1962.
‡ *Al Hayah,* Beirut, December 23, 1965.
§ *Middle East Economic Survey,* December 31, 1965.

ing of the Organization of Petroleum Exporting Countries, and would henceforth reap the benefits of that body's joint negotiations.

In the meantime, OPEC experts estimated that the Libyans had lost something like $100 million in taxes as a result of the petroleum act, which had been written for them by the oil companies' legal advisers. It was a figure that would be remembered by the young militants who were beginning to be active in student and army circles, and one day they would get it back with interest.

NOT THAT JOINING OPEC provided an instant cure for Libya's chronic maladministration and fiscal dishonesty; it was still the land of "the date palm and the greasy palm," as one American described it. In 1965 the Libyan government announced the opening of new areas in Tripolitania for oil development, and from all parts of the world, oil companies rushed in with their bids. Despite their complaints that Libya's new rules would drive them out of business, most of the independents already operating inside the country were among the bidders.

To the astonishment of the outside world, but not to those who listened to gossip in Tripoli, two of the choicest territories went not to any of the established companies but to an outsider, Occidental Petroleum Corporation of California. The company had been taken over eight years earlier by a formidable entrepreneur named Dr. Armand Hammer, whose expertise as a financial manipulator can be judged by the fact that while still a young man, he became a millionaire as a trade go-between in the Soviet Union. Grandson of a Russian emigrant to the United States, he went to the Soviet Union in 1921 as a young doctor, and there realized that the Russians were starving for want of grain, whereas America had a surplus of wheat. He had a million dollars he had inherited from his family (they made pharmaceutical products) and he lent it to the Russians to finance wheat deals. They paid him back in asbestos, timber and furs. He was twenty-three years old and his fortune was made. Later, in the United States, he became involved in a wide spectrum of activities demonstrating the breadth of his tastes and interests (ranging from hard liquor to art), and in 1957, after he bought a modest California company named Occidental, he began his preparations to enter the international petroleum market.

Dr. Hammer blandly ascribed his good fortune in winning the new Libyan concessions to the meticulous care with which Occidental's bid had been prepared, and to the fact that he had paid minute attention to Libyan susceptibilities and pride by tying up the bid in ribbons fashioned in Libya's national colors. But subsequent developments led the less naïve to believe that the success of the Occidental bids was due less to the color of ribbons than to the amount of dollar bills which went with them. As the *Wall Street Journal* was to put it later:

> Some little-noticed documents on file in federal district court here [New York] in connection with a suit against Occidental by Allen and Co. show some of the influences that may have been brought to bear in Occidental's winning of its bonanza. Involved are an agreement by Occidental to pay $200,000 to Ferdinand Galic, a bon vivant European business man and promoter; the financing of a documentary film written by Fuad Kabazi, Libya's former oil minister, and said to cost $100,000; alleged payments by Occidental to Taher Ogbi, the company's Libyan representative who became minister of labor and social affairs, and to "General de Rovin," a notorious international swindler whose real name is François Fortune Louis Pegulu; and the $100 million breach-of-contract suit filed against Occidental by Allen and Co., the Wall Street investment banking firm.*

Documents deposited with the federal district court in New York in 1972 suggest that Occidental's fantastic success may have begun in the summer of 1964 when two colorful characters met in Paris for the first time. One of them called himself General de Rovin; the other was Ferdinand Galic. The court documents reveal that De Rovin, then sixty-three years old, "was a swindler in

* *Wall Street Journal*, February 8, 1972.

Berlin, Vienna and elsewhere in the period prior to World War II; that he had dealings with the Nazis during the War for which a French court later sentenced him to death in absentia; that in the postwar period he traveled about South America and Canada, making a living by passing bad checks; that he eventually returned to France from Argentina under the phony name of 'de Rovin'; that he was employed by a French firm and promptly squandered its assets; that in February 1970 he was sentenced in absentia to a year in jail by a French court after a conviction of violation of foreign exchange controls . . ." *

Galic, a Czech married to an American and a well-known figure in Paris society, had no knowledge of General de Rovin's gaudy past when they first met. De Rovin had a proposition: if Galic could find an oil company willing to put up millions, he had the means, through a highly placed Libyan, of obtaining lucrative concessions in that country.

Galic at once telephoned a friend of his, Charles Allen, multimillionaire founder and general partner in Allen and Co., a New York brokerage firm. "Charley told me, 'Give me time, I will look around,'" Galic subsequently recounted in his deposition.

In September 1964 Dr. Armand Hammer, Herbert Allen, Sr. (Charles Allen's brother), Galic and De Rovin met in Claridge's in London with a Libyan businessman named Taher Ogbi. It soon became clear that Ogbi was the highly placed Libyan whom De Rovin had been talking about. Subsequently it was claimed that as a result of this meeting, certain agreements were made between Occidental and Galic, De Rovin and Ogbi in regard to Libyan oil concessions.† A separate agreement was said to have been made with Allen and Co.

At this stage Fuad Kabazi appeared on the scene. Oil Minister Kabazi, one of the strong men of King Idris' government, was introduced to Galic by Taher Ogbi. They became "friends from

* *Ibid.* Also quoted by *International Herald Tribune,* Paris, February 9, 1972.
† Occidental agrees that it later paid Galic $200,000, and a lesser sum to Ogbi, but declines to reveal what it paid to De Rovin.

the first day we met," Kabazi said in a deposition left with the federal district court, and he came to admire Galic "for his personality and his connections in the world of business." Thereafter he kept Galic fully informed of how the government felt about the concessions, and even pressed King Idris himself to grant Occidental the choicest territories.

Later the oil minister maintained that never for a moment did he take these unprecedented steps on Occidental's behalf because he was being bribed. He knew from the start why Galic was cultivating his acquaintance, he said—to secure concessions for Dr. Hammer's organization. "The whole purpose of his contact and close relation [with me] was to get this done," he testified, and though he knew that it was a crime punishable by a jail sentence for a Libyan official to disclose Cabinet secrets, he fed Galic information because the Czech represented Occidental and was "the man authorized to talk and the man to whom I should tell everything." But he believed that the actions he was taking were "for Libya's good"—and also, there now seems little doubt, because Galic had completely won him over. Galic had discovered that Kabazi, a poet and an intellectual, was keenly interested in books and movies, and they talked for hours about film techniques. "I think every time [Galic] came to Libya, the first person he came to see was myself," Kabazi testified further. "We also met in Europe. We corresponded approximately every fortnight—the time it takes for two letters to go off and back."

The two men employed a secret code in which Hammer was referred to as "Marteau" (the French word for "hammer"), and occasionally their meetings took on a cloak-and-dagger quality. Once, when the concession bidding was reaching a crucial stage, the two friends discovered that they were on the same plane together, but "we pretended not to know each other," Kabazi said.

By this time Galic had agreed to put up $100,000 to finance a film which had been written by the oil minister and was to be called *On the Crest of the Dune*. It was to be directed by an Italian, Guido Arata, and Kabazi's brother-in-law was to receive

90 percent of the profits from its earnings.* In his conferences with Kabazi over the script of the film, Arata testified, the minister could not concentrate because he was so worried about the problem of securing the concessions for Occidental. Certain influential people in the government wanted other companies to have the choicest areas, he told Arata. He also had doubts about whether Occidental had enough money to undertake the vast drilling projects that the concessions would entail, but Galic assured him that Allen and Co. would find the money.

By this time, Arata continued, Kabazi seemed to believe that he had a "commitment" to help Galic to get the concessions for Occidental. "There are many, many interests involved," he told Arata. "Imagine, it's as though there were a large dish filled with all little bones and around this dish are many, many dogs that are trying to edge each other out to grab a hold of the contents of the dish, but in view of the fact that Galic is a dog larger than the rest, he will eat the bone he has asked to eat." †

Early in 1965 Allen and Co. discovered that General de Rovin was not all he claimed to be, and they felt it their duty to inform Dr. Hammer and Occidental of the checkered nature of their go-between's past. The deadline for submitting bids for the new Libyan concessions was July 29, 1965, and it was not until two weeks before this date that Occidental sent telegrams to De Rovin and Galic telling them that an agreement Dr. Hammer had signed with them in London in September 1964 was canceled forthwith. At the same time a telegram reached Allen and Co. telling them that their agreement of December 1964 with Dr. Hammer was also canceled.

By this time it was too late for Galic and Allen and Co. to work out an arrangement with another oil company. Occidental's bid, gift-wrapped in the Libyan national colors, already recommended by Kabazi and backed by the Cabinet ("at the end,

* The movie was subsequently made but never shown commercially.

† Pretrial testimony of Guido Arata, quoted in the *Wall Street Journal,* February 8, 1972.

when I had the support of the prime minister," testified the oil minister), was deposited in Tripoli, and to the amazement of the outside world, accepted. Occidental was granted the two choicest areas in the new field, Concessions Nos. 102 and 103.

Fuad Kabazi is now in jail in Tripoli, charged with treason by the new Libyan Administration. Ferdinand Galic has pocketed his $200,000 from Occidental, but has associated himself with Allen and Co.'s suit for damages against the oil company. Occidental itself thrives. Lots Nos. 102 and 103 have proved to be two of the most lucrative oil fields in Libya, and in 1970 they produced 240.6 million barrels of high-grade crude. The small company, which had an operating revenue of $800,000 in 1957, when Dr. Hammer took over, is now a giant of the industry, with $403.5 million invested in oil property, and average gross earnings of nearly $1 billion a year. As the *Wall Street Journal* remarks: "There are, of course, complex dealings in many huge international business arrangements. And no one has even intimated that Occidental did anything illegal in its successful effort to gain the big oil concession." *

Moreover, Occidental contends that it was its superior bid which won the concessions, especially since it contained a promise to build an ammonia plant in Libya and to use 5 percent of its pretax profits for an agricultural project. "Nevertheless," says the *Wall Street Journal*, "Occidental's activities provide an insight into how huge companies sometimes operate in far-off lands." †

No doubt more details of those activities will be forthcoming when Allen and Co.'s lawsuit finally comes to trial. But when it does, it seems unlikely that the fast-talking, cosmopolitan General de Rovin will be giving testimony; at this writing his whereabouts continue to remain unknown.

* *Ibid.*
† *Ibid.*

CHAPTER TWENTY-FIVE

Coup d'État

SHEIK SHAKHBUT OF ABU DHABI was not only a miser but also the worst kind of snob. In 1965, as oil revenues flowed into the back-room annexe where he stored his wealth, and the sound of the rats and mice gnawing away at the piled notes became almost audible, his British adviser, Colonel Hugh Boustead, urged him to do something about the water shortage on the small island in the Persian Gulf which gave the desert sheikdom its name. Hundreds of technicians were at work in the desert oil fields and at sea on the continental shelf of the Persian Gulf. With a population of 15,000, Abu Dhabi could not possibly supply the labor force that the two major oil companies needed, so clerks were swarming in from India and Pakistan, and workmen from Yemen and all the poorer emirates along the shores of the Gulf. For this influx it was necessary to pipe in water from Buraimi Oasis because Abu Dhabi itself had only a few wells and could not cope with the thirst of the population explosion.

But the pipeline system would cost £1 million to build, according to the tender which the British firm of Paulings had submitted, and the figure made the sheik blanch. In 1965 Abu Dhabi Petroleum Company (a subsidiary of IPC) paid the sheik £20 million in oil revenues for its concessions, and Abu Dhabi Marine

Areas (a BP subsidiary) added another £10 million from its offshore concessions, but Shakhbut "simply could not bring himself to spend money, perhaps because he could not believe that his oil revenue would continue to pour in." * Finally, however, confronted by a sudden epidemic of dysentery on the island, and warned by Boustead that it could well be followed by cholera, he consented to discuss the cost of construction with Paulings, and the firm cabled that it was sending out its general manager at once. Shakhbut then asked the man's name. He was told it; it was that of a plain Englishman. At once he withdrew his offer to discuss the project; it would be beneath his dignity, he told his advisers, to discuss any such matter with someone who did not possess a title. When Paulings sent a lord, a genuine peer of the realm, to see him, only then would he grant an interview.

Many British firms have a baron or baronet on their board of directors. Paulings did not, but with a million pounds at stake, they soon acquired one. Lord Brentford was brought in, made temporary chairman of the company, and flown out to Abu Dhabi to get the pipeline scheme under way.

Shakhbut signed, but he bitterly resented spending the money, and took it out on Colonel Boustead. For month after month he kept him waiting for his pay until the adviser, due to go home at last on retirement, had to lose face by sending an emissary to ask for it. The following day two of Shakhbut's slaves arrived carrying two sacks, which they plumped on Boustead's desk, filled with small coins amounting to the exact sum of back pay due. The Englishman signed a receipt, handed it to the slaves and waited for them to depart. When they did not go he asked them why they were waiting. "For the sacks," one of them replied. "His Highness told us that under no circumstances were we to come back without the sacks."

FROM HIS DESERT STRONGHOLD in Buraimi Oasis, the ruler's younger brother, Sheik Zaid, watched Shakhbut's miserly antics

* Boustead, *The Wind of Morning*.

with growing distaste. Two things angered him in particular. One was the ruler's obstinate refusal to join OPEC—he feared the organization's interference in his affairs and was dead set against paying it any dues—and the other was his penurious attitude about improving the welfare of the people. Zaid's loyalty to his brother during the Buraimi crisis with Saudi Arabi had earned him no thanks. Instead, Shakhbut sneered at what he termed Zaid's softness in working to improve the lot of the Buraimi villagers, and whenever he came to Abu Dhabi the ruler would insult him by talking about him as if he were not in the room. Shakhbut must have been aware that this was risky, for he himself had come to power by slaying a relative. Not one of the past five rulers of Abu Dhabi had inherited the position; each had succeeded by a bloody coup, wiping out his predecessor with knife or gun. But for all his stinginess, Shakhbut was not without courage, and in any case he was a fatalist who believed that no human action could change the will of God. Allah would decide. In the meantime he would hold on to his money and eschew cheap popularity by wasting it on his people.

In such circumstances, a coup d'état was inevitable, and when it came it had the wholehearted approval of the populace. Zaid was popular everywhere, and especially in Buraimi, one of Abu Dhabi's main centers of population. "I was always astonished at the crowds who gathered around him wherever he went in Buraimi," wrote Boustead, "and who treated him with the sort of reverence and attention due to a minor saint. He invariably had a kind word for everybody, and was most generous with his money. I was immediately struck by all that had been done in Al Ain, his home town, and in the Buraimi area, for the benefit of the people. The Persian springs had been dug out to increase the water supply to the gardens, wells dug and pumps supplied and cemented baths for men and women had been built in the fallujahs. Everyone who visited Buraimi noticed the happiness of the people in the area." *

* *Ibid.*

Coup d'État

Under the circumstances, not many Abu Dhabians would have cared if Zaid had usurped his brother in the usual way, with a quick slice of the knife across the throat. In fact, some of his close associates did urge him to kill Shakhbut in order to save himself potential trouble in the future. In typical fashion, he replied, "If my brother ever finds sufficient people in Abu Dhabi to back him in a counterrevolution, I will willingly cede to him, because I shall have failed my country, anyway. Once he is gone, no one could possibly want him back—unless a tyrant has taken his place." *

Thus, Sheik Shakhbut became the first of his line for several generations to lose his throne but not his life. While his palace guard (British-trained and British-officered) looked the other way, Zaid and his followers swept into his brother's chambers and presented him with an ultimatum: abdication or his life. The moment he had been assured that he would not be left to starve in exile, Shakhbut accepted his dethronement with good grace and departed for Europe in an RAF plane the following day.

A few hours after the announcement of the coup, the first English newspaperman arrived in Abu Dhabi and was taken to the palace to see the new ruler. He found Sheik Zaid standing in the audience chamber, the little finger of his left hand intertwined with that of a burly young Arab named Ahmad Obaidli who had masterminded the coup.† The newspaperman, Ralph Izzard, a veteran of many rebellions from Bangkok to Berlin, was taken aback when Zaid's first words to him were:

"Tell me, Mr. Izzard, what would be the first thing *you* would do for Abu Dhabi if you were suddenly to become the ruler?"

"I would do two things," replied Izzard. "I would first announce that I am going to build schools and hospitals. Then I would build them. Too many rulers in this part of the world do the first but never get around to the second."

"We will try to prove ourself the exception," said Zaid.‡

* Quoted to the author by a member of Zaid's entourage.
† He is now one of Zaid's chief advisers.
‡ Recounted to the author by Ralph Izzard.

SHEIK ZAID SET ABOUT THE TASK of reforming his brother's disastrous Administration and shaping Abu Dhabi into a state capable of handling one of the biggest oil revenues in the Persian Gulf. He summoned agricultural experts, builders, teachers, and banking experts.* He also called in a Beirut-based Iraqi oil expert named Nadim Pachachi to revise Abu Dhabi's oil taxation laws to produce the additional revenue needed for the new building programs. He let OPEC know that once he had put the state's financial affairs in order, Abu Dhabi would join the organization and enable the oil-exporting countries to present a solidly united front to the companies operating in their midst.

Nineteen sixty-six was the year when the Middle East oil states had never seemed more bound together by their mutual interest in getting the most for their petroleum. As a bloc they controlled oil resources and output which made the rest of the world's supplies seem puny. A comparison with the oil industry in the United States gives a striking example of the Middle East's wealth. In 1966 the oil wells of the United States (one estimate was that there were nearly 400,000 of them) produced an average of 14.2 barrels each a day. In four of the major Middle East oil states (Iran, Iraq, Saudi Arabia plus the Neutral Zone and Kuwait) there was a fraction of that number of wells, but they each produced an average of more than 7,500 barrels a day. Iran's 184 wells alone averaged more than 11,586 barrels a day, and Kuwait's 503 wells 4,938 barrels a day.† This (1966) was the last year when U.S. output of oil exceeded that of the four countries named above (3.027,763 billion barrels against 2.908,548 billion barrels), and the figures for the Middle East did not include the mounting production in Abu Dhabi, Libya, Qatar and Dubai. Since then production has been increased to such an extent that

* Who tried—and failed—to turn the shredded notes in Shakhbut's back room into viable currency.

† The figures quoted are from Stocking, *Middle East Oil.*

the Middle East's figures far outweigh those of U.S. domestic oil fields.

"Once we are strong enough to shut down all the wells and shut off the pipelines," Abdullah Tariki was once reported to have said, "the companies will see a great light. The world cannot live without Middle East oil." * But in those days OPEC was still in swaddling clothes; now it had developed beyond the teething stage, and in several encounters with the oil companies it had shown the growing strength of its bite. Not many OPEC members contemplated an overall shutdown as a means of pressuring the companies, however, and they swallowed statements like Tariki's in silent embarrassment. But with Abu Dhabi as a member (Libya having already joined) by the end of 1966 OPEC felt that sufficient links had been formed between the oil states to make a chain that wouldn't break in any future crisis with the oil companies. They looked forward confidently to 1967 as a year of fruitful negotiations in which, beginning with Saudi Arabia, each oil state would improve its financial terms with the companies operating on its territory, all the others standing firm behind it during the negotiations. "1967 will be the year when OPEC comes of age," forecast a writer in a Baghdad newspaper.

But any birthday celebrations that were planned—as well as the series of oil negotiations—were interrupted by an event which rocked the Middle East to its foundations. In June 1967 the Arab-Israeli war broke out, and OPEC was one of those hit in the crossfire.

* Quoted in an article by Harold H. Martin in the *Saturday Evening Post* (February 17, 1962), though Tariki subsequently denied having said it.

Part Seven

END OF AN EPOCH?

PRECEDING PAGE: *King Faisal during a visit to a refinery.* PHOTO: ARAMCO

CHAPTER TWENTY-SIX

Boycott

ON JUNE 4, 1967, AS THE SIX-DAY WAR BEGAN, Gamal Abdel Nasser announced to the world that Britain and the United States had joined Israel in its attack upon Egypt and Jordan. It was not true, of course, but it was believed by the Arabs, most of whom were already convinced that the British and Americans were tied to Israel by an umbilical cord. In their rage their first instinct was to retaliate in the place where it would hurt the most: the Anglo-American oil enterprises. In Kuwait, in Saudi Arabia and along the Persian Gulf, bands of saboteurs assembled with explosives, ready to destroy the big companies' installations once and for all.

Fortunately for everybody, as it turned out, on the night of June 4 Iraq announced that it was taking control of IPC's operations and was shutting down all wells, in order to deny petroleum to the "imperialist aggressors." Foreign ministers from the other Arab countries, but not from Iran, hurried to Baghdad to confer, and then announced that they would do the same. All over the Middle East, Britons and Americans sighed with relief as troops moved in to occupy the oil fields and refineries. The soldiers' orders were to stop the flow of oil, but at least their presence would prevent sabotage.

By this speedy action the Arab oil states prevented permanent damage being done, but the boycott into which they had been coerced by Nasser's inaccurate propaganda was a body blow to their economy. As they all discovered within a few days, none of them had the financial reserves to carry on without the aid of revenue from the companies. Saudi Arabia was the first to feel the pinch acutely; on June 12, less than a week after the declaration of the boycott, King Faisal was informed by his finance minister that there was no more money in the till, and that for once Aramco was unable to help.

Then the oil wells must be started up again, said the king. After all, it was known by now that Nasser wasn't telling the truth.

No Arab would believe that, his minister answered. The Saudi government would be accused of sabotaging the Egyptian in his hour of defeat.

To which the king pointed out that Saudi Arabia could hardly do anything worse to Nasser than he had done to himself.*

At this point Ahmed Zaki Yamani, the oil minister, proposed that Aramco be told to restart operations, but to agree in writing to withhold all oil from "those states named by His Majesty's Government as having taken part in the aggression on Arab states." † Yamani pointed out that Aramco's oil went principally to the Far East; only a small proportion went to Britain and the United States, the "aggressor states," and while paying lip service to Arab antipathies, this device would do the least damage to the country's economy. The lifting of the general boycott on these terms was announced on June 13, but it was not until the end of the month that normal shipments were resumed by Aramco, and already Yamani calculated that Saudi Arabia had lost $30,-264,900 in revenue. Kuwait, which estimated its losses at just under $1 million a day, was the next to follow suit, and most of

* Information from a Saudi source who does not wish to be identified.

† Statement issued by the Saudi government, June 13, 1967.

the Arab oil states did likewise quickly afterward, with the exception of Iraq, which waited until the end of July.

By now the Arab oil states had begun to discover the weaknesses of OPEC, which only a few months before had been thought so strong and united. True, OPEC as a body had not been asked to institute an oil stoppage, but its two principal non-Arab members, Iran and Venezuela, might have been expected to aid their fellow members' boycott by limiting their own operations to the rate of shipments prevailing before the Six-Day War. Instead, they stepped up production to take advantage of all the shortages their Arab confreres and the closing of the Suez Canal had caused. For some time the shah of Iran had been agitating for a considerable increase in the output of oil from the consortium's fields; but the consortium, all of whose member companies had interests in other parts of the Middle East, had so far resisted,* on the grounds that taking more from Iran would mean taking less from Saudi Arabi, Abu Dhabi, Iraq and Kuwait, thus arousing Arab hostility. Now they did not need to exercise such restraint, since the Arabs had halted oil shipments from these countries. The result was a tremendous rise in Iranian sales to Britain and Germany,† the two main targets of the Arab boycott. Venezuela, whose usual principal market was the United States, also increased its shipments to Europe. At the same time King Idris of Libya, who was far from being pro-Egyptian and did not mourn Nasser's defeat at all, agreed to ban shipments to Britain but not to West Germany, and the increased supply thereupon sent to the Federal Republic was more than enough to divert the surplus from there to the United Kingdom.

On August 29, 1967, the Arab oil nations met in Khartoum to discuss the situation. By now most of them glumly realized that the worst sufferers from the boycott were the oil states them-

* Though they did sell some extra oil to NIOC (the National Iranian Oil Company).

† Which was considered to have adopted a pro-Israeli stand in 1968.

selves. Speakers valiantly spouted the theme of oil as "a weapon against the imperialists," but privately they admitted that the move had been an economic disaster for them. From the Arab point of view, the embargo had failed because:

1. The U.S., the principal offender in Arab eyes, was not hurt because of the insignificance of its Middle East imports (about 300,000 barrels a day). On the contrary, the emergency enabled U.S. companies with sizeable U.S., Venezuelan or North African production to make handsome profits;
2. The international oil industry did an outstanding job of making up the shortfall in supplies from other sources, despite the closure of the Suez Canal. (In this case, the focus of the world oil supply crisis shifted somewhat from the availability of crude oil at the source to a tight situation as regards [tanker] transportation.)
3. No quota ceilings were imposed on liftings of oil approved by the Arab governments concerned, which encouraged overlifting from Mediterranean and other ports. The excess oil no doubt found its way via West Germany to Britain.
4. There was no uniform interpretation of the coverage of the embargo. The North African countries did not in fact embargo West Germany.
5. Both Britain and West Germany had oil stocks of between 8 and 12 weeks which, with rationing, could be made to last them 5 to 6 months, even assuming that emergency supplies were unavailable. That is not to say that Britain and the rest of Europe got off lightly, but only that such discomfiture as they suffered was attributable more to the closure of the Suez Canal than to the partial embargo.*

The whole affair had been badly handled, everyone admitted. "Injudiciously used, the oil weapon loses much if not all of its importance and effectiveness," declared Ahmed Zaki Yamani. "If we do not use it properly, we are behaving like someone who fires a bullet into the air, missing the enemy and allowing it to rebound on himself." † As a sop to the pro-Nasser mobs, Saudi Arabia, Libya and Kuwait agreed to pay a yearly subsidy to

* Fuad Itayim, quoted in *Middle East Economic Survey* supplement, May 1970.
† Quoted in *Middle East Economic Survey,* July 21, 1967.

Egypt and Jordan of £135 million ($378 million), but all of them fervently hoped that no Middle East demagogue would ever maneuver them into such an unhappy and costly situation again.

Unfortunately, it soon seemed only too likely.

ONE NIGHT IN OCTOBER 1967, some four months after the end of the Six-Day War, a taxicab drew up outside a restaurant called Les Ambassadeurs, in Park Lane, London, and four young men emerged and went inside. Les Ambassadeurs is a private club much frequented by film stars, newspaper columnists and publicity men, and upstairs there is gambling at stakes high enough to seem impressive even by Monte Carlo or Las Vegas standards. At that time the owner was a burly ex-Polish army sergeant named John Mills who liked to keep the club "exclusive," and young men who were not recognizably famous and did not appear to be millionaires—and this quartet did not—were apt to be turned away.

Two of the young men, however, were Arabs; moreover, one of them, as his passport showed, was a sheik from a trucial emirate on the Persian Gulf. Many a London gambling promoter's fortune has been made by being polite to sheiks, and the four young men were promptly ushered inside. Later in the evening they found themselves upstairs, where they watched the gamblers at play. Suddenly the second Arab, a good-looking but awkward-moving Libyan, stopped at one of the tables and said, "But I know this man! Who is it he plays with?"

One of the English with him explained that the player he was asking about was a well-known Greek shipowner. The swarthy man who was his partner at chemin de fer had been recognized by the Libyan as a fellow countryman, one of the king's closest advisers. For the next hour he watched the two players, and it was an interval that was to change his life. In the time he stood there, the two men between them apparently lost approximately £150,000, and all the time the young Libyan stared down at them, his body rigid with tension, his eyes burning. When his companions tried to draw him away, he shook them off. "Let me watch them!" he said fiercely. "Carrion, carrion! First they rob us, then this is what they do with the money!"

It was the first and probably last time that the Libyan was in a London gambling club, and it confirmed every opinion he had formed, since arriving in England to take a military course, of the decadence of the country where he was learning to be a soldier. But it was the sight of the Greek and his fellow countryman gambling away thousands of pounds on the turn of a card which excited his hatred, for these were the two elements, he believed—Greek shipowners and Libyan ministers—who had profiteered most at the expense of the Arabs during the Six-Day War. One of these days, the young Libyan vowed, he was going to do something about such people.* His name was Muammar al Gaddafi, and two years later he would be master of his country.

It is true that the Greek shipowners had done well for themselves in the aftermath of the Six-Day War and the resulting closure of the Suez Canal, but no better than any other owner with a tanker to lease. The war came just too soon for the major oil companies to take delivery of the giant tankers they had ordered after the Canal had been closed down for the first time, during the Suez crisis of 1956, and they were still forced to rely on the charterers. One of the Greeks, Basil Mavroleon, admitted that he was leasing his tankers to the oil companies in the Gulf for the long journey around the Cape of Good Hope, which the Canal closure made necessary, at £5,000 a day on an 80,000-ton vessel. Almost simultaneously a Norwegian, Sigval Bergeson, was chartering one of his 80,000-ton vessels to Shell for £1 million for two journeys to and from the Gulf.†

Still, Gaddafi's spleen was misplaced. As Henri Deterding once

* This account of Gaddafi's reactions was told by Sheik Ahmad al Abah of Oman.
† Statements in the *Sunday Times*, London, July 2, 1967.

said, the tanker business is "the biggest floating crap game in the world," and a Greek was just as much a gambler when he leased his ship as when he played for high stakes in a London night club. He was just as likely to lose his shirt at either game, and he could hardly be blamed for chartering his tankers at $17.50 a ton during a crisis—the rate prevailing immediately after the Six-Day War—if the going price was likely to drop to $1.20 during a slump, as it did in 1972. (By that date, so many new tankers had been launched that many a Greek or Norwegian charterer was forced to put his ships in dry dock or try to adapt them for grain shipping.)

However, Gaddafi's rage at his fellow countrymen's excesses was only too well justified. By 1967 Libya had become an Arabian Klondike, with all the attendant evils that such a description implies. Tripoli was filled with wheelers and dealers from all parts of the Western world, accompanied by the cheap-jacks who tug at their purse strings. Ministers built themselves luxurious villas, opened numbered bank accounts in Switzerland, and went to London to whore and gamble. But there were no schools, few hospitals, and shoeless urchins in rags hung around the streets.

Whether King Idris was aware of the corruption is questionable; certainly he was too old, too weak and too complaisant to do anything about it. But to young Libyan army militants like Gaddafi, who to a man were pro-Nasserites, his most grievous sin was his reluctance to give support wholeheartedly to the Egyptian leader during the Six-Day War. Like his fellow monarch, Faisal of Saudi Arabia, King Idris feared the Egyptian leader for his socialistic and antiroyalist pretensions, and though Libya halted the shipment of oil during hostilities and for a time afterwards, this was because King Idris did not have much choice: the dockers had gone on strike and the oil field workers had stopped the flow of petroleum through the pipelines. To get them back to work and to demonstrate his "solidarity" with his neighbor, the king agreed to join Kuwait and Saudi Arabia in compensating

Egypt for its war losses, but his ministers pointed out that the millions of dollars this subvention would cost could only be provided if the oil fields were started up again.

Eventually the workers returned, and the oil companies were told to raise their production rate to take care of the increased demand which Libya anticipated in view of the cutting off of oil supplies from the Persian Gulf caused by the Suez Canal closure. At the same time, the companies were ordered to raise the posted price of Libyan oil by 80 U.S. cents a barrel. "There are three factors in the present situation which call for a review and adjustment of posted prices," the Libyan government announced. "1) The world-wide increase in prices of petroleum and petroleum products, particularly in European markets, caused by the Zionist aggression. 2) The more favorable geographical position of Libya for the supply of these markets in comparison with the Persian Gulf, owing to the closure of the Suez Canal. 3) The rise in freight rates which increases the favorable situation and demand for Libyan oil compared with oil from the Persian Gulf."*

Thereafter the boom was on. In the year that followed, Libya's revenue from oil production—and there was precious little income from any other source—almost doubled as the demands of a petroleum-starved Europe began pouring in. Not much of the increased revenue found its way into funds for the health and welfare of the Libyan people, and large amounts of it were siphoned into the pockets of the king's ministers and their hangers-on. One Libyan family alone, now safely in exile in Switzerland, was said to have made $4 million in the 1967–1969 oil boom. Only too well aware of the venality of their masters, Libyan civil servants, police, customs men and petty officials got into the act by demanding higher and higher bribes for their services. "Gimme Gimme Land" was the new name for Libya among the oilmen.

To the young army captain, Muammar al Gaddafi, the spectacle was sickening. Until the age of sixteen he had wandered the Tripolitanian desert with his family of Senussi tribesmen, and

* Statement at a press conference in London, August 9, 1967.

a goatskin tent had been his home. In his eyes, cities were decadent places in any case, but what he particularly despised was the growing evidence on every side of what he considered to be the degrading smear of effete Western civilization. Gaddafi was not only pro-Nasser and socialistic-minded, he was also a puritanical Muslim to a degree approaching fanaticism. Among his roster of heroes only two non-Arabs figured, Abraham Lincoln and General Bernard Montgomery—the first for ending slavery, the second for his ascetic private life and soldierly qualities. Otherwise he despised the modern West, and claimed that Arabs who aped European ways, associated with Western women and gambled or drank "made me vomit."

In 1969, when the Libyan oil bonanza was at its peak, an event occurred which must indeed have turned the young Libyan's stomach. In the spring of that year, the Occidental Petroleum Corporation completed construction of a pipeline from its oil field in Tripolitania to the port of Sirte on the Mediterranean. By this time Occidental was in the money—the Six-Day War had been a godsend to the company—and relations between its president, Dr. Armand Hammer, and the king and his ministers were all cordiality and mutual admiration. King Idris consented to officiate at the opening ceremony of the new pipeline, and when a guard of honor was picked from the Libyan army for the event, Captain Gaddafi was one of the officers chosen. It was quite an occasion. Dr. Hammer brought friends and members of his board from the United States. All the government ministers came, including Oil Minister Kabazi (though his much-admired friend, Galic, was not present), and the throng was ornamented by the presence of ladies from Tripoli's Italian colony. The young Libyan captain had a close view of his countrymen arriving in their Rolls Royces and Cadillacs, drinking champagne, smoking fat cigars, and fawning not only over their American guests but also over the Italians who had once turned Libyans into semislaves. The sight appears to have reinforced Gaddafi's feelings, and to have convinced him that his country was rapidly becoming a sink of Babylonian iniquity. Back in his barracks at Bab Azizia outside Tripoli, he talked long into the night with a clique of young associates. A few weeks later he had convinced them that a revolution was feasible, and had even composed the slogan to rally the people: "Poor and barefooted, but with Nasser!"

On September 1, 1969, while King Idris was on an official visit to Turkey, Gaddafi and his group quietly moved companies of faithful troops to strategic positions throughout the kingdom and took over the regime in a bloodless coup d'état. Within hours, several ministers—including Kabazi and the gambler whose excesses had so shocked Gaddafi in London—were under arrest, and the desert kingdom which many Arabs had called "the Camel with the Feet of Gold" was no more.

A few nights later a young man wearing a dark cape over his shoulders slipped into a popular Tripoli cabaret and stood at the bar sipping an orange juice and staring at the crowded tables of guests loudly cheering a belly dancer. Suddenly he whipped off his cape, revealing his army uniform, and blew a whistle. Troops burst into the room, and one hundred and fifty guests and show girls were shepherded into trucks and carted off to jail. By the next day all cabarets in Libya were closed down, never to open again. Newspapers covered their front pages with denunciations of their immorality, and of the decadence infecting Libyans as a result of contact with the West.

Not long afterward the same young man, dressed this time in a simple Senussi *gallabiah,* stood in the outpatients' clinic of a Tripoli hospital while a group of nurses and interns chatted together. Finally he timidly approached them. "Come back tomorrow," said one of the nurses abruptly.

"When I come back tomorrow," the young man said, "you will no longer be here. You will be in jail."

Soon all Libya was talking about the nocturnal peregrinations of the young wolf of the Libyan revolution. Gaddafi had become a modern Haroun al Raschid, liable to appear anywhere among his people, watching, listening and then acting. Slowly but inexorably, the purge of Western influences began. Each time British and American oilmen came on leave to Tripoli, they found more evidence that life was changing. Bars and liquor stores had closed down; street names were in Arabic only; urchins no longer importuned for money; and hotel servants were no longer subservient—or even willing to provide service. The xenophobia in the air was palpable, and the new rectitude of the government was tangible. An Esso International official returned to his office to declare in astonishment, "I actually got a paper signed without having to pay baksheesh for it."

But soon Gaddafi would find other ways of making the oil companies pay. To begin with, however, he busied himself with preparations for his first big challenge to the Western influences which had dominated Libya for so long. A few months after taking over, once the people had been prepared by propaganda groundwork, he ordered the U.S. Air Force to quit its base at Wheelus Airfield and leave the country. The Arab world waited, and the militants held their breath. How would the United States react to this impudent challenge to their strategic policy in the Middle East from a group of young upstarts? From Saudi Arabia, Faisal is said to have told the exiled King Idris, "Don't worry. You will soon be back; the Americans will see to it. Those misguided young men have gone too far." *

But neither Faisal nor many another Arab incumbent realized how things had changed for the big powers in their confrontations with the new regimes rising out of the ashes of the Six-Day War. Henceforth, arrogance and defiance on the part of the Arabs was the order of the day; the era of subservience was over. The U.S. Air Force packed up its supplies and obediently flew away, to be followed not long afterward by the British from their bases in Tobruk and Benghazi. True, the British did not go quite so willingly, but a few riots in the streets, some Britons beaten up and nurses raped, speedily brought the situation to the point where the British had either to evacuate or to bring in the modern equivalent

* Report in *Al Siyasha*, Kuwait, April 15, 1970.

of a gunboat. But as Suez had proved in 1956, the days of gunboat diplomacy were over. If the United States couldn't defy the revolutionaries of Libya, how could a weakened Britain dare to?

The foreign troops departed and down came the Pepsi posters and all the other signs of Cola colonization.* How long, the oilmen began to wonder, would it be before the young wolf got around to them?

ACTUALLY, the home offices of the big oil companies in Britain and the United States were not worrying much about the sword of Damocles now hanging over their operations in Libya. If it had been 1967 or 1968, the threat of Colonel Gaddafi's interference in the oil fields would have given them ulcers, because it was Libya's oil which had kept Europe's wheels turning after the Six-Day War. But by 1970 the Anglo-Americans had the situation in hand. The giant tankers ordered in the aftermath of the 1956 Suez crisis were now being launched, and freight charges from the Persian Gulf were starting to drop. Fuel stocks were building, and soon the shortage would be over. In these circumstances, what had they to fear? The Libyan leader was in no position to threaten them. It was only a question of presenting a united front against his demands and refusing to knuckle under. In a crunch, it was Libya that would suffer most. The oil companies had alternate sources of supply—in the Gulf, in Iran, in Venezuela and in their reserves—to keep their customers supplied. Even if the Libyan fields shut down completely, none of them need lose a cent in sales or profits. On the other hand, Gaddafi would lose his oil revenues just at the moment when he needed them for new programs and to replace the millions which venal ministers had smuggled out of the country.

Veterans of tough negotiations in Iran, Iraq and Venezuela,

* Coca-Cola itself had already been banned by all Arab countries in retaliation for its commercial activities in Israel. For the same reason, the Ford Motor Company and Alka-Seltzer were also under Arab proscription.

with countries ten times the size of Libya and dictators twice his age, the oilmen were convinced they could handle the young colonel and his Revolutionary Command Council. In the eyes of many an oilman, they were just a bunch of naïve fanatics with sand running out of their ears, and the word went out to stand up to the new regime and not take any nonsense.

As they were to discover to their cost, Muammar al Gaddafi is not so naïve as he looks and sounds, and in his first encounter with the combines he outwitted them. Not for the first time, he consulted his second in command in the Revolutionary Command Council on the strategy to be adopted, and Major Abdessalam Jalloud did not let him down. Among this strange pack of Islam fanatics dedicated to the precepts of the Koran, the glory of the Arabs and the downfall of the Zionists, Jalloud stuck out like a pine tree in a desert. He was a townsman rather than a child of the desert, and he had been known to sigh over what his fellow rebels had done to Tripoli in pursuit of faith and asceticism. A stay in the West, far from producing a revulsion against modern civilization, as it had in the case of Gaddafi, had given him a taste for European culture, an affinity for jazz, and an avid taste for good food, wine and intelligent female companionship. Some of the beautiful young members of the Italian colony in Tripoli and Benghazi had come to know him well, and he had continued his friendship for them some weeks after the revolution, and he had given them up only when it became evident that Gaddafi would make them the next target for expulsion.

When the militants first took over the reins of government, they all lived together in the barracks at Bab Azizia, an arrangement that Jalloud found constricting and monastic,* and shortly afterward he moved to a luxurious villa some distance out of Tripoli which had been taken over from a departed minister. The young major did not enjoy its sybaritic pleasures for long, because about a week later he drove out one morning to discover

* Though, in fact, none of the militants has ever deprived himself of wives or other female companions.

that someone had been at work during the night; on the garden wall of the villa had been painted: "WHO BRIBED YOU WITH THIS VILLA?"

Jalloud did not need to ask who had ordered the insulting graffito, and the next day he said good-bye to his wine cellar, beach, swimming pool, air-conditioned bedroom and his girl friends, and moved back to the communal life of the military barracks. It is a measure of his popularity with Gaddafi, and of his usefulness to the new regime, that nothing was ever said to him about his attempt to break out of the monastery. Plenty of Libyans had ended up in jail for much less.

Jalloud's conviviality was not expended simply in the pursuit of personal pleasure; his general air of bonhomie made him popular with oilmen, and he picked up information about the way the oil companies were thinking which the junta might otherwise have missed. He knew that these men had expected Gaddafi to show his contempt for them by nationalizing the old operations in Libya, and certainly it had been a move that the young military leader was known to favor. Jalloud persuaded him not to make that mistake, for of course Libya would never at that time be able to run the oil fields on its own. He urged Gaddafi to make the companies pay an immediate increase of 50 cents in the posted price of Libyan oil, and 60 percent instead of 50 percent in Libya's share of the sales.

Gaddafi wanted to make it a straight $1 raise, and to turn the confrontation into a trial of strength. "We must show we are the masters here," he said.

Jalloud told his chief that he had already demonstrated this well enough by kicking out the American and British armies, and warned him against empty political victories at a time when Libya needed something more important—money. What Jalloud saw as his main problem was how to split the ranks of the oil companies. If they remained united, he knew he could not beat them. But if one of them broke, the others would have to break too.

Boycott

Word spread through the oil fields that the crisis was coming, and the major companies got ready to face up to it, united in their resolve not to give way or pay out any more money. They waited for the summons, but it did not come. It turned out that Jalloud had called in the representative of only one company: Occidental Petroleum. Of all the independents operating in Libya, Occidental was the most vulnerable to an ultimatum, for it had no alternative sources of supply. Its fortunes had been based on its success in Libya, $450 million of its profits had been invested in new developments inside the country, and it could not afford to have its operations closed down. It could not even face a cutback in production rates, and any hesitation it may have had about accepting the Libyan demands was hastily forgotten when troops moved into the fields and ordered an immediate cut of 30 percent in the company's output.

On September 14, 1970, Occidental announced that it was accepting a compromise agreement with the Libyan government and was immediately increasing its posted price of oil by 30 U.S. cents per barrel, and increasing its tax level from 50 percent to 58 percent. All the other independents hurriedly announced that they were following suit. By the end of the month the major oil companies had come into line. The united front disintegrated, and one by one the concessionaires were summoned to sign new agreements. The major companies had been outsmarted; it was a clear-cut victory for the Revolutionary Command Council.

As one cynical oilman put it: "At this rate, Gaddafi will soon be getting ideas above his nation."

CHAPTER TWENTY-SEVEN

Tidying Up

THE BRITISH ARE A CONSCIENTIOUS PEOPLE, and when they relinquish sovereignty in someone else's country they always try to leave it clean and tidy. They are not proud of the mess they left behind them when they quit India in 1947 and Palestine a year later, but on both occasions the departure came at the end of a long and exhausting war, and the inhabitants of the lands they were leaving made it abundantly clear that they preferred any dirty linen which the English might leave behind to a prolongation of their presence. Since those two regretted lapses, Whitehall had contrived to do better, and the lowering of the Union Jack over Malaya, Singapore, the West Indies and the several states of East and West Africa had been carried out amid handshakes and mutual expressions of good will. More important, reasonably stable native administrations had been trained to carry on.

Now it was the turn of the Trucial States of the Persian Gulf to gain their independence, and since Britain would be retaining some of her most valuable overseas investments—mainly her petroleum concessions—in these countries, it was particularly important to hand over the reins to regimes strong enough to control an unruly and backward populace, but supple and enlightened enough to cope with demands for improvement.

Tidying Up

There were some heavy sighs and doubtful shakings of the head from old colonials when it was announced that Britain would quit the Gulf on December 1, 1971. For more than a hundred years the Gulf had been a British lake controlling the passage to the India and Far East, and for most of that period Britain would have fought a war rather than allow any other nation to gain military influence there.

In 1900, when he was viceroy of India, Lord Curzon and his American wife, Mary, sailed into the Gulf on Her Majesty's ship *Hardinge* and descended on the sultan of Muscat and Oman, who, the viceroy had heard, was thinking of accepting a large bribe from the Russians to allow them to establish a coaling station in the area. Curzon had already expressed his view that any Briton who allowed a foreign power to infiltrate this vital area of British influence should be impeached and hanged. Now, with the guns of Her Majesty's ships trained on them, Curzon and his lady stepped ashore to sip sherbet and exchange welcoming speeches with the ruler.

In his remarks the sultan referred to Mary Curzon as a charming beauty, a veritable "pearl." Curzon replied by saying that the Persian Gulf was also a pearl, and with a certain emphasis in his voice, that it was a pearl which Britain considered beyond price. The point was noted and more sherbet was consumed.

"The morning after we left Muscat," wrote Mary Curzon in her diary, "we cruised all day among the islands and fiords of the most barren description. The land is uninhabited save for a few fishermen, and its main interest is a strategic one, as the bays and inlets afford anchorage for a fleet, and as the land is No Man's Land, Russia or France could take advantage of the harbourage in the event of war." *

It must never be allowed to happen, Curzon vowed, and during his lifetime it did not. But times had changed; Britain no longer had either the military strength or the political influence to maintain herself in the Gulf against local hostility—and ever

* Quoted from the author's biography of Lord Curzon, *The Glorious Fault.*

since the Suez crisis there had been plenty of that. Their experience in Aden, where a raffish set of left-wing brigands had seized power when Britain left, had taught the British that staying too long could produce a more radical local government than if they departed sooner and more willingly. Aden was a mistake to be avoided in the Gulf, so Whitehall dispatched its best Arabist troubleshooter, Sir William Luce, to make sure that when the colonial administration quit, no wild revolutionaries would immediately take over.

British foreign service officers come in all shapes and sizes, but in some respects they are always the same. Luce was a chip off the same block that had produced Sir Percy Cox, whose stern paternal attitudes toward Arab leaders had shaped and delimited the boundaries of Arabia after World War I. The son of an admiral, down from Cambridge with a good degree and a "blue" in rugby, a devotee of riding, shooting and sailing (he still mentions the three as his recreations in *Who's Who*), Luce had climbed the ladder of the colonial service by way of the Sudan. He spoke fluent Arabic and had been dealing with Muslim tribes of one sort or another for most of his adult life. From 1960 to 1966 he had been British political resident in the Gulf, with headquarters in Bahrain, and during this time had followed the time-honored British policy, where Arabs were concerned, of never interfering with the local rulers except to encourage rivalries between them—this tactic being sufficient to keep them divided and therefore unlikely to combine together to kick the British out.

Now Luce was suddenly called upon to reverse his role. With Britain departing, and taking with her the gunboats and troops with which she had protected both the sheiks and the petroleum routes, the small states along the Arabian shore must be left in a condition to defend themselves. Not against attacks from one another, however, but from the incursions of more powerful neighbors, or from the revolutionary elements which were starting to fan across the deserts on the heels of the British withdrawal. Therefore, instead of encouraging the Gulf states to glower at one another across their borders, Luce must now try to persuade them to forget their rivalries and work together. *Federation* was the magic word he brought to the sheiks, a word that would work as many miracles for them as *oil*! If they combined together in a federation of Arabian states with a joint army, air force and navy, they would be able to withstand assaults on their independence from all but the major powers. Most important of all, they would be able to protect their oil fields from infiltration and sabotage.

To emphasize the support that Britain was prepared to give to such a federation, Sir William told the various rulers that his government was prepared to sign treaties of friendship with them all; and to help them form combined defense forces, Britain would assign them the locally raised levies which had been used to maintain colonial order. Their British officers would be left behind to lead them—though of course they would first resign their British commissions and sign on as "advisers."

While the emirs and sheiks were brooding over Sir William's offer, the British envoy was planning another move which was the antithesis of all he had been taught as a good colonial servant; he was preparing to interfere in the domestic affairs of one of the Gulf states.

If a federation did come about, its members must have administrations which could rely on the support of the majority of their subjects. By 1970 it could be said that most of the British-controlled Gulf states—especially three of the four key states, Bahrain, Abu Dhabi and Dubai *—had rulers whose regimes were reasonably stable and against whom a revolt was unlikely. But to the south of Abu Dhabi and Dubai, her frontiers inextricably mixed up with those of the participants in the proposed federation, was the Sultanate of Muscat and Oman. Oman was not being asked to become a member because it was so large and unwieldy that it would have unbalanced the union. But it was so closely involved geographically, socially and economically with its neigh-

* Kuwait had been granted its independence in 1961.

bors that what happened to it would inevitably affect the others. And Oman was the rotten apple in the barrel, a hotbed of misery and discontent.

The British believed that the main reason for Oman's unsettled state was the despotism of its ruler, Sultan Said bin Taimur, whose tyrannical methods were fanning the flames of revolt all over the southern deserts. Before federation came and the new independent states were left on their own, something had to be done about this man. Otherwise the diseases his misrule provoked might spread the infection far beyond his own borders.

IN 1970 OMAN was hardly an oil state in the same class as Abu Dhabi or Kuwait, but its fortunes had improved considerably in the years since Wendell Phillips was given the concession for the Omani dependency of Dhofar. Admittedly, that concession had proved a severe disappointment to the independent U.S. companies which had taken it over, both because the wells gave off high-sulfur crude oil not suitable for shipping abroad, and because rebellious tribesmen had made subsequent exploration extremely hazardous. But farther north in Oman proper, a company called Petroleum Development (Oman) Limited * had begun exploiting reasonably productive fields in the desert at Al Huwaisa, Yibal and Natin Fahud, and sending oil by pipeline from a collection point at Izki to a tanker-loading port on the Gulf of Oman near the capital city of Muscat. The revenue from this operation was sufficient for PDO to pay Sultan bin Taimur $75 million a year in taxes, which should have been enough to bring about some amelioration in the lot of the Omani people. But the sultan was a stubborn man who clung to his belief in a feudal state dedicated to faith in Allah and the teachings of the Koran. His people not only lacked hospitals and schools; they were also oppressed by the rigid rules of a theocratic kingdom

* In which Shell held 85 percent of the shares, Compagnie Française des Pétroles 10 percent, and the Gulbenkian interests 5 percent.

where minor vices or indulgences were regarded as crimes punishable by flogging, amputation or death.

Young Omanis who were rich enough to be sent abroad for their education, returned with stories of the wonders of Cairo, Beirut and Europe, and their discontent with conditions in their homeland had begun to permeate the youth of the country. They listened avidly to the propaganda broadcasts from South Yemen supporting the rebels operating in the Jabal Akhdar region of Dhofar. They were sympathetic; anything was better than the slough of despair which Oman had become under the sultan.

On June 11, 1970, members of the SAF (the Sultan's Armed Forces, the British-officered defense force which kept law and order in the kingdom) were attacked in one of their camps at Izki near the pipeline collecting point of PDO. Mines were placed along the barbed-wire perimeter of the camp, and when they exploded, a group of guerrillas burst through, throwing hand grenades and aiming bursts of machine-gun fire at the SAF tents. It was dark when the attack was launched, and there was great confusion. Fortunately, however, the infiltrators seemed to be just as confused as the defenders, and after a sharp exchange of fire they were all rounded up and either killed or captured.

Under subsequent questioning, the guerrillas revealed themselves to be a Communist organization known as PFLOAG (the Popular Front for the Liberation of Oman and the Arabian Gulf). It was a shock to learn that they were operating this far north, for until then their activities had been confined to Dhofar. There were more unpleasant surprises to come. Further questioning at the hands of the sultan's own executioners revealed the whereabouts of several caches of arms smuggled in from Russia and a lot of revolutionary literature, most of it printed in China. Some of the caches were uncovered in Muscat itself. Finally, the names of the co-conspirators and secret sympathizers of the rebels, including officials close to the sultan's court, were extracted during the interrogations.

Taimur ordered an immediate purge, and the fetid old Portuguese fort in Muscat, which he used as a jail, was soon full of suspects, most of whom he intended to keep incarcerated until they rotted. Indeed, he was so alarmed by the extent of the conspiracy that he ordered his own son and heir, Prince Qaboos, to be confined to his house near Muscat. The sultan was taking no chances; having usurped his own father, he was determined that his son would never get the chance to do the same.

But Prince Qaboos, an amiable young man whose outlook on life was completely opposite to his father's, had friends. Though trained at Sandhurst in England, he had not confined his studies to military tactics, and he had ideas for Oman whose liberality and worldliness would have outraged the sultan. He had often discussed these ideas with Wendell Phillips whose six-shooter he proudly wore, and with the two senior British officers of the SAF, General Philip Graham and Colonel Hugh Oldham. These two soldiers had become admirers of Qaboos and had little doubt that he would make an infinitely better ruler of Oman than his reactionary father. The sultan had ordered the SAF to keep Prince Qaboos under tight guard and prevent him from seeing anyone who might conspire with him. But since Graham and Oldham were in command of the force, there was no such restriction on their own contacts with the prince. After the Izki guerrilla attack, their meetings with Qaboos were almost as frequent as their conferences with the British envoy, Sir William Luce.

By now rumors were ricocheting around the Gulf that Sultan bin Taimur had decided to abdicate in favor of his son. Who had started these reports no one could say, though they were persistent enough to be repeated on July 19, 1970, in a dispatch from the Cairo correspondent of the Paris newspaper *Le Monde.* The British blandly denied knowing anything about it, and when approached, Sir William pointed out that Britain never interfered in the domestic affairs of the Gulf states. The sultan himself vehemently repudiated any intention of abdicating, and after ordering General Graham to reinforce the guard on his son's residence, he departed in a Royal Air Force plane for his palace at Salala, eight hundred miles south of Muscat, where he imagined he would be well out of the way of any conspirators. The governor of the province was an old fighting companion, and Said bin Taimur trusted him implicitly.

But on the afternoon of July 23, the governor was visiting the island of Masira, halfway up the coast on the Arabian Sea, where the RAF had a base. Normally he would have turned down the invitation to be a guest of the British airmen, for General Graham was coming to Salala that day for a conference with Colonel Oldham, and on such occasions the governor liked to be present to receive the greetings of the commander in chief of the Sultan's Armed Forces. On this occasion, however, someone had neglected to inform him that the general was coming. In fact, someone also neglected to inform the sultan himself of General Graham's presence in the town. On the other hand, the commander of Taimur's bodyguard, Colonel Alastair Turnhill, certainly knew about the general's visit, for he was on the airstrip with Colonel Oldham to greet Graham when he touched down. The three officers immediately drove down the coast road out of Salala, on what they later described as a sightseeing tour of the Portuguese forts which dot the coast in these parts. By the time they returned, things had changed.

The climate in and around Salala is one of the most pleasant on the Arabian peninsula, and the sultan's palace catches any cool breeze which wafts across the Indian Ocean. Standing on the seashore overlooking a series of white beaches, it dominates the town, as James Morris describes it, "Like the Castle at Windsor or the Mormon Tabernacle in Salt Lake City." Morris continues: "All these seaside capitals of Eastern Arabia had their big fortress palaces in which the ruler reclined among his dependents and sycophants, his outlook, like his creature comforts, generally depending upon the size of his oil royalties. Kuwait's was the grandest, Bahrain's the most stylish, Doha's the most horrible, and Salala's probably the nicest. It [is] a long crenellated building surrounded by high walls and complicated by connecting courtyards and alleyways. A wide avenue of palm-trees led to the double gate-tower at its entrance . . . Every Tuesday morning a slim young Englishman in uniform marched out of the palace after his weekly conference with the ruler—he was the commander of the Dhofar Force, the sultan's local army. Muscular, well-armed slaves guarded the gateway to the palace." *

But on this day the commander was sightseeing with his general, and the guards drowsed in the afternoon sun. At least they must have been drowsing, for how else could they have missed and failed to challenge a small body of Omanis in the uniform of the sultan's defense force which passed under the gate tower and jogged on the double into the palace?

Said bin Taimur was in his study, a long room overlooking the sea, a copy of the Koran in his lap as he squatted on the carpet, his back resting against an ornately upholstered camel saddle. Two slaves stood in shadowy corners of the room, watching over him, but neither heard the faint shuffle of soft shoes outside as the guard on duty was overpowered and carried away.

Suddenly the large door of the chamber opened and seven men slid inside. One of them the sultan recognized at once as Braikh, the young son of the governor of Salala, and he angrily asked him why he had burst in so abruptly. Braikh replied that he had not come to harm the sultan, but was acting on instructions to take him away. If he would consent to come willingly and at once, there would be no trouble. The young man spoke with great respect, and repeatedly called the sultan by his royal title, Jalalath (Majesty).†

But Said bin Taimur was neither impressed by the intruder's deference nor frightened by the posse of men standing behind him with daggers in their belts and fingering the triggers of their

* Morris, *Sultan in Oman.*

† The various parties involved in these events, British and Omani alike, have been extremely discreet about them. This account comes from various sources, including the sultan's own as recounted to a number of his friends, including Dr. Wendell Phillips.

rifles. As he was to say later, he would have been ashamed to give up without a fight, and he had no intention of doing so. Reaching under some cushions near the camel saddle, he pulled out a revolver—a six-shooter of exactly the same type as Wendell Phillips had given to his son—and opened fire at Braikh. The young man ducked just in time, and the bullet hit one of the men behind him. At the same time the two huge African slaves emerged from the shadows, unsheathing the great swords which they wore around their waists. Flailing them above their heads, they waded into battle. One crashed to the floor, shot in the stomach, but the other's murderous blade, whirling like a windmill sail, effectively held the intruders at bay while his master dashed to the end of the room and disappeared behind some curtains.

There ensued what might have been a scene from an old-fashioned chase film. Besides the many corridors that lead from building to building in the palace at Salala, its walls have hidden doors, sliding panels and dark peepholes in which those inside them can see without being seen. The sultan used all these to the full, and might not have been discovered for a considerable time had not young Sheik Braikh, by this time frantic at the way things were going, started shooting wildly into the paneled walls. Fortunately for him, he scored a hit; from behind one wall came a sudden cry and then a moan for help, and Sultan Said bin Taimur was brought out, bleeding badly from a wound in the shoulder. Patched up by an RAF doctor—who appeared so quickly that he might almost have been awaiting the summons—Taimur had recovered sufficiently by evening to consent to abdicate in his son's favor. But to ensure his own safety, he insisted that he would sign the papers only in the presence of Colonel Turnhill, the commander of his bodyguard, and with a guarantee of safe-conduct from General Graham. The two Britons in question, by this time back from their sightseeing trip, were immediately available to give him the assurances he demanded. The sultan signed at once, and left for England in an RAF plane the following day.*

* He died in exile in London in 1972.

Great was the rejoicing in Oman at the news. There was dancing in the streets of Muscat, lights stayed on after dark, the curfew was abandoned, and music from Radio Cairo and Kuwait blared out. Even women walked in the streets, and a few of the younger ones dared lift their veils. Everything that had been forbidden for so long (except perhaps the imbibing of alcoholic liquor) was done in the next few days, and to those relatives and progressive Omanis who had been driven into exile by Taimur's despotism, Qaboos sent reassurances that all was forgiven.

One of the first to send telegrams of felicitation to the new sultan was the chairman of the Shell Trading Company, whose subsidiary, PDO, added a message expressing their joy at the "historic event." In London the Foreign Office was more restrained, declaring that it "would not dream of embarrassing Sultan Qaboos" by commenting on a purely internal event. Colonel Oldham assured newspaper correspondents now arriving in Muscat that he was "prepared to swear on my honor that I knew nothing about what was taking place," and General Graham in turn declared that he had been in Salala on the day of the coup "purely by accident" and had been "an involuntary witness" of what had taken place. He added, "But as a good Omani officer, disciplined and respectful of authority, I could not do other than submit to the orders of Sultan Qaboos, to whom I have already given my oath of allegiance." *

The only person who made no comment at all was Sir William Luce, the British envoy in the Gulf. But he did not need to; the rotten apple in the barrel had been disposed of, and he could now get on with the job of strengthening the federation.

THE FEDERATION which Sir William and the British had originally envisaged was to consist of nine states in all: Bahrain, Qatar, Abu Dhabi, Dubai, Sharjah, Al Fujairah, Ajman, Umm al Qaiwain and Ras al Khaimah. But first Bahrain and then Qatar

* These quotations come from *Le Monde*, Paris, May 27, 1971.

dropped out, both of them arrogantly confident that they were rich and secure enough not to need the federation's protective umbrella. That left seven. Then the ruler of the tiny state of Ras al Khaimah, Sheik Sakr bin Mohammed al Qasimi, let the wind out of Luce's sails by announcing that he too was staying out.

Sheik Sakr is one of the wiliest characters in Arabia, and his combination of guile and truculence can be attributed to inheritance. His ancestors were responsible for the British installing themselves in the Gulf in the first place, for until 1820, Ras al Khaimah was the center of the bloodthirsty buccaneering industry which had gained for this part of Arabia the name of the Pirate Coast. Ras al Khaimah sticks out like a sore thumb into the waters of the Gulf, and with the two islands which it owns called Greater and Lesser Tunb, controls the thirty-mile Strait of Hormuz between Arabia and Iran through which all ships must pass to reach the Indian Ocean. From its ports and islands, the pirate fleets of Sheik Sakr's ancestors sallied forth to pillage the merchantmen passing through on their way from Mesopotamia, Persia, and the Gulf ports of Arabia. When the Royal Navy, exasperated by the pillage and the harassment of its lines of communication with their Indian empire, at last descended upon Ras al Khaimah, they found sixty-three warships and eight hundred smaller vessels at anchor, all of them engaged in buccaneering. The dockside sheds were stuffed with booty, and life in Ras al Khaimah town was gaudy in the extreme.

Nostalgic about the adventurous days of his forebears, Sakr dreamed of restoring his country's colorful glories, if not the mayhem, once his country had succeeded in finding oil, like its coastal neighbors. He had handed over the concession to the Union Oil Company of California, and never has a bunch of American oilmen worked harder to fulfill a sheik's hopes as well as their own. Surveys of the offshore concessions had revealed "promising" evidence, and at enormous expense rigs had been brought from California and drilling started. Each day Sheik Sakr sailed out in one of the boats of the Ras al Khaimah navy to see how work was progressing. The navy was nothing like the mighty pirate fleet of his ancestors; it consisted solely of three rubber dinghies with outboard motors and one small pinnace, but it displayed the Ras al Khaimah flag around the islands of Tunb and flapped it defiantly under the cliffs of mighty Iran on the far shore. Sheik Sakr knew that the shah of Iran, fearful of the blackmail to which the Arabs might one day subject his tankers on their way from Abadan and Kharg Island to the outside world, was determined to gain control of the islands of Tunb, for whoever controlled the Tunbs controlled the strait.

In the course of his negotiations, Sir William Luce had been to see the shah, and afterward he had suggested that Sheik Sakr accept the same compromise which he and the shah had worked out with the ruler of Sharjah, which owned Abu Musa, another small strategic island in this part of the Gulf. Around Abu Musa's shores there were promising signs of oil, and Sir William had negotiated an agreement in which the ruler of Sharjah accepted an Iranian military presence on the island, agreed to split both sovereignty over the island with Iran and all oil revenues discovered around it, and would, until such time as oil was found, accept from Iran a yearly subsidy of £1.5 million.

This comfortable compromise could be arranged between Iran and Sheik Sakr over the islands of Tunb, Sir William suggested, if only the ruler would give his consent. But he would not. The drills were descending into the sea bed off Ras al Khaimah, and at any moment Sakr was sure that the Union Oil Company would hit a rich field of petroleum or gas. Then Ras al Khaimah would be wealthy again, and by staying out of the federation and holding on to the islands of Tunb, would become a great power in the Gulf, once more to be held in respect by all who sailed through her waters.

Soon very few of the 54,000 people in Ras al Khaimah were unaware of the fact that their ruler had opted for independence, for sovereignty over their islands, and for a gloriously prosperous future lubricated by oil. In the meantime they lived on the profits raked in by the Lebanese croupiers at Ras al Khaimah's casino (the only one in the Gulf), from the sale of vegetables to neighboring emirates, and on hope.

In the summer of 1971 their dreams were dashed. John Turk, Union's general manager, drove to the ruler's palace to tell him that "heavy shoals" of gas had been struck in one location at 17,900 feet, and that "high-quality oil" had been found in another location at 16,000 feet.

"But this is good news!" said Sheik Sakr.

Unfortunately, Turk went on, neither the flow of gas nor of oil had lived up to expectations. Only 1,200 barrels a day were coming out of the well. Union Oil had already spent $15 million on operations, apart from the $200,000 annual subsidy they gave the ruler, and the future looked "real bumpy," as Turk put it. Of course work would continue, and oil would be found and pumped, but the bonanza which everyone had hoped for seemed a long way off.

When news reached Sir William Luce of the waning hopes of Sheik Sakr, the indefatigable envoy * set off at once for Ras al Kaimah to tell the ruler that it was not too late for him to change his mind. A place in the Federation of Arabian Emirates, as the new body would be called, was still open. So, it was quietly suggested, was a joint arrangement with Iran for control of the islands of Tunb.

Unfortunately, Arab "face" was now involved. Sheik Sakr was a proud man, and he had told his people that their country would be independent. How could he now go back to them and say that he had been forced to knuckle under to rich Abu Dhabi and Dubai in the federation, and to allow Iranian soldiers to establish themselves on the barren shores of the Tunbs? It could not be done; the people would jeer at him, calling him a weakling and a fool. He would have to stand by his decision; there was no other way out.

* He had now been traveling back and forth around the Gulf, and thence to Teheran and London, for over two years.

But of course there was. When you have able colonial service advisers of Sir William Luce's caliber at your elbow, there is always an acceptable alternative.

BRITISH CONTROL OF THE TRUCIAL STATES was due to end on December 1, 1971. Until then, Britain was still technically the "protector" of all the emirates of the Gulf, and bound by treaty to come to the aid of any of them attacked by a foreign power.

On November 30, twenty-four hours before the treaty ended, Iranian naval vessels anchored off the islands of Abu Musa and the Greater and Lesser Tunbs, and troops were ferried ashore. On Abu Musa they were welcomed by a delegation from the ruler of Sharjah, with whom, thanks to Sir William Luce, they had made a treaty. But when the force went ashore on Greater Tunb—the only one of the two islands that was inhabited—a Ras al Khaimah police post opened fire, and in the subsequent skirmish three Iranians and one Arab policeman were killed. However, a few hours later the islands were flying the Iranian flag, and units of the armed forces were settling in. The Strait of Hormuz was under Iranian control, and the petroleum tankers which passed through it every ten minutes, day and night, from the great oil ports of the Gulf would henceforth be under Iranian supervision.

Cries of indignation rose from all the centers of the Arab world, and Iran was attacked as an aggressive power seeking to impose her will upon the Arabs. But to no one's surprise in London, least of all Sir William Luce's, the main barrage of Arab indignation was turned on Britain, which was savagely accused of breaking her treaty obligations in the Gulf by failing to come to Ras al Khaimah's aid.

It was all just what Sir William and Sheik Sakr had expected. Far from losing face over the loss of the Tunbs, the ruler of Ras al Khaimah suddenly became the hero of the hour. He had dared to fight in a situation where Britain had cringed and run away. He was hailed by the people not only of Ras al Khaimah but of the other states on the Gulf. Moreover, when their rulers met in Abu Dhabi on December 2, 1971 to declare themselves formally members of the Federation of Arabian Emirates, many of the sheiks expressed deep regret that the "heroic" Sakr was not among them. So it was decided to invite him all over again, and to accord his state the privileges in the new union which his brave stand had merited. Two weeks later he accepted.

Sir William Luce and the British government were content. What did it matter that the windows of some British offices were stoned, and that mobs in Kuwait and Bahrain had burned Union Jacks? The rulers understood what had happened, even if the mobs did not. And the object of the exercise had been achieved: Iran had the all-important islands of Tunb, and Ras al Khaimah was a member of the federation.

AS IT TURNED OUT, one leading Arab had not in any way "understood" what had occurred behind the scenes in the occupation of the Tunbs, and that was Muammar al Gaddafi of Libya. Apparently he was unaware that the mild denunciations which were issuing from government offices in Cairo, Kuwait, Baghdad and Riyadh came from regimes which had no intentions of doing anything about the situation, and their embarrassment was considerable when Gaddafi urged them to band together and throw the Iranian aggressor back into the sea. Most embarrassed of all were the rulers of the new Federation of Arabian Emirates, especially when Gaddafi offered them direct aid for what he presumed would be their imminent counterattack against the invader.

On December 5, six days after Iran's occupation of the islands, OPEC met in Abu Dhabi for one of the organization's regular conferences. Libya had announced that it was boycotting the meeting; it refused to sit down at the same table as the Iranians. But OPEC's discussions were to be followed by a meeting of OAPEC (the Organization of Arabian Petroleum Exporting

Countries),* and to this the Libyans had sent its usual delegation.

While OPEC settled down to the discussion of its agenda, the Libyan delegation went up the coast to Ras al Khaimah. There they burst in upon the astonished Sheik Sakr, and to his considerable confusion, handed him a message from Gaddafi offering him Libyan troops and arms ready to fight and expel the Iranians. Though he had no intention of taking it seriously, the ruler thanked the delegates warmly for their offer and gave them a signed photograph of himself to take back to their leader.

Back in Abu Dhabi, the Libyans informed the oil ministers of Abu Dhabi and Dubai of the nature of their visit to Ras al Khaimah, and told them that in the event Libyan troops were sent, passage would be needed for them through their territories. The ministers promised to convey this information to their rulers, who must have been considerably more alarmed at the prospect of having Libyan troops in Abu Dhabi and Dubai than of having Iranian forces on the Tunbs. Fortunately, Sheik Sakr tactfully informed the Libyans that he could rely upon his Arab brothers in the Gulf for any aid he required, and that Libyan troops would not be needed.

Frustrated in his hopes of retaliating, Gaddafi turned on the perfidious British, who in any case had been responsible for the whole affair. At least in his own country he did not require anyone's by-your-leave. So on December 7, 1971, he seized Britain's oil installations in Libya and announced the nationalization of British Petroleum.

"HE'S CLOSING US DOWN!" said a BP engineer in Tripoli when news of Gaddafi's action reached the head office in Libya. "I wonder if he knows what he's doing?"

"Let's not tell him, anyway," said one of his colleagues.

The crude oil of the Libyan desert has qualities which make it particularly desirable. It has an extremely low sulfur content, and

* OAPEC had been formed by the Arab oil states in 1967 after the Six-Day War.

is particularly suitable for conversion to gas, for producing chemical feeding stocks, and for high-grade petroleum.* But it has one element in it which can cause problems: the amount of wax in the oil. Petroleum engineers use mechanical gadgets called "go devils" to keep the oil from clogging the pipelines on its way from the wells to the ports, and these are particularly vital in the Libyan fields because the waxy content of the crude oil can clog the pipes within hours.

For two or three days after the seizure of their assets in Libya, followed by the withdrawal of personnel and the closing down of field operations, BP officials kept to themselves the thought of what was going to happen to Concession 65 if action wasn't taken quickly. The wells would begin to clog up, the crude in the tank farms would begin to coagulate, and the pipeline to Tobruk, the loading port, would soon become what one engineer described as "a bloody great wax candle, fit for nothing except chopping up into sections and putting on your dinner table."

Under the circumstances, some of the BP staff can perhaps be forgiven for the irresponsible pleasure this prospect gave them. One of those who did not share their delight was an American named Nelson Bunker Hunt. Hunt is one of the three sons of the Texas oil billionaire, H. L. Hunt, and he and his brother Lamar owned and operated their own oil company, Hunt International Petroleum Company. Hunt International was a part owner with BP of Concession 65, and the Americans had their assets seized by the Libyans at the same time as those of their British partner. Furious, Bunker Hunt flew in from Texas to Tripoli to protest to Gaddafi's right-hand man, Major Jalloud. He was not British, he pointed out, he had no connections with Iran and no operations in the Gulf. Why was he being held up for ransom? Hadn't he always been a good friend of the Arabs?

When this plea failed to have any effect, Hunt took off for Algiers, where he and his brother also had operations, and from

* Oil from the Gulf, for instance, is apt to be high in sulfur, resulting in pollution problems at the refineries and in tankers.

Colonel Houari Boumedienne, the Algerian premier, he got both sympathy and help. Boumedienne got in touch with Gaddafi at once to point out that because of Hunt's friendly attitude and his "unswerving sympathy for the Arab cause," he should not be penalized by Gaddafi's quarrel with the British. The appeal was effective and the seizure of Hunt International's holdings was canceled.

Gaddafi's gesture could not have been more fortunate for Libya. The moment he was reinstated, Bunker Hunt looked over Concession 65 and warned that unless something was done quickly, the oil would solidify in the pipes and approximately $10 million worth of damage would be done to one of the most prosperous oil fields in the Western Desert. Within twenty-four hours he had been given permission to fly in thirty-five of his own experts from Texas, and under their supervision the field was put back into operation and Concession 65 was saved.*

It is doubtful whether the news of Bunker Hunt's last-minute intervention raised any cheers in Britannic House, BP's headquarters in the City of London. Knowing how short Libya was on first-class oil technicians, the British company had counted on this to force the Libyans to the negotiating table. Now this hope had been dashed by their American partner, and their £100 million investment in Libya was back in operation, but out of

* If Bunker Hunt expected gratitude and a privileged position for his Libyan operations in return for his help, he was mistaken. On October 4, 1972, the Libyan Oil Ministry formally submitted a list of demands on Hunt, the most important of which were: 1) 50 percent of the profits on the oil his company had sold from his half-share of the Sarir field (Concession 65) since the nationalization of BP, and 2) a 50 percent participation by the Libyan state in his half-share of the field. Otherwise Libya threatened to embargo Hunt's lifting of Sarir crude oil. "As you know, Bunker Hunt's concession in Libya is a purely personal or should I say family affair," said Oil Minister Izz al-din al Mabruk. "And frankly, such personal or family concessions are against the principles of our Libyan revolution . . . Bunker Hunt is doing nothing at all for us, and therefore we don't need him at all . . . He should adhere to his agreement with us; otherwise he will lose not just 50 percent of his interest but the whole lot." (*Middle East Economic Survey,* November 3, 1972). On December 14, 1972, it was announced that Hunt had rejected the Libyan request for a 50 percent participation, probably with the backing of other U.S. companies in Libya, fearful that if Hunt accepted, the others would have to follow.

their hands. It was just like 1951 all over again, when Mossadegh had nationalized their oil operations in Iran, only this time they could not hope that the major U.S. oil companies would come to their rescue. The climate had changed, and so had the nature of the oil industry. For one thing, there were too many independents around for the majors to enforce a boycott. And why should they, in any case? By helping BP in Iran, they had gained a sizable stake in Iranian oil, but in Libya they already had their concessions, so what could they gain?

Hence, BP had to content itself with protesting the seizure to the Libyan government, condemning it as a clear violation of international law, and calling for arbitration. In the meantime the company took space in the world's newspapers warning any would-be purchasers of oil from Concession 65 that they would be buying "pirate" oil. "The company has also reminded the Libyan Government," BP's statement concluded, "that the government's wrongful acts are under international law incapable of depriving the company of its rights under the agreement. Accordingly, the attention of all those who may be concerned with these developments, whether as purchasers of oil or otherwise, is drawn to the continuance of the company's rights. It is the intention of the company to assert those rights wherever and whenever necessary against those who infringe them." *

It was the same action that BP had taken during the dispute with Mossadegh, and as in the case of Iran, the company planned to prosecute anyone who tried to buy or sell oil from Concession 65.

In 1951 the threat had worked; no one had dared to buy, and Iran had been starved into submission. But this was twenty years later, and unlike Iran, Libya was a relatively rich country, with $5 billion in the bank, ready to be used in just such an emergency.

* Part of an announcement by BP Exploration Company (Libya) Limited in, among other newspapers, the *International Herald Tribune*, Paris, December 23, 1971.

Part Eight
CRUDE FACTS

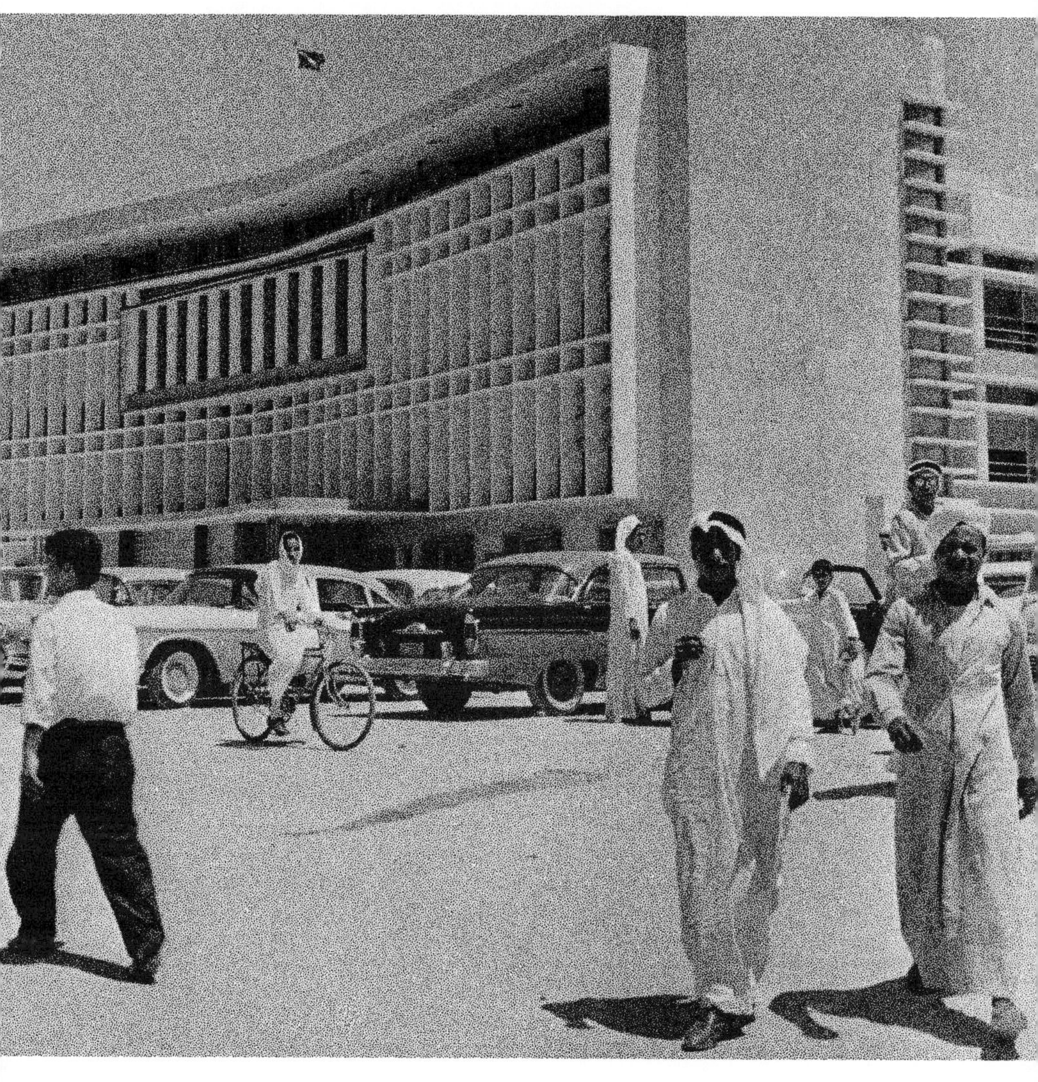

PRECEDING PAGE: *One of the new schools in Kuwait built with the profits from oil.*
PHOTO: G.A. JOHNSON/CAMERA PRESS LTD.

CHAPTER TWENTY-EIGHT

Private Club

IN FEBRUARY 1971, members of the Organization of Petroleum Exporting Countries met the major Western oil companies in a series of confrontations in Teheran, and by bluff, threats and smart maneuvering squeezed huge increases in revenues out of the companies. It was probably the most expensive commercial negotiation ever; by 1975, it would cost the major combines the astronomical sum of $30 billion in extra payments to the Arabs and Iranians. Kuwait alone, a state of only 5,800 square miles and with a population of around 700,000, had its projected 1972 income from oil more than doubled, from $700 million to $1.450 billion, which is not bad going for such a small country.

Among oilmen in Kuwait there is a story that when he kneels down to pray five times a day, the ruler, Sheik Sabah al Salem al Sabah, inevitably ends by saying, "Allah, tell me and my people what we are to do with all our money!"

With the possible exception of Abu Dhabi, which has an even smaller population, Kuwait is the richest state in the world. The tiny pearl- and prawn-fishing village of thirty years ago is now a land of broad highways, modern hotels and fantastic villas, of shops stuffed with expensive imported clothes, food, jewelry, watches and perfume. It has fourteen hospitals with a bed for every eighty Kuwaitis, and their equipment is the envy of the medical world. It has two hundred and thirty public and forty private schools, with a teacher for every ten pupils, and lessons are taught by the latest audio-visual aids. Leisure is encouraged by tennis courts, swimming pools and playgrounds in every school. All of this is free.

One in three Kuwaiti citizens now owns a car, and that includes babies and women, only a few of whom drive. The cars are generally big American ones, and Kuwaitis vie with the Japanese as the most dangerous drivers in the world. In the desert sands alongside the highways leading out of the city, brand-new Cadillacs, Chryslers and Plymouths lie like beer cans, crashed at break-neck speed, burned out for lack of oil or merely abandoned because of a blown-out tire. Their owners prefer to buy new ones rather than take them into a garage.

Land in the center of Kuwait is now $420 a square foot; yet hotels and office buildings continue to rise, as do new schools, parliament buildings, exhibition halls and ministries. The state has built the biggest desalinization plant in the world, and water is used lavishly for gardens and small parks and to quench the thirst of the oleanders which line most of the highways—and this in a land which has one of the hottest climates in the world (it averages 115 degrees Fahrenheit in the shade in August). The telephone is free, citizens pay no taxes on their incomes, houses, basic rates for electricity and gas in their homes, or for gasoline in their cars. Any Kuwaiti can borrow 25,000 dinars ($75,000) free of interest from the government to build himself a home, and pay it back at a rate of not more than 5 percent of his salary over twenty-five or thirty years. These children of the desert have constructed air-conditioned villas on the sands where once they grazed their camels and lived in goatskin tents, and on summer evenings they come out of them, trailing TV sets, fans and refrigerators on long extension wires, and squat in the sands as of yore, sipping ice-cold Pepsi-Cola and watching *Bonanza* or the latest film from Cairo.

But these incredible benefits of a super welfare state are the privilege of Kuwaiti citizens only, and it is not easy to become a member of the club. Of the state's 700,000 inhabitants, fewer than 400,000 are Kuwaitis. The rest are Palestinians (about 100,-000), Egyptians, Iranians, Iraqis, Jordanians and Saudis who have swarmed into this desert El Dorado in search of work. They have no trouble in finding it, for why should a Kuwaiti work when he has all that money? But jobs do not bring with them the privileges that a citizen automatically enjoys, and only those who have some special skill to give the state are granted citizenship. The rest do most of the work to keep Kuwait going, and watch their rich masters taking it easy.

Out in the desert, just beyond the confines of the city, are the shantytowns of the non-Kuwaitis who will never achieve the special skills to qualify them for citizenship, and who have come from all points of the Middle East to wait for the crumbs from the rich man's table. They live in an atmosphere laden with the stench of oil, and at night their squalid huts are lit by the flare of burning gas. They have no air conditioning to ameliorate the appalling heat of summer, and little protection from the violent sandstorms for which Kuwait is notorious. In winter they are whipped by storms, and their settlements turned into brown seas of mud by torrential rain. They are fellow Arabs or Iranians who, by an accident of geography, do not qualify for a share of the riches, and perhaps they would not resent their misfortune so much if the Kuwaitis were not so contemptuous of them.

The Kuwaiti government pays some 60 million dinars (about $180 million) in annual subsidies to Egypt and Jordan, and is prompt in paying its dues to the Arab League, but in individual encounters with a brother Muslim who has come to work for him, the average Kuwaiti is apt to treat him like a servant. "It is bad enough to see the young men flinging their money around, wasting it on useless extravagances," said one Palestinian bitterly, "but even worse is their arrogance. I always thought that one of the faults of us Palestinians was our superior manner, but compared with Kuwaitis we are humble people. They will not allow us to buy land, or go into business without a Kuwaiti partner, or get medical attention without paying a fee—but we could stomach all this if only they would acknowledge the contribution we make to the running of their state. Without us it would stop functioning. They go to college but come away with easy art or sociology degrees, and they don't know a bolt from a spanner. So they sit around in offices where the government has invented jobs for them, and pass out nonsensical orders that we take no notice of. If we did, everything would go wrong! And they have the impudence to look down on us just because they have money!"

One of these days, some prophets darkly forecast, a solution will be found to the Israeli-Arab situation, and then the Palestinians in Kuwait will need a new cause to excite their passions. It may well be a campaign to acquire a more equitable share of Kuwait's riches for those who keep the wheels of state turning. Perhaps it is the vision of this unhappy day, and of the knowledge that they are envied and even hated by most of their neighbors which makes Kuwaiti men the haughty, unstable characters that they are. Oil wealth seems to have brought them only uneasiness, suspicion and xenophobia. A *New York Times* reporter surveying the Kuwaiti scene in 1971 quoted a psychiatrist's observation that outpatients in the city's mental hospitals had risen from 450 to 18,000. The psychiatrist stated: "Many of [their] neuroses are related to the way the Kuwaiti has come to consider himself an elite person. It is not important in his eyes whether he works or not. If he does something wrong he feels it can be excused. He looks down on naturalized citizens as second class and non-Kuwaitis as quite something else again. These people come to me and say they can't sleep, can't work and can't enjoy life."

The dispatch continued: "Some of Kuwait's leaders have begun to feel that the perpetual onward and upward financial spiral has brought no rewards in terms of happiness and they are calling for a slowdown, a time for Kuwaitis to catch up emotionally and culturally with their new economic status." *

It does not help the stability of the Kuwaitis to find themselves constantly looking over their shoulders at their neighbors, by whom they are regarded as the spoiled brats of the Persian Gulf, and all of whom have territorial claims against the country. The vague border agreements made by Sir Percy Cox at Oqair in 1922 satisfy neither the Saudi Arabians nor the Iraqis, and Iran has strong objections to the extent of Kuwaiti territorial control over the offshore regions of this area of the Gulf. Since the considerable oil revenues of these three nations fall far short of their budgets, they can be forgiven for the covetous looks they cast at a city-state which is perplexed about what to do with all its money.

TWENTY-FIVE MILES SOUTH of Al Kuwait is Ahmadi, the headquarters of the Kuwait Oil Company, an Anglo-American combine whose 692 wells keep half the world's industrial complexes supplied with power. When people in the Western world use such glib words as "rapacious" or "capitalist exploiters" when talking about the Anglo-American oil companies in the Middle East, they obviously have no conception of places like Ahmadi in Kuwait, Dhahran and Al Khobar in Saudi Arabia, and several company towns in Iran where millions of dollars have been spent on people.

Ahmadi is a company town; in one way or another all of its 30,000 people are connected with the operations of the Kuwait Oil Company. As in Al Kuwait, there are houses for them to live in, swimming pools, tennis courts and recreation grounds where they can play, theaters, cinemas, and still another first-class hospital with first-rate doctors. But here there is no discrimination against the non-Kuwaitis, and the 300 British, 44 Americans, 1,400 Kuwaitis, 280 Indians, 160 Pakistanis and 1,600 non-Ku-

* Dana Adams Schmidt in the *New York Times,* August 28, 1971.

waiti Arabs all have the same privileges and attention. So do their dependents, as well as the Kuwaiti members of the fire department, police, postal services and maintenance facilities that help to keep the town going. All of these services are a financial drain on the oil company.

Ahmadi is a green town. British housewives have encouraged the Kuwaiti and Indian women to cultivate their gardens, and massive amounts of desalinated water spray the grass and flowers in the backyards of company houses. In March and April they are a sight to see, green oases in which riots of colorful flowers bloom, only to die in the onslaught of heat at the beginning of May. Each weekend, cars stream out from Kuwait packed with families come to see the soothing green and to picnic in the streets. When the police try to move them on, they refuse to budge; "It is our town," they say.

Indeed it is, though it is paid for and administered by the Kuwait Oil Company. Sometimes a weary American or British administrator of the town, squeezing municipal affairs in between his normal work, expresses the hope that one day the Kuwaitis will take over the place. But "each time we suggest it, they shrug their shoulders and say that as long as it runs well, why should they bother? I wish they would take the same view of the oil operations."

TEN MILES OUT into the shallow waters of the Persian Gulf, a 66-inch underwater pipeline—wide enough for a Volkswagen to drive through it—pumps 30,000 barrels of crude oil an hour into the bowels of the 250,000-ton tankers anchored at Sea Island, KOC's loading point. But sometimes storms blow up in the Gulf, and a great wind rides across the desert, mixing sand and oil smoke in a choking fog. Then the empty tankers are forced to sail out to deeper waters, where they cannot be blown ashore, and the marine masters anxiously watch all the weather signs, calculating how long they must wait before calling the tankers back in to link up with the pipelines. If they wait too long, the tank farms fill up to overflowing, and then instructions must go out to the feeding points and wellheads to close down.

No oil company official is more alarmed than the ruler of Kuwait when this situation threatens. In the spring of 1972 a storm blew up which raged along the shores of the Gulf for nearly eight days, forcing tankers to put back to sea, feeding operations to stop, and wells to be shut down. No one informed the people of Kuwait about what was happening, because the volatile population of the capital might have panicked. With good reason, for Kuwait still runs on gas from the oil fields and the state would grind to a standstill without it.

If the storm of 1972 had not abated when it did, the city would have been without power in another four hours. Afterward a KOC official said, "In a way, it's a comfort for us to know that this can happen. And for them to know it too. When the Kuwaitis are feeling particularly bloody-minded—usually because of some action quite apart from our operations, something to do with American or British policy towards Israel or Vietnam, say—their militants start threatening us. 'We'll show you who's master here,' they say. 'With one turn of a spigot, we can deny the West the oil it needs from us. We'll call a general strike and close down the oil fields.' At which the government in Kuwait rushes out, wringing its hands, and says, 'For Allah's sake, no. Don't you realize what would happen? No gas. No power to give us light or air conditioning or to pump our water. We'll all die of heat prostration—or if we survive that, of disease.' There's one thing we can be sure of: they can starve us, beat us, shoot us, make life unpleasant in the extreme. But for their own sakes, they can never close us down."

CHAPTER TWENTY-NINE

Participation

ON A CRISP AFTERNOON EARLY IN JANUARY 1972 I drove through Riyadh, the Saudi Arabian capital, to a meeting with Ahmed Zaki Yamani, the king's minister of petroleum and mineral resources, when we would talk about the future of foreign oil companies in the Middle East.

At that moment nothing could have seemed more remote. It was the end of a bitterly cold day of the kind that always surprises the visitor to central Arabia and makes him wish he had brought warmer clothing. The sun was setting and the sky, which had been a brilliant blue all day, was now turning crimson, its glow washing over the walls of the palaces, ministries and mosques lining the broad streets of the city's modern quarter. Everything was stained red by the sun, even the black, round, cheerful face of my Bedouin driver, and reddest of all were the uneven crags and shallow defiles of the desert, rising and falling toward the horizon at the end of every street. From several mosques within earshot, the loudspeakers in the minarets were calling the faithful to the fifth and final prayer of the day. In the time of Ibn Saud, everything would have stopped at the call of the muezzin, but now, though groups of men standing outside the government offices began to fall on their knees, the traffic—big American cars, heavily laden trucks, an occasional bus, and not a camel or donkey in sight—went right on rolling down the streets.

Riyadh is the capital of the richest oil state in the world, but it doesn't look the part. It is a small, clean sandstone town dominated by a giant water tower—floodlit by night and visible for miles around—whose slender stem and mushroom cap catch every glint of light. Some see in its shape a symbol of Saudi Arabian virility, but to the traveler glimpsing it across the desert, the earthly delights it suggests are not likely to be fulfilled when he reaches the city. When the U.S. boxer Muhammad Ali visited the city, he was heard to remark, "Doesn't anyone ever sing or smile in this town?"

The answer is that only the muezzins sing, and there really is not much to smile about. Riyadh has always had the reputation of being Arabia's most inhospitable city—"saturnine and sanctimonious" was what Ameen Rihani called it—and even Saudis from other parts of the country find it unbending. It is not simply that there are no movies, theaters or other places of entertainment (these are forbidden in other parts of the country too), but that there is a pervasive air of gray sanctity about the place, a hint of prudishness in the air, which strangers find inhibiting and seems to make the natives glum. Arabs who know both places say that Mecca is much more cheerful. As for Western infidels, most of them live not in Riyadh but on either side of Saudi Arabia's two coasts—in Dhahran or Al Khobar on the Persian Gulf if they are in oil, in Jidda on the Red Sea if they are in shipping or trade *—and make short visits to the capital only when necessary.

Ahmed Zaki Yamani is an unlikely figure to find in such a restrictive atmosphere, for in manner, appearance and outlook he gives every sign of worldly sophistication. Of all the oil min-

* Or in diplomacy. All foreign embassies are in Jidda, as is the Saudi Ministry of Foreign Affairs. Envoys come to Riyadh only when summoned there by the king.

isters with whom Western negotiators have to deal, Yamani is the one with whom they seem to get along best. Howard Page, who faced him several times across the table on behalf of Standard of New Jersey, said recently of him, "I know damn well that there are occasions when Yamani will insist on doing something which he knows is not in his country's best interests, but that's because he's got himself hooked politically."

This ability to persuade Western oilmen that he is doing his best to meet them halfway, though under pressure, while at the same time convincing his militant Arab and Iranian colleagues that he is squeezing the last drop of blood out of the companies, is Yamani's principal asset as a negotiator. But he has others. "He's very clever," said Page. "I remember when we were working on negotiations with OPEC in Saudi Arabia. Things weren't going well, and Yamani needed an inspiration to get the negotiations unclogged. He took me and the other Western delegates to see King Faisal. The parent companies of Aramco were arguing about some tax situation with the Saudis. Well, Yamani put me right up in front when we got to the palace, closer to the king than you are to me, and it turned out that he had Faisal well briefed. The king hauled me over the coals for not agreeing to some of the proposals about taxes that OPEC had made. The hell of it was that I was perfectly willing to agree to the proposals, but the two other fellers squatting there on the cushions [from two of the other parent companies] were not, so of course I couldn't say anything. But I think Yamani knew the score, and what he was trying to do was break me down so that I would say to Faisal, 'I agree. It's only these two fellers who are holding it up.' He was trying to put them on the spot and make them back down."

For his private home in Riyadh, the Saudi oil minister has annexed a wing of the capital's newest hotel and surrounded it with a garden lush with green grass, palms, oleanders and mallows. Inside, concealed lights glow on vivid silk carpets hanging on the woolen-padded walls. The carpets are deep and soft underfoot, and strewn with low-slung damask chairs. A hi-fi system plays tapes from Western musicals and films, so that a conversation about oil or Arab politics is apt to be carried on against a background of the theme music from *Doctor Zhivago.* At times a rustle of silk comes through the filigreed partition at the end of the room, and the faint scent of French perfume, and one remembers the smart, good-looking woman who is sometimes seen with Yamani on his trips to Beirut, Teheran and San Francisco—modern and unveiled on those occasions but relegated to her own quarters in the restricted conditions of the capital.

YAMANI HAS A NIGHTMARE which he fears may yet come true: that one day, in a fit of nationalistic rage, probably as a result of an exacerbation of the Israeli-Arab situation, all the Middle East oil states will turn on the Western oil companies and nationalize the lot of them, just as Algeria has done to the French, and Libya and Iraq to the British. The policy of annexation preached by all local militants ("Arab oil belongs to the Arab people") could, Yamani believes, do nothing but harm to the oil states—economically, sociologically and politically.

When he talks about this, three terms come into his conversation, as they do into the talk of most oilmen. The first is "upstream," which means all the activities of the oil companies in the country where the crude oil is pumped out of the wells. The second, "downstream," refers to the outside world, where the oil is actually marketed. The third term is "off-takers"; these are companies which have no part in the "upstream" production of oil, but only in the "downstream" business of selling it. Yamani elaborated on this theme in a speech he gave a few years ago:

"Under the present setup, the majors make the bulk of their profits in the producing end of the business. In a country like Saudi Arabia, for instance, they produce the oil at a cost of, say, 15 cents a barrel, perhaps less; and they pay Saudi Arabia perhaps 90 cents a barrel in royalty and income tax. Now Arabian light crude, for example, has a posted price of $1.80 per barrel, but its realized price in the market may be only $1.40.* So if the majors then feed this oil into their refineries at the posted price, this means that they do not actually make any profit in their downstream operations, and in some markets may even register losses. Thus virtually all their profits . . . are made upstream at the production stage." †

For this reason, Yamani pointed out, no major oil company operating under today's conditions is really interested in bringing down the price of oil. On the contrary, the major oil companies have a vested interest in keeping up the price, and besides, "one might say that they are now really the only available bulwark of stability in the world market." On the other hand, the producing countries should welcome the existing setup because they then can count on a regular and stable income, enabling them to plan ahead.

"Now, against this background, let us look at nationalization and see what will happen if we nationalize the producing operations in our countries . . . Nationalization of their upstream operations would inevitably deprive the majors of any further interest in maintaining crude-oil price levels. They would then become mere offtakers buying the crude oil from the producing countries and moving it to their markets in Europe, Japan and the rest of the world. In other words, their present integrated profit structure, whereby the bulk of their profits are concentrated in the producing end, would be totally transformed. With the elimination of their present profit margin of, say, 40 cents a barrel from production operations, the majors would have to make this up by shifting their profit focus downstream to their refining and product-marketing operations. Consequently, their interest would be identical with that of the consumers—namely, to buy crude oil at the cheapest possible price. They would put their full weight behind efforts to drive down crude-oil prices, and in this they would undoubtedly succeed."

There could be only one result from such a situation, Yamani maintained:

* The figures given are for 1969. The posted price is, of course, now very much higher, reinforcing Yamani's point.

† Quoted from a paper presented at the third seminar on the Economics of the Petroleum Industry, held at the American University of Beirut, spring 1969.

". . . a dramatic collapse of the price structure, with each of the producing countries trying to maintain its budgeted income requirements in the face of the declining prices by moving larger volumes of oil to the market. I need hardly emphasize the disastrous effects of such an eventuality on the economic and political life of the producing countries. Financial instability would inevitably lead to political instability."*

Yamani admitted that for a time, the consuming countries might enjoy a glorious period of cheap oil at their power plants and gas stations, but he believes that the subsequent chaotic conditions would in the end outweigh any economic advantages.

The chief apostle of nationalization of the Middle East oil fields is Yamani's predecessor as Saudi oil minister and his bitter rival, Abdullah Tariki. Today Tariki operates as one of that new breed, a petroleum adviser, and in that capacity gives advice on tactics and strategy to the oil administrations in Algeria, Libya, Iraq and Kuwait. In the first three countries he has succeeded in persuading his clients to seize some or all of the oil operations of the major companies, but no one in the oil world is in any doubt that Tariki's main target for nationalization is in a country for which he has no mandate to act as an adviser: Aramco of Saudi Arabia. He has always been antipathetic toward the great American combine, and though Saudi Arabia is his native land, he has a consuming hatred for the present Administration, which has driven him into exile, which is trying to persuade friendly governments to ban him entirely from the Middle East, and whose ruler is his implacable enemy.

Tariki was in Baghdad when the Iraqi government nationalized the northern territory operations of the Iraq Petroleum Company on June 1, 1972; and those who heard Major General Ahmed Hassan Bakr, president of the Revolutionary Command Council, announce the takeover guessed that the Saudi adviser had not only masterminded the act but had also written the words, which exactly echoed his own fiery sentiments:

* *Ibid.*

"You know that the oil companies are the dangerous tools which represent imperialist logic, the logic of plunder and monopolistic exploitation and the impoverishment of the masses. They have always continued to represent imperialist domination, both symbolically and in fact, so that it has become clear and obvious that any true national liberation is incomplete unless the requirements of national sovereignty are imposed on these companies which have acted in accordance with their imperialist nature, contrary to the interests of our masses, the spirit of the times and the progress of history, and have considered themselves a state within a state. It is clear to the struggling masses from their experiences that putting an end to the domination of the monopolistic oil companies is the only way to ensure national sovereignty and the economic independence which is the tangible essence of political independence."*

This was genuine Tariki rhetoric, full of the references he loves to make to imperialism and plentifully sown with favorite phrases like "struggling masses," "economic independence" and "national liberation."

Tariki was not allowed to proceed to Beirut for the emergency meeting of OPEC which was called on June 9–10 to discuss the Iraqi nationalization because at the request of the Saudis, the Lebanese government has declared him *persona non grata.* But his clients among the delegates came well primed with his arguments, and rumors in the OPEC lobbies had it that the Iraqis had been told by Tariki to rest assured that within a year of IPC's nationalization, that of Aramco would follow. Thereafter, the rest of the oil concessions of the Middle East would automatically be "liberated."

It was a difficult OPEC conference for Ahmed Zaki Yamani, for eventually it endorsed the Iraqi nationalization of IPC and promised the full support of the oil-producing nations to its fellow member in the event of any trouble with the Anglo-American-French-Dutch directorate of the seized company. By voting to help make Iraq's annexation a success, Yamani was not only aiding an enemy of himself and his state; he was also sabo-

* Extract from President Bakr's speech on Radio Baghdad, June 1, 1972, translated by *Middle East Economic Survey,* Beirut, June 2, 1972.

taging his own plans for solving the crisis which faces the oil states and the companies in the present decade.

But Yamani's objections to nationalization are not simply negative. He has been the most active supporter among the ministers of the oil states for what he believes is the more sensible alternative to annexation: participation by the producing countries in the operations of the majors. Ever since Enrico Mattei started the trend in Iran in 1958, new concessions awarded by Middle East oil states have had a clause in the contract stipulating that once petroleum was discovered, the concessionaire would make the host country an active partner in the operations. Yamani, for example, is now an active director of the Arabian Oil Company (the Japanese combine which has discovered rich fields offshore Saudi Arabia and Kuwait), and participates in all the higher policy decisions made by the company.

With the Anglo-American companies, however, the concessions (even those which have been revised) are of the old type, operated by foreigners whose principal responsibility is to their parent companies or shareholders in Europe or the United States, and in whose plans and policies the host countries have no say whatsoever.

"Take Aramco, for example," says Yamani as we made ourselves comfortable over a cup of coffee in his sumptuous living room. "Since Aramco is incorporated in this country [Saudi Arabia] I, as oil minister, have been made a member of the board. But that's just show business—it is meaningless. Aramco has what it calls its executive committee, and that is where they plan the real business, where they get instructions from their parent companies in America. *I want to be there.*" He spoke with great emphasis. "I want to know what is going on and I want to have a say in it. And that is what frightens them most."

Yamani has been trying to make a participation deal with Aramco ever since 1963, but so far (fall of 1972) has been fobbed off repeatedly by a succession of U.S. presidents of the company. "It isn't the financial aspects of participation which frightens the company—and all the other companies—the most, though they insist that this is the biggest problem," he went on. "What they are worried about is having Arabs or Iranians taking active part in the direction of the company. They are afraid that we might not have the know-how, might overestimate our own ability, or might interfere. Take Aramco again. I wanted them to relax on this point, so in 1963 I started asking the company to enter into a joint venture with us on a small scale. They said, 'Why?' I said, 'Just to know each other, nothing more. To do business together.' We even found a third party to join with us to make it easier for them. They refused."

In 1968 Yamani had tried again. "There is an oil structure lying partly inside Aramco's concession and partly in the Saudi government's own concession. I went to them and said, 'How about developing this oil structure together? If we find oil, we can go into it together.' By this time they couldn't say no, because by 1968, foreign companies couldn't afford to be rude to host governments any longer. So they agreed, and then they went off to America, obviously to consult with their parent companies. Then they returned, not saying no again, but you know how people behave when they don't want to do something but don't want to refuse outright—playing for time and so on. Then we announced an invitation to any comers to go into our area with us, and when we did, they came running to us and said, 'Let's do it now.' But by then it was too late."

By 1971 the oil states of the Middle East were beginning to separate into militants and moderates, and Tariki had become the thorn in Yamani's flesh. All around him he could see signs that Tariki's activist policy of nationalization was gaining ground with Arab leaders in some countries, and with the Arab and Iranian masses in all of them. But apart from the National Iranian Oil Company, which had special relations with the Anglo-American consortium, not a single national oil company of any of the oil-exporting countries in the Middle East was successful. The prime example of failure was in that phenomenally rich little state, Kuwait, where conditions were such that it would have seemed impossible to achieve a loss; yet the local national oil company had managed to do so. Its brand-new refinery had failed to cope adequately with the free oil handed over to it by the Kuwait Oil Company, and its marketing plans had gone so awry that Russia was buying up Kuwait National's crude oil cheap for sale to its clients in Asia and selling refined oil in Europe to the company at high prices for its markets there, entailing losses all round. Saudi Arabia's national company was having trouble with its petrochemical division at home and its marketing abroad, and almost exactly the same situation prevailed in Iraq, where the Iraq National Oil Company (INOC) had had a particularly bad year.

Under these circumstances it was not difficult to convince reasonable men that for at least a decade, nationalization was the worst possible solution to the eventual goal of self-control of its oil by each country. For Yamani and his fellow moderates, however, the difficulty was that the Iranian and Arabian masses were not reasonable, because militants like Tariki and the masters he served—Gaddafi of Libya, Bakr of Iraq and Boumedienne of Algeria—insisted on portraying the major oil companies as the enemy they could defeat only by seizing their concessions.

"Lord Balfour once said that, pity though it is, most colonial people prefer self-government to good government," a British oilman in Iran said. "It's the same with these people. It's foolish of them, but the Iranians and Arabs prefer to run their own oil fields badly rather than have them run well by outsiders, particularly British and Americans."

Yamani's task was to try to convince the governments of the oil states (he knew he had no chance of convincing the masses, whipped up by Tariki-type rhetoric) that participation was the better way because it would achieve control by gradual means without interrupting the flow of oil "upstream" or its profitable sale "downstream" at a stable price. His arguments, his passionate sincerity and the wealth of documentation which he produced so convinced Middle East leaders that by mid-1971 he had the votes of all the member states in OPEC for a joint demand for participation. The producer states would approach the major Western concessions and ask for an immediate 20 percent share in their operations, this percentage to be increased as time went on and the Middle East gained more technical expertise.

Tariki fought hard to persuade his client states to veto this proposal, but failed. Libya, Algeria and Iraq indicated that they would make their own policies, but otherwise were sympathetic to the goals of the seven other members.

But now the task was to convince the major oil companies that they would have to make a gesture to the moderates and sacrifice some of their sovereignty to the states in which they were operating. Yamani was well aware that this was going to be the hardest part of his struggle, and that Western oilmen didn't believe a word of it when he insisted that the sacrifice would be for their own good. As he had said in Beirut two years before:

> "Unfortunately, the majors—at least some of them in their public statements—seem to be obsessed with the empire they have built. It is so vast and it took them so many decades to achieve. And now they see these newcomers—these national oil companies in the producing countries—wanting to come and take a piece of their cake, which is the last thing they want to happen."

Later in the same speech he made an indirect appeal for co-operation by the majors:

> "Participation serves so many purposes. First and foremost it will maintain the stability of prices. It has some common features with nationalization, but without the drawbacks. From our point of view, it will give us [the oil states] a position of influence in the market, instead of being pawns in the hands of other people. It will be a good thing for the oil companies because it will save them from nationalization and provide them with an enduring link with the producing countries. The majors' concessions will be starting to expire in twenty years or so and, unless they have this bond with the producers, their very existence will be in jeopardy later on. They will have nothing at the producing end, and their position in the market will be weak."*

* Paper on the Economics of the Petroleum Industry, *op. cit.*

The reaction of Aramco to this had been blunt, Yamani recalled. "Aramco did not have a negative attitude toward participation," he said wryly. "They simply said no." Then he added crisply, "But we will make them say yes."

Anyone traveling around the Middle East oil states at the beginning of 1972 could not fail to be aware that a crisis was imminent. Algeria had already taken over the Sahara oil fields from France. Libya had nationalized BP, and there were strong rumors that Standard of New Jersey's concessions were next on the list. In Iraq the Baathist government was increasing its pressure on IPC. In Iran the shah had made several statements attacking the "trickery" and "unfair economic pressure" which had been employed by the Anglo-American consortium to make his people give in to their demands in 1954, and he had gone on to threaten reprisals when the concession came to an end in 1979.

Ahmed Zaki Yamani admitted to being disturbed. Although he had rallied the moderate OPEC states behind him against nationalization and in favor of participation, there had been no gesture in response by the major oil companies. Time was running out, and the militants were gaining ground. "If I were a consumer in the West," Yamani remarked, "and looked down the road to 1980, I would feel rather scared that sometime in the eighties I would not find enough energy for my industry. I would also be worrying about the nationalist trend which is prevailing now all over the world, mainly in the evolving countries—the national pride they have vis-à-vis their natural resources. The only way I would feel secure about my sources of oil supply would be if I had accepted the principle of participation and my interests were tied up with those of the producing countries."

It was January 3, 1972; Yamani was speaking almost immediately after a meeting with Aramco officials and was evidently depressed by the way his discussions had gone. In two weeks, OPEC would be meeting the majors in Geneva to force a decision from them on participation, and it was evident that in his preliminary talks he had failed to persuade the American directors of Aramco that the situation was critical. "The trouble with the Americans is that they don't look at it the way we or the consumers in the non-American world do," he said. "They think they have enough reserves in Alaska. Already they import half of the oil they consume, but they have Venezuela next door, and they think that with the help of their major oil companies they can satisfy their needs. But I think they are wrong. I think that by 1979 there will be a drastic change in the oil industry, especially here in the Middle East. The Iranians have made it very clear that they will not live with the Anglo-American consortium after that date. Then, even though the concessions are supposed to last until the year 2000 and longer, I think that the oil companies are fooling themselves if they believe that they can continue with their agreements in Kuwait, Saudi Arabia, Abu Dhabi and elsewhere. If Libya goes ahead and nationalizes all its oil, and if Iraq follows suit, responsible ministers in places like Kuwait and Saudi Arabia will have to take cognizance of Arab pride, and we won't be able to keep quiet or remain moderate for too long. There will have to be changes."

The conversation went into neutral gear while a servant padded across the carpets and deftly poured our fifth cup of scented Yemeni coffee. My host sipped morosely, listening to the music oozing out of the speaker above him.

"It is always hard to be a moderate," Yamani continued, "but it is particularly hard with American oil people because they are apt to mistake moderation for weakness. They know that we in Saudi Arabia and Kuwait can't be and don't want to be as violent as the Libyans. Gaddafi just draws his oil royalties and puts the money in the bank, or uses it to suborn people. We spend it on public works, on education, on people, and I think ours is the better way. I believe that those leaders with large bank accounts are becoming not only too rich but too arrogant. It is having a bad effect on them. They think only of the way they can use their hoarded gold to destroy, and we who are moderates suffer from their arrogance—especially if those toward whom we are being moderate misunderstand and refuse to co-operate."

Yamani was asked what would happen if the majors continued to follow Aramco's lead and turned down the principle of a participation agreement at the forthcoming conference. He shook his coffee cup to indicate to the servant that he wanted no more, but there was something about the convulsiveness of the gesture which suggested that for a moment, the cup he was shaking was an obstinate American oilman.

"I don't expect to finish the negotiations at one session at Geneva," he replied. "But if the Anglo-Americans are still as stubborn as they have been until now—if they give us a flat no to our participation proposals—then OPEC will meet and decide what to do. And what I think we will do is force the companies to give us shares by legislation. It would cause trouble, though probably more trouble for them than for us. But we cannot afford to play into the hands of the militants; we have to show that we mean business."

Yamani was asked whether the major companies realized how crucial the situation was. He shook his head and said, "In July 1967 I had a meeting in a small town in New York State with the four companies which own Aramco. They had just reduced the posted price without warning, completely unbalancing the Saudi budget, and the situation was desperate. I went all the way to New York to meet them and advised them to do something to alleviate our financial situation. I told them, 'If you don't do anything for us, you will have to pay for it in the end. So do something quickly. I'm warning you.' Well, perhaps they laughed at me; at any rate, they didn't do anything, and the result was that they got OPEC and have been paying the price ever since —at least double the price, I think, that they would have had to pay if they had been more amenable."

Yamani got up and paced for a while across the carpets. Finally, squatting beside me, he said, "I think you will be seeing the Aramco people in Dhahran in a day or two. You can convey a message to them from me. It is this: '*I am warning you again. Do something about participation. If you do not, the price you will have to pay will be heavier than ever before.*' "

He rose again, this time to accompany me to the door. It was dark over Riyadh now, and except for the glow of the floodlit water tower, there was no light to be seen. It was very silent.

"I hope they receive the message properly," said Ahmed Zaki Yamani. "But I am afraid that sometimes, when oilmen think only from a money angle, they get blind. Salaam aleikum."

ON JANUARY 7 I flew from Jidda across the dun-colored wastes of the Empty Quarter to Dhahran on the Persian Gulf. It was cloudy and stormy most of the way, and for the last half-hour of the journey, from Al Hasa Oasis to the coast, the dunes and crags of the desert were crisscrossed with pipelines and smudged with the smoke and flame of gas flares. In 1971 alone, those pipelines had carried 1.641,615 billion barrels of Aramco oil to the tank farms of Ras Tanura, the loading port on the Gulf. When the oil that was pumped by the Japanese-owned Arabian Oil Company (65.348 million barrels) and by the wells of the Getty Oil Company (33.762 million barrels) is added to this, it brings Saudi Arabia's output for 1971 to 1.741 billion barrels, or 4.769,-863 million barrels a day, making the country the greatest producer of oil in the Middle East, well ahead of its two nearest rivals, Iran and Kuwait. Out in the Gulf, like great whales breasting the cresting waves, tankers were lined up for the loading points; there they would take on 300,000 tons of crude oil in thirty hours, and sail off to the markets at the rate of one ship every forty-five minutes.

Dhahran looks like a small residential town in a Southern California valley—full of neat, air-conditioned bungalows with green lawns running down to the sidewalks, street lamps out of a Pasadena side street, and oleander hedges. American, British, Saudi and Arabian officials and their families live side by side, and get together for evening parties. There are schools, hospitals, clubs, playing fields and swimming pools, and a cosmopolitan commissary where the choices run from hamburgers to curry and marzipan cakes to pecan pie. Schools, education and medical attention are free, and any Aramco employee can get an interest-free loan to build a house, after he has been with the organization for a year.

A Saudi Arabian official of Aramco (trained at Syracuse, passionately anti-Nixon but fervently pro-company) showed me around the College of Petroleum which, with generous help from Aramco of nearly $15 million, has become the best training school for oil technology in the Middle East. The grounds lie in the shadow of the old derrick of Well No. 7, on a hill now known officially as Jabal Steineke, in memory of the U.S. geologist who first drilled oil there and founded Aramco's fortunes. Statistics about Aramco rolled off the Saudi official's lips:

> Of the company's 10,353 employees, 8,630 are Saudi Arabian and only 883 are American;
>
> nearly 7,000 of the Saudis have been working for Aramco for more than fifteen years;
>
> the average salary of the Saudis in the company is $3,900 a year;*
>
> 7,578 Saudis have built their houses through Aramco's home-loan program;
>
> at company expense, 200 Saudis are sent every two years to the United States for postgraduate courses;
>
> the forty-four schools the Saudis' children attend have cost the company $35.4 million.

"We pump more oil than any other company in the Middle East," said the official, "and market it more efficiently than any other company. We pay more to our government than any company in any other country,† and we are Saudi Arabian, registered and incorporated inside the state."

* Ahmed Zaki Yamani told me of an official survey made by an independent investigation team of the priorities of Aramco's Saudi employees. "First, they wanted the company to give them more education. Second, they wanted the company to give their children a proper education. Third, they wanted raises in salaries. They are the only workers in the Middle East oil industry who put education before more money."

† $1.866 million in 1971.

How does one tell a company which has "Saudized" itself so thoroughly, and runs so efficiently, that henceforth it must cede 20 percent of its share—and eventually more—to the state? That it must hand over not only a fifth of its action inside the Saudi state, but all the way "downstream" to the gas stations on the freeways of the world?

"Ahmed Zaki Yamani can call it anything he likes, participation or whatever," said an American official of the company who prefers to remain anonymous, "but it still sounds like nationalization to us. Twenty percent nationalization to begin with, and then a gradual step-up of the takeover, year by year, until they have control of the whole operation. It just isn't in the cards. Even if Aramco accepted the proposition, can you see our parent companies agreeing? Can you see Jersey, Socal, Texaco or Mobil—especially Mobil—letting them in on their operations?"

But what Yamani and other participants were envisaging was an arrangement more complicated but somewhat less drastic than this American official was anticipating. The Saudi oil minister had put it in these words:

> "You see, we do not want to get the oil merely for its own sake. We do not want just to obtain, say, 20 percent of Aramco's production, then 30 percent another year, and so on. What would we do with all that oil? If we went and dumped it on the market, we would be committing suicide. So we have to have an arrangement with the Aramco owners, jointly or severally, whereby:
>
> "1. Part of our entitlement of oil production would be marketed jointly with them through joint ventures downstream; and this is a business matter.
>
> "2. Another part would be marketed by them on our behalf on a commission basis.
>
> "3. A third part would be sold by us acting alone through the various ways and means at our disposal.
>
> "So I think all these elements should go together if we are to have a healthy, stable situation."*

The American shook his head when this point was made to him. "No matter how you slice it," he said, "it's still nationaliza-

* Paper on the Economics of the Petroleum Industry, *op. cit.*

tion, and that's something we don't believe in. We're Americans, remember? Free enterprise, and all that. We'll never accept a government, even the American government, as our partner."

In that case, it appeared that Aramco and the other majors were in for a bumpy ride when the Geneva negotiations on participation started in a few days, for the oil ministers, with Yamani as their spokesman, were determined not to take no for an answer.

THE NIGHT BEFORE I FLEW OUT OF ARABIA, a senior Saudi official of Aramco asked me and my wife to a party at his house in Dhahran. Gailan al Bassani had spent his early childhood wandering the fringes of the Empty Quarter with his father's Bedouin tribe, and he is so steeped in desert lore that those of his American fellow workers whose favorite weekend pastime is picnicking and exploring the secrets of the vast Arabian sand sea are delighted when he consents to come along. He holds degrees in engineering and geology from U.S. universities, speaks fluent English and French, and is regarded as one of the most promising members of the company's executive staff.

Bassani had invited a cross section of Aramco's senior staff to meet us: a young American vice-president of Tapline; an exuberant Bahraini who works for Bapco, the subsidiary of two of Aramco's parent companies, in the refineries on Bahrain Island; a Saudi geologist and his English wife; and a sprinkling of American, Lebanese and European secretaries. In the middle of dinner the president of Aramco, Frank Jungers, arrived with his wife, and what had started as a social evening almost immediately became more businesslike, for Jungers wanted to hear in detail what Ahmed Zaki Yamani had told me about OPEC's attitude toward participation.

It seemed surprising that evidently there was such poor liaison between Aramco and the Saudi government that the gravity of the participation question had to be conveyed to the Americans through a visiting writer. Jungers seemed genuinely startled when I mentioned the determination of Yamani and his OPEC colleagues to push the participation question to the limit, and that in the event of a rejection by the major oil companies, the oil-producing countries would legislate. But he hastened to reply that Aramco was by no means as intransigent as some of his junior officials had suggested. He was flying to Geneva for the meeting with OPEC representatives the following day, he said, and "I can tell you that we have no intention of turning them down. You will find that we will be going a long way toward meeting them."

But not, it turned out, in the direction that the oil states wanted to go. Not at that time, anyway.

THE GREATEST WEAKNESSES of the Middle East oil states have always been their inability to stand together in an emergency, and their susceptibility to bribery and corruption. Even the formation of OPEC has not eliminated the tendency of member states to revert to wily Oriental tactics when a quick and immediate profit presents itself. This was the case, for instance, after the Six-Day War, in 1967, when the Arab oil states voted to boycott certain markets, and their fellow member, Iran, immediately stepped up production to meet the shortage, to its considerable profit. Throughout the history of Middle East oil, one state or another has been played off against its rivals by the skillful manipulations of the majors.

But throughout the participation negotiations, which went on all through 1972, the oil states stood firm in the face of considerable temptation. Ironically, considering the posturing of the militant Arab states whenever oil talks are in progress, it was the attitude of two of the most autocratic of the old regimes which kept the oil states united, and it was King Faisal of Saudi Arabia and the shah of Iran who propelled the negotiations toward a successful conclusion, even if it took time to achieve it.

However, the expected confrontation failed to take place in

Geneva, for at the last moment Ahmed Zaki Yamani skillfully postponed it. The delegates from OPEC and the representatives of the majors had gathered in Switzerland for two main purposes: to decide at what rate the oil states should be compensated for the revenue they had lost through the devaluation of the U.S. dollar in 1971; and to make a decision about participation. The dollar question was efficiently dealt with; the majors, having gone to Geneva expecting to pay more, quickly agreed to terms.

But the resigned amiability which they displayed over the money question—they would, after all, simply pass the increases over to their consumers—changed to outright hostility when the participation question reached the agenda. The Americans, particularly the four companies owning Aramco (Standard of California, Standard of New Jersey, Texaco and Mobil), made it clear to their more dovish British, Dutch and French colleagues that they were determined to take a hard line. It was a situation made for the troublemakers among OPEC's more militant members, who could easily exploit American intransigence and force a public showdown from which, for political reasons, it might be impossible for the oil states to retreat. With the help of OPEC's secretary-general, a flexible Iraqi named Nadim Pachachi, and the lobbying of the Iranian delegate, Mansoor Froozan, Yamani maneuvered the member states into agreeing to let him conduct the negotiations on his own. Simultaneously, the British, Dutch and French agreed to allow the four stubborn U.S. companies to represent them in the talks with Yamani. Thus it had become a straight Saudi-Aramco confrontation. Yamani set the date for the talks between the two as February 1, 1972, and as a location chose Jidda, in Saudi Arabia, where close government control was kept over the press and no details of the negotiations could leak out except what he wanted to appear.

It was at this stage that the U.S. companies, using Frank Jungers of Aramco as their front man, decided to try a last trick to break the solid OPEC line-up on participation.

JIDDA is the most tawdry of Saudi Arabia's cities. Its old wooden houses with jutting latticed balconies are jostled by tall office buildings which squat beside them on the shores of the Red Sea. February is the month when the pilgrims to the great *hajj* stream back into the port from Mecca, forty miles up the road, and the bazaars along the waterfront are thick with Muslims from the far corners of the earth, buying up souvenirs before taking their chartered planes and ships back home. One of the small pleasures of attending the Aramco-Saudi confrontation was to see the preoccupied, businesslike expressions on the faces of the American oilmen soften slightly as they pushed their way to the morning conferences through merry hordes of giggling, frizzy-haired girls from Nigeria, fuzzy-wuzzies from the Sudan, bare-breasted old ladies from Abidjan, and smiling women in saris from Bali, the Philippines, Malaysia and India. Here were human beings at peace with themselves, happy and relaxed, and even momentary contact with them was a relief from the tense business at hand.

For the Americans had decided on a bluff, and they were nervous about whether they could get away with it. What was the best way to tempt Saudi Arabia to break the solid ranks against them? The American delegates had planned and argued long into the warm nights in search of an answer, and eventually they had come up with a proposition which they believed would get them off the hook. At their meeting with Ahmed Zaki Yamani, on February 15, the four companies offered him not 20 percent but 50 percent in future participation projects with them. The only snag was that the offer was not for Aramco's present functioning operations, but for developing and operating certain areas of their concession where the presence of oil had been proved but not yet developed. Even so, for the Saudis it must have been a tempting offer, for Aramco was sitting on vast reservoirs of untapped oil which only its enormously rich organization had the capital to develop. Moreover, the U.S. delegates made it plain to Yamani that if he accepted the offer on behalf of Saudi Arabia, they would make a similar approach to Iran, where the Anglo-American consortium also controlled great stretches of proved but unexploited reserves. After all, had not Yamani himself approached Aramco on two separate occasions with proposals for 50–50 participation deals in Saudi Arabia? But on those occasions he was not representing all the OPEC states; nor was there any danger, if they had gone through, that participation deals in other oil states would be compromised. The fact was that with the exception of Iran, none of the other states had further viable areas which could be exploited under a 50–50 deal; their only hope for participation lay in the OPEC demand for a 20 percent share of *existing* operations, and acceptance of this new Aramco offer would sabotage that.

So Yamani rejected the American offer. "It may be reckoned a tribute to OPEC solidarity that even the potential beneficiaries rejected the proposal out of hand," the *Middle East Economic Survey* reported.

The moment when a boxer throws his master punch and misses, he is at his most vulnerable, and so it was with the Americans. On the evening of February 15, a few hours after he had rejected Aramco's offer out of hand, Ahmed Zaki Yamani drove out of the old town of Jidda to the modest palace King Faisal has built for himself along the flat, humid shores of the Red Sea. Iran already knew of his rejection of the majors' offer, and the shah had let it be known that he approved. Now Yamani decided that this was the moment for him to throw a master punch of his own. He had a long meeting with the Saudi monarch and came away well satisfied with the results of his audience. Once before (as Howard Page of Standard of New Jersey had cause to remember), Yamani had used King Faisal to back up his tactics; now he was about to play the king again.

That night the U.S. delegates were informed that they would be expected to attend an extraordinary session the following morning, and when they were assembled, the Saudi oil minister wasted few words. He reminded them that the purpose of the conference in Jidda was to find a solution to the crisis between the oil states and the companies over participation, and that no progress had been made. But the deadlock must be broken, and to emphasize the urgency of the situation, he read a message to the delegates from His Majesty the King:

"Gentlemen: The implementation of effective participation is imperative, and we expect the companies to co-operate with us with a view to reaching a satisfactory settlement. They should not oblige us to take measures in order to put into effect the implementation of participation."*

It was another way of saying what Yamani had warned six weeks earlier: "Give in, or we will legislate." Only this time the ultimatum was given by the king himself, and his prestige, as well as that of his government, was now involved.

Nevertheless, it took another three weeks for the Americans to come around, and Frank Jungers had to use his most persuasive arguments to convince some of the more recalcitrant of his parent companies that it was the end of the road. But after a succession of emergency conferences in Beirut, New York, San Francisco and Houston, Ahmed Zaki Yamani announced on March 11, 1972:

"Last night I received a letter from Aramco indicating the acceptance by Aramco and its shareholders of the principle of participation to the extent of a minimum of 20 percent. The details of this question will be subject to further negotiations which will start in a week's time in Riyadh, Saudi Arabia. The OPEC extraordinary conference will be informed today of this step taken by Aramco and its shareholders."†

It was done. Aramco had accepted and the rest of the companies would follow. Admittedly, it was only the principle of participation which had been accepted so far, but everyone knew that there could be no turning back. One anonymous oil expert

* Quoted by the official Saudi News Agency (SNA) from Jidda, February 18, 1972.
† Quoted by *Middle East Economic Survey* in issue dated March 10, 1972.

in the Middle East commented: "One should not, of course, underestimate the formidable problems that still remain to be settled—notably the price to be paid for the governments' 20 percent share * . . . Nevertheless, basically the Rubicon has been crossed, and it can safely be affirmed that the final movement towards host-government participation in the world's largest concentration of low-cost oil reserves has been well and truly launched." †

Things would never be the same again in the Middle East oil business, and in the eyes of some veteran American and British oilmen, it was the beginning of the end. But it was also "a well-deserved personal triumph for Shaikh Ahmed Zaki Yamani, who for many years has been the foremost champion of the idea of participation both in Saudi Arabia and in the general framework of OPEC." ‡

Now he had to face the next stage: how much to pay the majors in compensation for letting the host countries take over an increasing share of the action. Billions of dollars were involved, and the arguments about compensation—whether to pay merely the book value of the concessions, which is what the host countries were offering, or the potential value of the oil in the ground, which is what the companies were asking—and such complicated matters as "buy-back" prices would go on all through the summer and fall of 1972.§ But despite the vast sums involved, it was something of an anticlimax compared with the principle accepted by the majors on March 10, 1972.

* It eventually turned out to be 25 percent.

† *Ibid.*

‡ *Ibid.*

§ An agreement on compensation was finally reached on October 5, 1972, but details were not immediately revealed. An agreement on "buy-back" prices was reached five weeks later.

CHAPTER THIRTY

The Bottom of the Barrel?

SO WE REACH THE PRESENT DAY in the history of oil in the Middle East. What happens next? It is not likely that the successors of Frank Holmes, Knox D'Arcy, Gulbenkian, Getty, Max Steineke and Enrico Mattei will be as flamboyant as their predecessors, but the events in which they find themselves involved will almost certainly be as spectacular, hazardous and unexpected. The Middle East oil industry is changing fast, and no matter what happens, the changes are not going to make life comfortable for the Western oilmen who run the concessions and have invested billions of dollars in their future.

Once upon a time the United States used to exploit the oil wells of Arabia for the money to be made out of them; nowadays it has become more interested in the oil itself. "In 1970, the United States consumed 710 million tons (30 percent of world consumption), or 15 million barrels a day," says a recent U.S. report. "By 1980 it will demand between 20 and 25 million barrels a day, but U.S. production, now around 10 million, will rise to only 11 million from presently known reserves. Alaska could provide 2,000,000 barrels a day if the currently planned (and hotly contested) pipeline is built; newly found offshore deposits could swell output if environmental objections could be met. Even so, industry sources predict that there will be a gap." *

Until new sources of energy have been found, this gap can be filled only by buying more oil, principally from the Middle East, and the price there is going up, both monetarily and politically. To fill America's needs, Ahmed Zaki Yamani of Saudi Arabia has proposed a preferential oil agreement between the two countries whereby priority would be given to America's fuel requirements over all other customers in return for an "open door" for Saudi investment of its oil profits in the United States. This might well solve America's fuel problem, though at the price of allowing Arab money (and with it, Arab influence) to flood into American trade and industry. On the other hand, it would create grave problems for the rest of the free world. Today most oil from Arab and Iranian fields goes to Western Europe or Asia.† At a moment when several Arab states—Libya and Kuwait in particular—are deliberately cutting their production to hold on to diminishing reserves, oilmen fear that the incursion of the United States on the Middle East market as a *consumer* can only drive up the price which everyone will have to pay. "Our rates of consumption are so large that we can see the bottom of the barrel," says David Freeman, a former energy expert for President Nixon.‡

Admittedly, there are other experts in the Middle East who would not agree that the oil barrel is liable to be scraped in the immediate future. Kuwait and Libya may have fewer reserves than they believed a year or two ago, and Western oilmen hope that the knowledge may somewhat chasten the arrogant regimes in those countries. But the potential reserves in both Iran and Saudi Arabia grow in size each time new areas are surveyed. Iran

* Quoted in *Time,* June 12, 1972.

† Exports of crude oil from Saudi Arabia in 1971, for instance, were as follows: 48 percent to Western Europe, 37 percent to Asia, Africa and Australia, 10 percent to the Western hemisphere, and 5 percent elsewhere.

‡ *Time, op. cit.*

alone, which produced 52 million tons of crude oil in 1960, 191.7 million tons in 1970 and 227 million tons in 1971, is planning to increase production to 399 million tons in the next ten years. Saudi Arabia has even greater expansion plans. Ahmed Zaki Yamani announced in New York in October 1972 that his country plans to drive production up by 1980 to an astounding 20 million barrels a day, or 1 billion tons a year.

What most worries Western strategists when they contemplate the part the Middle East is destined to play in the world fuel situation is neither the expense nor the shortage of oil, but the increasing tendency of Arab militants to attempt to use oil as a political weapon. As long as there are factories to be fueled, automobiles to be lubricated and houses to be heated, the West will go on buying oil, no matter how high OPEC forces up the price. Petroleum has become as much the drug of Western civilization as cigarettes and alcohol, and the addicts will go on paying. But are their governments prepared to pay the political price which may one day be demanded of them?

It was not simply antipathy toward the Israelis that persuaded General de Gaulle to swing France's policy behind the Arabs in the last years of his regime, and has impelled his successor, President Georges Pompidou, to curry Arab favors. Both leaders were prepared to toe the Arab line in the hope that one day it would bring dividends in the form of oil concessions, and in the summer of 1972 this policy paid off. When Iraq nationalized IPC's holdings in June 1972, the confiscation was directed at the American, British and Dutch partners in the combine. The French partner, Compagnie Française des Pétroles, was singled out for special exemption, and invited to discuss ways and means of taking over from its erstwhile confreres. France was being rewarded with a "privileged position," said Saddam Hussein, vice-president of the Iraqi Revolutionary Command Council, because of the "just policy that France follows concerning Arab causes, specifically the Palestinian cause against Israel." He added that any country which adopted a similar attitude "will be able to profit from a privileged situation, not only in Iraq but also in all the Arab countries." *

Hussein's words must have aroused dismay among Western oilmen throughout the Middle East, and it echoed what one of many U.S. oil executives had said during interviews, "Why do we have to keep backing Israel, anyway? What does it get us, except the Jewish vote in New York? Why don't we behave like France and subordinate our Middle East policy to our self-interest? A pro-Arab attitude in the State Department would make all the difference."

There is no doubt that a change in policy could buy time for the majors and stave off the day when complete control of their fields is taken over by national Kuwaiti, Saudi Arabian, Abu Dhabian, and Iraqi oil companies. But oil executives in Britain and the United States, Foreign Office strategists, State Department planners, and U.S. senators and congressmen are almost certainly fooling themselves if they believe that a pro-Arab—and its necessary concomitant, an anti-Israel—policy will help them more than temporarily.

Iran, which demands no anti-Israel or pro-Arab line from its foreign oil concessionaires, is nonetheless determined to take over the Anglo-American consortium and run the whole enterprise herself within a foreseeable period of years, for no other reason than that it is now run by foreigners and the Iranians want to run it themselves. In March 1972, the shah of Iran gave an interview to a French oil journal in which he was asked: "Do you intend eventually to take your oil affairs into your own hands and sell your oil products directly?"

"There is no doubt that is our final goal," he replied. "We are currently studying this question, notably with our neighbors and friends in Kuwait and Saudi Arabia. As far as we Iranians are concerned, we are too sophisticated to act purely and simply as nationalists. Twenty years ago, we suffered a very serious reverse.

* Statement made at a press conference given in Paris, June 15, 1972, following a meeting between Hussein and President Pompidou.

It was a hard and lasting lesson. We still have abundant oil for a good fifty years, while other countries see their reserves drying up. We are also trying to set up long-term plans. Already we are thinking of 1979, the date of the end of the first period of the [Anglo-American] oil agreement. Obviously, we could accept another three further five-year periods [as set out in the agreement], but that would be the easy way out." *

The shah has made it clear that though he approved of Ahmed Zaki Yamani's campaign for participation in the operations of the majors, such an arrangement is of no direct concern to Iran, which secured a participation arrangement (albeit an unsatisfactory one) from the consortium under the agreement of 1954. He regards participation as no more than a temporary expedient to make use of the foreign companies and their facilities while building up a technical and sales staff, and marketing expertise. "I am thinking of the forty million Iranians who will be alive in the year 2000," the shah went on. "That is why I say that the ideal final solution is for us to take our oil into our own hands, by-passing the intermediaries [the majors] to an ever and greater degree. Today this might come as a shock, but I do not think that when it comes the producers and the hundreds of millions of consumers will complain about it." †

If the ruler of a land with close political and economic ties with the United States and Western Europe is nevertheless preparing eventually to nationalize the concessions in his country, is it likely that similar plans can be staved off in other Middle East oil states, even by the most servile policy toward the Arabs?

Experts would probably reply by pointing out that the Anglo-American concession in Iran expires in 1979, whereas those in Kuwait, Saudi Arabia and the Gulf emirates run to the end of the century or even longer. To Western oilmen for whom a contract is sacred, that may be a comforting fact, but there are few Arab governments today which consider themselves bound by

* Quoted from *Bulletin de l'Industrie Pétrolière,* Paris (March 20, 1972).
† *Ibid.*

the concession agreements they signed in the years before World War II, and if they hold to them for the moment it is only because there are not enough technicians and marketing experts. Secretary-General Nadim Pachachi expressed the view of most members of OPEC when he said:

"As you are aware, the basis of a valid contract is the free will of the two parties. Most of the oil concessions granted in the Middle East before and after World War Two were concluded by states under the mandate or influence of a colonial power. When the IPC agreement was concluded in 1925, Iraq was under British mandate. It is a well-known fact that the frontier commission appointed by the League of Nations to settle the dispute between Iraq and Turkey over the vilayet [province] of Mosul refused to give a decision in favor of Iraq before an oil concession was concluded with the IPC. The British government would not allow the ratification of the new constitution before the oil concession was granted."*

Therefore Pachachi maintains there is nothing to prevent a state from repudiating a concession or altering it whenever there is a "strong and valid reason for the revision." But what constitutes "a strong and valid reason"? Muammar Gaddafi nationalized British Petroleum's concession in Libya because Britain failed to prevent Iran from seizing the offshore islands in the Persian Gulf. Iraq nationalized IPC's northern holdings because, she claimed, the company was refusing to increase production.

Yet curiously enough, to date not a single Middle East oil state has nationalized any U.S. oil company, despite the fact that American policy in support of Israel continues to bedevil relations between the Arab world and Washington. If British inaction in the Persian Gulf is sufficient reason for seizing BP's assets in Libya, why hasn't Gaddafi seized Esso's even bigger concession in retaliation for the sale of U.S. Phantoms to Israel? Could it be that Gaddafi thought he had enough technicians to handle BP's operations, but lacks the experts to handle Esso as well? Or sim-

* Quoted from a lecture delivered by Dr. Pachachi on "The Role of OPEC in the Emergence of New Patterns in Government-Company Relationships" at the Royal Institute of International Affairs, Chatham House, London, on May 19, 1972.

ply that he can gain easy propaganda returns from twisting the tail of a comparatively helpless British lion, but finds the United States too big an opponent, too useful as a market and too good a provider of technical expertise to wish to take that step?

There are students of the Middle East oil situation who believe that it would be better for everyone, particularly the consumers, if the OPEC states did nationalize all oil operations inside their countries, and sooner rather than later. James Akin, director of the Office of Fuels and Energy in the U.S. State Department, discussed the implications of such a step in a speech he made on June 2, 1972, to the Eighth Arab Petroleum Congress in Algiers. He made the point that nationalization might not necessarily be a disaster for the United States:

"Investment in production is about one-third of the total of [U.S.] investment in foreign oil; the remainder is in tankers, refineries and distribution. If the entire investment in production were lost—everything in Canada, Venezuela, Australia, non-OPEC Africa along with the Middle East—the loss to [our] balance of payments . . . would be only about $700 million."

What would the Arabs gain?

"The Arab income from oil in 1970 was $4.8 billion. Based on increases in taxes already agreed by OPEC and the companies, and on published figures for increases in production, this will rise to over $12 billion by 1975. And by 1980, assuming further price increases, the annual income has been estimated at somewhere between $30 billion and $50 billion. I should venture that if production of all the major companies in the Arab world were nationalized, the additional benefit to the Arabs would be well under $1 billion a year. And if the consumers started crash programs to develop other energy forms, it could be substantially less."

And what would the consumer lose? He might even gain, Akin pointed out, and mentioned those

". . . who argue that nationalization should be encouraged. Their argument is that once the oil companies' buffer is removed, OPEC's solid front will disappear and the national companies will compete among each other for a greater share of the markets, and prices will go down to one dollar a barrel in the Gulf."*

To the major companies themselves, of course, Akin's remarks bring no comfort, and they view the prospect of the nationalization of their enterprises not only as an unmitigated disaster but as a fate they do not deserve. They cite the enormous capital investment they have made (and are planning to make) in the Middle East, and insist that no compensation offered by a confiscating government could possibly match the effort and expenditure they have put into their concessions.

Indeed their effort has been vast and unremitting, as these pages have shown. But it has also been rewarding. The majors have recovered their investments many times over, and they continue to make magnificent profits. The U.S. Department of Commerce, for instance, reporting on direct U.S investments in 1970, pointed out that the net assets of the petroleum industry in the Middle East amounted to $1.466 billion. The yield from those assets in 1970 alone was $1.161 billion, representing an annual return of 79.2 percent. In contrast, U.S. investments in overseas mining and smelting industries yielded 13.5 percent in 1970, and in the same year the yield from overseas manufacturing companies was 10.2 percent.

Nationalization would get the United States and the West off the hook politically in the Middle East. It would depoliticize petroleum and restore it as an article of merchandise to be sold at the going price, and it would remove the oil fields from the Arab-Israel arena. Moreover, any militant who thereupon suggested that the oil states controlling their own oil should immediately begin a boycott of petroleum to those Western nations pursuing "anti-Arab" policies would crash with a considerable impact against the facts of commercial life. In business on their own, unbolstered by the majors' posted-price system, no oil state would dare to initiate a boycott against any consumer nation

* Quoted in the *Middle East Economic Survey,* June 16, 1972.

with the money to buy. In all of the Middle East there are not enough storage tanks to hold the petroleum that would immediately become surplus if such a boycott were attempted. The deserts would start to run with crude, the gas stoves of Kuwait and eastern Arabia would sputter and go out, the desalinization tanks would cease to function and the water tanks would run dry. And those militants advocating a massive cutback in production in order to create the "seller's" market that a political boycott would necessitate would then be faced with one of the harsh facts about oil fields: you can diminish the flow from the wells or shut them down completely, but if you want to keep them healthy you can close them down for only a limited time. A well that is capped for too long may not only be impossible to reactivate; it may also play havoc with the rest of the field.

Not even the pleasure of seeing the United States squirm under a boycott is worth the destruction of a billion-dollar oil field.

CERTAINLY NATIONALIZATION of the majors' concessions in the OPEC states would effectively destroy the international oil cartel. It would also bring to an end the artificial price structure for petroleum which between them the cartel and OPEC countries have created. The companies would become "off-takers" searching the world for the cheapest oil they could buy for their markets; the OPEC nations would lost the buffer organization which sells their product for them and maintains the artificial price. Only the consumers would gain.

It is for this reason that, despite Gaddafi, despite the Iraqis, despite the militants' cry of "Arab oil for the Arabs," the international cartel is not likely to be put to death yet by the oil states. As long as it has a role to fulfill, the cartel will operate and the majors will continue to make their heady profits. In forcing them to accept participation, Yamani has saved them from extinction. One wonders why they ever objected to it, for as Maurice Chevalier once said of those who grumble over old age, "Consider the nature of the alternative."

To be sure, in the days to come, the great Anglo-American combines will find themselves handing over larger and larger shares to their hosts, and paying bigger royalties and higher taxes. But judging from their past performance, they will simply pass the increases on to the consumer. Perhaps at the end of the present decade, or at the end of this century, power will have passed out of their hands into those of the producing countries. Then Gulf, Esso, BP, Shell and Mobil will still be names in neon lights along the world's highways, but their names will have disappeared from the wellheads in favor of NIOC, INOC, Petromin, KNOC, and ADNOC.* The directors of the ousted companies will bewail the lack of gratitude of the governments in whose deserts they pioneered and sweated, and probably their shareholders will grumble at the smallness of the compensation. But those who have read this story will no doubt come to the conclusion that, all things considered, they have had a pretty good run for our money.

* National Iranian Oil Company, Iraq National Oil Company, Saudi Arabian Oil Company, Kuwait National Oil Company, Abu Dhabi National Oil Company.

APPENDIX

PRECEDING PAGE: *A well in the Zakum field offshore Abu Dhabi.* PHOTO: BP COMPANY LTD.

APPENDIX

PRECEDING PAGE: *A well in the Zakum field offshore Abu Dhabi.* PHOTO: BP COMPANY LTD.

The Oil States of the Middle East: Some Facts and Figures

COUNTRY:	**Saudi Arabia**
LAND AREA:	1,350,000 square miles
POPULATION:	7,100,000 (est. 1968; a census is being planned)
OIL OUTPUT:	5.6 million barrels daily (January 1972 *)
OIL INCOME:	$2.571 billion (est. for 1972–73)
RULER:	King Faisal ibn Abdul Aziz
CAPITALS:	Riyadh and Mecca
GOVERNMENT:	Authoritarian and theocratic

OIL COMPANIES: *Arabian American Oil Company* (*Aramco*), onshore, offshore

(Owned by Standard Oil of California 30%; Standard Oil of New Jersey 30%; Texaco Inc. 30%; Mobil Oil Co. 10%.) Aramco produces 90% of Saudi Arabia's oil. Since January 1973 the Saudi government has a 25% "participation" in the operation, which will increase to 51% by 1980.

Saudi Arabian Oil Co. (*Saoc*), onshore

(Owned by AGIP of Italy 50%, Phillips Petroleum Co. 50%, Petromin, Saudi state company, which assigned the concession to the first two, gets profits from half-share participation.)

* Includes approximately 300,000 b/d (half-share with Kuwait) from the Neutral Zone.

Sun-Natomas, onshore-offshore, Red Sea

(Owned by Sun Oil Co. 60%; Natomas Co. 30%; Oil and Gas Development Co., Pakistan, 10%; and Petromin under same terms as above.)

Auxerap, onshore-offshore, Red Sea

(Owned by French government company, Erap, 66⅔%; Tenneco Oil Co. 33⅓%; and Petromin.)

Aminoil-Getty, operating in the Neutral Zone shared by Saudi Arabia and Kuwait

(Saudi Arabian half owned by Getty Oil Co.)

Arabian Oil Co. Ltd., Neutral Zone

(Owned by assorted Japanese consortium under the name of Japan Petroleum Trading Co. Ltd. 80%; Kuwaiti state 10%; Saudi state 10%.)

OIL RESERVES: 88.063 billion (est. 1970, since which time important discoveries have been made).

During a speech in Washington at the beginning of October 1972, Oil Minister Ahmed Zaki Yamani revealed that Saudi Arabia plans to raise its crude-oil production from the present 5.6 million barrels daily to what the *Middle East Economic Survey* called the "astounding" figure of 20 million barrels daily by 1980. He implied that the expansion would be carried out in association with U.S. interests.

POLITICAL PROSPECTS: Stable as long as King Faisal manages to stay in control (there have been two abortive moves to unseat him). But he is a sick man and his death could act as a signal for the unleashing of forces long resentful of the rigid, puritanical, authoritarian regime.

A revolutionary regime would almost certainly take over Aramco's vast holdings in the country. But a more moderate regime would be unlikely to upset the present development of "participation," in view of Aramco's increasing efforts to "Saudize" its operations and its ability to call up from the United States the vast financial reserves which will be needed if the ambitious plans for the development of newly discovered oil fields are to be carried through.

Some Facts and Figures

COUNTRY:	**Iran**
LAND AREA:	634,000 square miles
POPULATION:	20,149,000
OIL OUTPUT:	4.738 million barrels a day (January 1972)
OIL INCOME:	$2.260 billion (est. 1972–73)
RULER:	Mohammed Reza Shah
CAPITAL:	Teheran
GOVERNMENT:	Royalist and authoritarian

OIL COMPANIES: *Iranian Oil Exploration and Producing Co. (IOEPC)*, otherwise known as the Consortium. Onshore

(Owned by Iranian Oil Participants Ltd., whose member companies are: British Petroleum 40%; Netherlands Iranian Oil Trading Co., a wholly owned subsidiary of Shell, 14%; Mobil Oil Co. 7%; Standard Oil of California 7%; Standard Oil of New Jersey 7%; Texaco 7%; Compagnie Française des Pétroles 6%; and the remaining 5% split between six different U.S. "independents" in different proportions.) The Consortium produces 80% of Iranian oil.

Arepi, onshore-offshore

(Owned by Erap-Elf of France 32%; ENI of Italy 28%; Hispanoil of Spain 20%; Petrofina of Belgium 15%; OMV of Austria 5%.)

Continental Oil of Iran, onshore

(Owned 100% by Conoco of U.S.)

Elf-Iran, onshore-offshore

(Owned by Erap-Elf of France 80% and Aquitaine-Iran of France, in which Erap-Elf has a controlling interest, 20%.)

Farsi Petroleum Co. (FPC), offshore

(Owned by Société Française des Pétroles d'Iran 50% and National Iranian Oil Co. 50%.)

Iran-Pakistani Oil Co. (IPOC), offshore

(Owned by NIOC 50% and Pakistani State 50%.)

Iran–Pan American Oil Co. (IPAC), offshore

(Owned by Amoco, a wholly owned subsidiary of Standard Oil Company of Indiana, 50%; and NIOC of Iran 50%.)

Iranian Marine International Oil Company (IMINOCO), offshore

(Owned by AGIP of Italy 16⅔%; Hydrocarbons India Ltd. of India 16⅔%; Phillips Petroleum Co. 16⅔%; and NIOC 50%.)

Iranian Offshore Petroleum Co. (IROPCO)

(Owned by International Oil Co. of Spain 11¾%; Atlantic Richfield Co. of U.S. 6¾%; Cities Service Co. of U.S. 6¾%; Kerr-McGee Corp. of U.S. 6¾%; Skelley Oil Co., in which Getty Oil Co. holds controlling interest, 6¾%; Sun Oil Co. 6¾%; Superior Oil Co. of U.S. 6¾%; and NIOC 50%.)

Lavan Petroleum Co. (LAPCO), offshore

(Owned by Atlantic Richfield Co. 12.5%; Murphy Oil Co. of U.S. 12.5%; Sun Oil Co. 12.5%; Union Oil Co. of California 12.5%; and NIOC 50%.)

National Iranian Oil Co. (NIOC), onshore-offshore

(Owned 100% by the Iranian State.)

Persian Gulf Petroleum Co. (Pegupco), offshore

(Owned by Texaco 10%; Gelsenberg A.G. of Germany 10%; Wintershall Aktiengesellschaft of Germany 10%; Pressag A.G., owned by the West German State 6%; Veba-Chemie A.G. of Germany 6%; Gewerkschaft Elwerath, a German subsidiary of Shell and Standard Oil of New Jersey, 5%; Deutsche Schachbau- unter Tiefbohrgesellschaft m.b.H. of Germany 3%; and NIOC 50%.)

Société Irano-Italienne des Pétroles (SIRIP), onshore-offshore

(Owned by AGIP of Italy 50% and NIOC 50%.)

OIL RESERVES: 80 billion barrels.

The shah of Iran maintains that his country has ample oil reserves for the next fifty years, sufficient to justify an annual increase in production of 15 percent until a figure of 8 million barrels a day is reached.

POLITICAL PROSPECTS: The shah's regime has many enemies, notably among students, intellectuals, and Iranian émigrés. His control over the armed forces, and his popularity with them, is such that a coup d'état is unlikely, but he is a target for assassination, and a successful one could bring chaos to Iran. However, his regime is a shrewd mixture of superficial tolerance toward the young (miniskirts, discotheques, student forums) and a ruthlessness toward any serious opposition. Of foreign reports of anti-imperialist guerrillas operating in Iran he says dismissively: "They could be dealt with by the assistant cooks of the Imperial Iranian Army." He has ambitions to plug the gap left by the departure of the British from the Persian Gulf, with Iranian military and political influence, and is strengthening his armed forces to that end. They could probably lick any other forces in the area, including those of his bitter rivals, the Iraqis, but would be no match for a major power.

The shah's attitude toward foreign oil companies is to promise them secure access to oil supplies in return for a steadily rising price for the oil they extract. The Consortium's concession ends in 1979, and in January 1973 the shah offered the member companies two alternatives: 1) An "as-is" situation until the expiration of the concession, after which the Consortium would have to join "the long queue" to buy Iranian oil with no preferential treatment; or 2) a new long-term agreement now, which would hand over complete control to Iran immediately—but with secure long-term purchase contracts for the next 20–25 years.

COUNTRY: **Kuwait**

LAND AREA: 5,800 square miles (Neutral Zone: 2,500 square miles)

POPULATION: 733,196 (1970), of which 387,298 are non-Kuwaiti

OIL OUTPUT: 3.787 million barrels a day (including approximately 300,000 b/d from the Neutral Zone)

OIL INCOME: $1.63 billion (est. 1972–73)

RULER: Sheik Sabah al Salem al Sabah

CAPITAL: Al Kuwait

GOVERNMENT: "A nice little oligarchy" was how one observer once described the regime.

OIL COMPANIES: *Kuwait Oil Co. (KOC)*, onshore

(Owned by British Petroleum Ltd. 50% and Gulf Oil Co. of U.S. 50%.) Produces 90% of Kuwait oil. The Kuwait government now has a 25% "participation" in the operation, rising to 51% by 1980.

Kuwait National Petroleum Co. (KNPC)

(Owned by the Kuwait state 60% and private Kuwait interests 40%.)

Kuwait Shell Petroleum Development Co. (KSPD), offshore

(Owned 100% by Shell.)

Kuwait-Spanish Petroleum Co. (KSPC), onshore

(Owned by KNPC 51% and Hispanoil of Spain 49%.)

Aminoil

(Operates in conjunction with the Getty Oil Co. in the Neutral Zone and holds the Kuwaiti half interest: is owned by R. J. Reynolds Industries, a U.S. "independent" consortium.)

Arabian Oil Co. Ltd.

(Operates in the Neutral Zone. See under Saudi Arabia.)

OIL RESERVES: 74.5 billion barrels was the estimated reserve in 1969, but since then the figure has been drastically down-scaled and some experts calculate that at the present rate of extraction, Kuwait's oil will be exhausted by the end of the century.

POLITICAL PROSPECTS: Politics here is almost a family affair. Of the 730,000 inhabitants, fewer than 40,000 have a vote, and all non-Kuwaitis, all women and males under twenty-one are excluded. There is no party system, but about 15 percent of the members elected to the national assembly form a vociferous if not very effective oppositon; it is they who are most antagonistic toward foreign oil interests inside Kuwait. But the real threat to the regime comes from the non-Kuwaiti majority, a large number of them intelligent and resentful Palestinians laboring under a sense of grievance at their lack of privileges and civic rights.

Sheik Sabah al Salem is a member of a family which has ruled in Kuwait for two hundred years, and he sees to it that his close relatives fill many of the important jobs. He is a mild constitutional ruler leaning toward moderation, but there are more xenophobic influences hovering in the background. If they were to form an alliance with the Palestinians, the first to suffer would undoubtedly be the Anglo-American interests in the Kuwait Oil Co.

COUNTRY: Libya

LAND AREA: 679,358 square miles

POPULATION: 1,830,000 (est.)

Some Facts and Figures

OIL OUTPUT: 2.455 million barrels a day (January 1972)
OIL INCOME: $1.1 billion (est. 1972–73)
RULER: Colonel Muammar al Gaddafi, chairman of the junta
PREMIER: Major Abdul Salem Jalloud
CAPITAL: Tripoli
GOVERNMENT: Revolutionary and theocratic

OIL COMPANIES: *Esso International*

(Owned by Standard Oil of New Jersey.)

Mobil Libya and Amoseas

(Owned by Standard Oil of California 50%; Texaco 50%.)

Oasis

(Owned by Continental Oil Co. of U.S. 50% and Marathon Oil Co. of U.S. 50%.)

Occidental Oil Co.

(Owned by Occidental of California.)

Libya National Oil Co.

(Owned 100% by the Libyan state.)

Aquitaine

(Owned by Hispanoil of Spain 42%; Murphy Oil Co. of U.S. 16%; Erap-Elf of France 42%.)

British Petroleum Exploration Co. (Libya)

(Owned by BP 50% and Nelson Bunker Hunt of U.S. 50%.) Note: This company was nationalized by the Libyan government in 1971, but Hunt's share of the operation was subsequently freed from the nationalization decree. BP's share of the company was taken over by the Libyan state-owned company *Arabian Gulf Exploration Co. (Libya)*.

AGIP Libya

(Owned 100% by ENI of Italy.) In 1973 ENI granted 50% "participation" to the Libyan government.

OIL RESERVES: 33 billion barrels. The Libyan revolutionary junta believes that the reserves will run out in twenty years at the present rate of extraction, and is forcing foreign companies to make drastic cuts in production. Oil extraction was down 16.8% in 1971 from the rate in 1970, and further cuts were being made in 1972–1973.

POLITICAL PROSPECTS: Gaddafi's ambition is to nationalize all foreign oil interests in Libya. He took over BP's assets in 1971 and planned to follow with a seizure of Esso International shortly afterward, but was dissuaded by his more supple-minded revolutionary partner, Abdul Salem Jalloud, who recognizes Libya's need of the big oil companies as sales outlets. The companies are offering "participation" along similar lines to that granted to the Persian Gulf States, but Gaddafi is demanding more and has already extracted 50% "participation" from the Italians. How things will work out is anyone's guess. Gaddafi is unpredictable, with enough money in the bank to be able to put political militancy before financial prudence. As long as he remains the strongest member of the revolutionary junta, the future of foreign oil companies in Libya is uncertain. "We will stay on as long as there is a buck to be made" is the way one U.S. oilman put it.

COUNTRY:	**Iraq**
LAND AREA:	170,000 square miles
POPULATION:	6,650,000 (est.)
OIL OUTPUT:	1.923 million barrels a day (January 1972)
OIL INCOME:	$500 million (est. 1972–73)
PRESIDENT:	Ahmed Hassan Bakr
VICE-PRESIDENT:	Saddam Husain (strong man of the Revolutionary Command Council)
CAPITAL:	Baghdad
GOVERNMENT:	Revolutionary

OIL COMPANIES: *Iraq Petroleum Co. Ltd. (IPC)*, onshore

(Owned by British Petroleum 23¾%; Compagnie Française des Pétroles 23¾%; Near East Development Corp., in which Mobil Oil Co. and Standard of New Jersey have a half-share each, 23¾%; Shell Petroleum Co. Ltd. 23¾%; PARTEX, owned by the Gulbenkian Foundation, 5%.) Note: This company, which produces about 55% of Iraq's oil, was nationalized in 1972 and taken over by the state-owned Iraq Company for Oil Operations (ICOO). The Compagnie Française des Pétroles was allowed to retain its 23¾% holding in the company as a reward for the pro-Arab policies of France.

Basrah Petroleum Co. Ltd. (BPC), onshore

(Owned by the same companies as IPC.) Note: This company was not nationalized when IPC was taken over.

Mosul Petroleum Co. Ltd. (MPC), onshore

(Owned by the same companies as IPC.) Note: This company was not nationalized when IPC was taken over.

Erap-Elf of France, onshore-offshore

(Owned 100% by the French state.)

Iraq National Oil Co. (INOC), onshore-offshore

(Owned 100% by the Iraq state.)

OIL RESERVES: 32 billion barrels.

Iraq believes she has ample reserves to take her into the next century. Her complaint against IPC was that it did not extract sufficient oil from its concession. The government withdrew the North Rumaila fields from IPC some years earlier for the same reason, and has now begun to exploit them through INOC with Russian aid, selling the product mostly to East European–bloc markets pending settlement of its quarrels with IPC. New fields, potentially rich, have been discovered in the south.

POLITICAL PROSPECTS: By strengthening her ties with Russia and the Communist bloc, Iraq does not make operations easier for the Western oil companies. Iraq's revolutionary government has always wished to connect oil operations with politics, and to give concessions only to those oil companies whose home governments support its policy, particularly against Israel. At the same time, Iraq depends on Western oil revenues to keep the population fed, and this probably rules out a policy of wholesale nationalization. But Iraq has bitter memories of past oil-company exploitation.

COUNTRY: Abu Dhabi

LAND AREA: 50,000 square miles (including approximately 200 offshore islands)

POPULATION: 52,000 (est. 1970)

OIL OUTPUT: 1.047 million barrels a day (January 1972)

OIL INCOME: $450 million (est. 1972–73)

RULER: Sheik Zaid bin Sultan al Nahayan

CAPITAL: Abu Dhabi

GOVERNMENT: Authoritarian-paternalistic

OIL COMPANIES: *Abu Dhabi Marine Areas Ltd. (ADMA)*, offshore

(Owned by British Petroleum 66⅔% and Compagnie Française des Pétroles 33⅓%.) Note: In 1973 BP began negotiations to sell part of its interests in ADMA to a Japanese consortium. The Abu Dhabi government has a 25% "participation," rising to 51% by 1980.

Abu Dhabi Petroleum Co. Ltd. (ADPC), onshore

(Owned by same companies as IPC: see Iraq.) Note: The two above companies are responsible for the exploitation of 85% of Abu Dhabi's oil. The Abu Dhabi government now has a 50% "participation," rising to 51% by 1980.

Abu Dhabi Oil Co. (ADOCO), offshore

(Owned by Daikyo Oil Co. Ltd. of Japan 23⅓%; Maruzen Oil Co. Ltd., of which one-fifth share is held by Union Oil Co. of California, the rest Japanese, 23⅓%; Nippon Mining Co. Ltd. 23⅓%; Japan Petroleum Development Corp. 20%; Qatar Oil Co., a Japanese combine, 10%.) The Abu Dhabi government has notified ADOCO that it will exercise its option to take over 50% participation in May 1973.

Middle East Oil Co. (MECO), onshore

(Owned by Mitsubishi group of Japan 54.3%; Japan Petroleum Development Co. 42.1%; NYK Shipping Line of Japan 2.2%; and Tokyo Kaijo Kasai Hoken, marine and fire insurance group, 1.4%.)

Pan Ocean–Syracuse–Wingate, offshore

(Owned by the Pan Ocean Oil Corp. 60%; Syracuse Oils Ltd. 20%; Wingate Enterprises Ltd. 20%.)

Phillips–AGIP–Aminoil, onshore

(Owned by AGIP of Italy 41⅔%; Phillips Petroleum Co. 41⅔%; American Independent Oil Company, group operating in Kuwait Neutral Zone, 16⅔%.)

Abu Dhabi Petroleum Co.

(100% state-owned.)

OIL RESERVES: 14 billion barrels.

Experts calculate that Abu Dhabi's potentialities have not yet been fully realized, and that when certain areas still in dispute with

Saudi Arabia are developed, production could be more than doubled in the next ten years.

POLITICAL PROSPECTS: The last four rulers were either assassinated or usurped by their successors, and this could happen again, but it seems unlikely. Sheik Zaid is popular with his people, and unlike the brother he overturned, is anxious to use the oil revenue of the state for public works rather than hide it under his bed. Because of its wealth, the strength of its armed forces (6,000 British-led troops and a jet air force), and its outward-looking policy, Abu Dhabi has become the most influential element in the newly formed Federation of Arab Emirates. Its policy toward foreign oil companies is practical, particularly if the companies are interested in participation deals.

COUNTRY: **Qatar**

LAND AREA: 6,000 square miles

POPULATION: 130,000 (est.)

OIL OUTPUT: 452,000 barrels a day (January 1972)

OIL INCOME: $240 million (est. 1972–73)

RULER: Sheik Khalifa bin Hamad al Thani

CAPITAL: Doha

GOVERNMENT: Family oligarchy

OIL COMPANIES: *Qatar Petroleum Co. Ltd. (QPC)*, onshore

(Owned by IPC; see Iraq.) The Qatar government now has a 25% "participation," rising to 51% by 1980.

Bunduq Oil Co., offshore

(Owned by British Petroleum 33⅓%; Compagnie Française des Pétroles 33⅓%; United Petroleum Development, a combine of ADOCO of Abu Dhabi, Alaska Oil Co., North Slope Oil Co. and Qatar Oil Company, 33⅓%.) Plus government participation.

Qatar Oil Company (QOC)

(Owned by six Japanese oil companies 35.3%; three Japanese electric power companies 26.5%; Sumitomo group of Japan 15.9%; two Japanese iron and steel companies 5.9%; and Hitachi Shipbuilding Co. of Japan and associates 16.4%.) Also government participation.

Shell Co. of Qatar Ltd. (SCQ), offshore

(Owned 100% by Shell.)

OIL RESERVES: 5 billion barrels.

Qatar's prospects are similar to those of Abu Dhabi (above).

POLITICAL PROSPECTS: In February 1972, Sheik Ahmad bin Ali was overthrown by his cousin, Sheik Khalifa, in a bloodless coup. This did not visibly alter the political situation, since Khalifa was already prime minister and responsible for running the country while his cousin was away (which was often).

Khalifa has indicated that he will curb the privileges and perquisites of the ruling family, which is said to number no less than 12,000, all demanding jobs or subsidies from the state. In 1970 it was estimated that $64 million of the year's $110 million oil revenues went to the ruler and his family.

Khalifa is demanding increasing participation in foreign oil-company operations and greater utilization of Qatarians in the operations. But the locals lack know-how and face the same problem in miniature as Kuwait; large numbers of Palestinians have had to be imported to keep light, heat, fuel, air conditioning and Cadillacs operating.

COUNTRY: **Oman**

LAND AREA: 60,000 square miles (est.)
POPULATION: 65,000 (est.)
OIL OUTPUT: 280,000 barrels a day (January 1972)
OIL INCOME: $130 million (est. 1972–73)
RULER: Sultan Qaboos bin Sultan
CAPITAL: Muscat
GOVERNMENT: Authoritarian

OIL COMPANIES: *Petroleum Development (Oman) Ltd. (PDO)*, onshore including Dhofar, offshore Gulf of Oman

(Owned by Shell group 85%; Compagnie Française des Pétroles 10%; PARTEX, Gulbenkian Foundation 5%.)

Wintershall, offshore Gulf of Oman

(Owned by Wintershall of Germany 25%; Shell group 20%; Union Carbide Petroleum Co. of U.S. 20%; Compagnie Française des Pétroles 10%; Gelsenberg A.G. of Germany 10%; DST of Germany 10%; PARTEX 5%.)

Some Facts and Figures

OIL RESERVES: 2 billion barrels.

POLITICAL PROSPECTS: Sultan Qaboos is more popular with the Omanis than his father was and tries to do more for the welfare of the people of this backward, feudal land. But his oil revenues are swallowed up in maintaining a British-led force to beat back guerrilla incursions from South Yemen and the activities of armed Communist activists of PFLOAG, the so-called Popular Front for the Liberation of the Occupied Arabian Gulf.

Oil operations and explorations by foreign companies are encouraged, and there are believed to be rich fields in Dhofar province, especially in Jabal Akhdar (the Green Mountains). But guerrilla activities make operations there a particularly hazardous experience and hardly worth the huge amounts of money involved. Oman was the only Middle East oil state (except for Libya, where it was deliberately cut by state order) whose oil production dropped between 1970 and 1971 (by 13.6%). Those of Saudi Arabia, Iran, Kuwait and Abu Dhabi went up by 26.7%, 18.4%, 6.9% and 34.6% respectively.

COUNTRY: **Dubai**

LAND AREA: 1,500 square miles

POPULATION: 59,000

OIL OUTPUT: 130,000 barrels a day (January 1972)

OIL INCOME: $70 million (est. 1972–73)

RULER: Sheik Rashid bin Said al Makhtoum

CAPITAL: Dubai

GOVERNMENT: Paternal

OIL COMPANIES: *Dubai Marine Areas (DUMA)*, offshore

(Owned by Continental Oil Co. 30%; Compagnie Française des Pétroles 25%; Hispanoil of Spain, in which Marathon Oil of U.S. has just under a third-share, 25%; Texaco 10%; Delfzee of Germany 5%; and Sun Oil Co. 5%.) Plus government participation of 25%, rising to 51% by 1980.

Buttes-Clayco, onshore-offshore

(Owned by Buttes Gas and Oil Co. and Clayco Petroleum Co. of U.S.) Plus government participation.

Dubai Petroleum Co. (*DUPETCO*), onshore

(Owned by Continental Oil Co. 55%; Texaco 22½%; Sun Oil Co. 22½%.) Plus government participation.

OIL RESERVES: 1 billion barrels.

POLITICAL PROSPECTS: Sheik Rashid is one of the shrewdest rulers in the Persian Gulf and has made Dubai rich by encouraging trade, particularly in gold (which is imported legally at a tax of 2% and then smuggled into Pakistan and India) and oil. He hopes to turn Dubai into the commercial hub of the new Federation of Arab Emirates.

COUNTRY: Bahrain

LAND AREA: 598 square miles
POPULATION: 200,000
OIL OUTPUT: 74,000 barrels a day (January 1972)
OIL INCOME: $28 million (est. 1972–73)
RULER: Sheik Isa al Khalifa
CAPITAL: Manama
GOVERNMENT: Benevolently authoritarian

OIL COMPANIES: *Bahrain Petroleum Co.* (*BAPCO*), onshore-offshore

(Owned by Caltex Petroleum Co., in which Texaco and Standard Oil of California have half shares.) Plus government participation of 25%, rising to 51% by 1980.

Superior Oil International, Inc.

(Owned 100% by Superior Oil Co. of U.S.) Plus government participation.

OIL RESERVES: 500 million barrels.

POLITICAL PROSPECTS: This was the first American foothold in the Gulf, but the island's oil supplies are now dwindling, and Bahrain's future lies in the exploitation of its separate gas reserves, small industries, and the development of the island's airport for commercial flights between the West and Asia. The dockyard which once refueled and serviced the Royal Navy is also being enlarged as an eventual depot for tankers taking Gulf oil to the markets.

Some Facts and Figures

Besides Abu Dhabi and Dubai, several members of the Federation of Arab Emirates have oil operations, but for the moment none of them has made major discoveries and thus far the rich incomes enjoyed by the subjects of Sheik Zaid and Sheik Rashid have escaped them. Occidental Petroleum Co. of the U.S. is operating in Ajman. Shell-Bomin (an Anglo-Dutch-German combination) is at work on the Musandam peninsula of Al Fujairah and in its territorial waters in the Gulf of Oman. The Union Oil Co. of California has been operating in the offshore waters of Ras al Khaimah but must now come to an arrangement with the Iranian government, which has taken over the offshore islands, and with Buttes Gas and Oil Co., which also has claims in the area and has already made its peace with Teheran. With its partner, Clayco Petroleum Co., Buttes Gas and Oil is already operating offshore from Sharjah, where Shell and Shell-Bomin also have concessions, onshore and offshore. Sharjah had an attempted coup d'état in 1972, when Nasserite rivals of the ruler attempted to take over the state. But though the ruler was killed,* the coup was swiftly put down by a fire-brigade squad of troops dispatched from Abu Dhabi, the first overt demonstration of co-operation and cohesion between members of the new federation. The remaining member of the federation, Umm al Qaiwain, has two hopeful oil operations in progress, one run by Occidental, the other by Shell.

Other Middle East states have smaller oil operations, the largest of them being Egypt, which produces around 300,000 barrels a day from its Red Sea and Western Desert fields. Most of this is used domestically, and Egypt does not yet qualify to join either OPEC or OAPEC as an oil-exporting country. The same is true of Syria, which produced 124,000 barrels a day in 1971. The high political and financial price Syria is demanding to transport Iraqi and Saudi oil by pipeline across her territory is gradually turning that once efficient system of transportation into a too-costly operation compared with the relatively low rates for giant oil tankerage.

* He was succeeded by his brother, the present ruler, Sheik Sakr bin Mohammed.

COUNTRY: **Israel**

LAND AREA: 7,992 square miles (Israel proper; does not include occupied territory)

POPULATION: 3,500,000 (est., including occupied territory)

OIL OUTPUT: 130,000 barrels a day (including occupied territory)

OIL INCOME: No figures

PREMIER: Mrs. Golda Meier

CAPITAL: Jerusalem

GOVERNMENT: Democratic

OIL COMPANIES: *Israel Oil Prospectors* (Israeli-U.S.-Swiss)
Tri-Continental Drilling (Negev; Israeli-Canadian)
Pan-Israel Oil (Negev; Israeli-Canadian)
Israel Mediterranean Petroleum (Negev; Israeli-Canadian)
Sharon Oil Company
Israel-American Oil
Yellow Knife Power Company (Canadian)
Matsada (jointly owned by Lapidoth Co. and Israel Oil Prospects. Operating in Gaza)
Petrocana (offshore; Canadian)
Ashar Oil Company of Delaware (offshore)
Livingstone Oil Company (U.S.)

OIL RESERVES: No firm estimates.

POLITICAL PROSPECTS: Israel's domestic oil production was modest until after the 1968 Six-Day War and was mostly confined to operations around Gaza, Tiberias and the Dead Sea. Output in 1965 was about 6,000 barrels a day, mostly from a field near Gaza originally drilled by the Iraq Petroleum Company.

After the Six-Day War, Israel found herself in possession of the Egyptian oil fields in the Sinai Desert area. There were six main fields: Belayim Offshore, Belayim Onshore, Rudeis, Sidri, Sudr and Asl, operated on a partnership basis by the Italian company COPE (an ENI subsidiary) and the Egyptian national company, GPC. In 1966, these fields—mainly the Belayim offshore and onshore operations—produced 89,286 barrels a day. By the end of 1971 Israel had upped this output to 120,000 barrels a day and also uncovered considerable reserves. The future of these operations, which have solved Israel's domestic fuel problems, is bound to figure largely in any future Israeli-Arab settlement.

Israel has also been active in the establishment of pipelines. Originally a 16-inch pipeline was built from Eilat on the Red Sea via Beersheba and Ashdod Yam to Haifa, on the Mediterranean, and it was pumping 100,000 barrels a day by 1972, mainly to the old IPC refinery at Haifa (now owned by the Israeli state and Isaac Wolfson interests) for conversion to Israeli domestic use.

A second pipeline, in operation by 1972, is much larger, 42 inches, and capable of pumping 1 million barrels a day. It was built at a cost of $117 million and, since the closing of the Suez Canal, has become a useful means of eliminating the long haul of oil from the Middle East, around the Cape of Good Hope, and thence to the Mediterranean countries and Eastern Europe. Since Arab countries bar trade with Israel, and since she has no need to use the pipeline herself, it seems certain that the bulk of the crude oil passing through it comes from Iran, probably under an arrangement between the Israelis and the Iranian government. The Consortium in Iran (most of whose member companies have interests in Arab countries) is unlikely to supply the oil, but on the other hand, the National Iranian Oil Company is under no such inhibition. By an agreement signed between Iran and the Consortium in 1967, NIOC has the right to take 20 million tons of crude oil from the Consortium in the next five years. NIOC has signed barter agreements with Rumania, Bulgaria, Czechoslovakia, Poland, Hungary and Yugoslavia, and in 1971 lifted 35 billion barrels of crude from the Consortium's terminal on Kharg Island in order to supply these countries. It is almost certain that this oil was sent via the Israeli pipeline to the Mediterranean and thence by tanker to Eastern Europe. The southern terminal of the pipeline at Eilat, on the Red Sea, can accommodate tankers up to 125,000 tons; the Mediterranean terminal can accommodate 75,000-ton tankers, but this capacity is being raised to 125,000 tons.

Israel hopes that if there ever is an Arab-Israeli settlement, her 42-inch line would be used to supplement supplies passing through the (reopened) Suez Canal and via the SUMED (Suez-Mediterranean) pipeline now being projected by Egypt and a number of Western interests.

CRUDE-OIL PRODUCTION IN MIDDLE EAST AND NORTH AFRICA 1971

(Thousand Barrels Daily)

	1971	*1970*	*% Change*
Saudi Arabia *	4,771	3,797	+ 25.7
Iran	4,567	3,856	+ 18.4
Kuwait †	3,198	2,989	+ 7.0
Libya	2,758	3,313	− 16.8
Iraq	1,705	1,559	+ 9.4
Abu Dhabi	934	694	+ 34.6
Algeria	780	1,022	− 23.7
Qatar	430	362	+ 18.8
Egypt	306	328	− 6.7
Oman	287	332	− 13.6
Dubai	125	84	+ 48.8
Syria	124	81	+ 53.1
Tunisia	89	88	+ 1.1
Bahrain	75	77	− 2.6
Total	20,149	18,582	+ 8.4

* Includes 273,000 b/d from the Neutral Zone for 1971 (181,000 b/d from Arabian Oil and 92,000 b/d from Getty Oil) and 248,000 b/d for 1970 (172,000 b/d from Arabian Oil and 76,000 b/d from Getty Oil).

† Includes 273,000 b/d from the Neutral Zone for 1971 (181,000 b/d from Arabian Oil and 92,000 b/d from Aminoil) and 254,000 b/d for 1970 (172,000 from Arabian Oil and 82,000 b/d from Aminoil).

NOTE: 1971 figures for Algeria, Egypt, Syria and Tunisia are MEES estimates.

Bibliography

Arfa, Hassan, *Under Five Shahs*. London: John Murray, 1964 (New York: Morrow, 1965).

Assa, Ahmed, *Miracle dans les sables*. Paris: Librairie d'Amérique et d'Orient, 1969.

Barrows, Gordon H., *The International Petroleum Industry*. New York: International Petroleum Institute, 1965.

Beeby-Thompson, Arthur, *Oil Pioneer*. London: Sidgwick & Jackson, 1961 (*Black Gold: The Story of an Oil Pioneer;* New York: Doubleday, 1961).

Boustead, Sir Hugh, *The Wind of Morning*. London: Chatto & Windus, 1971.

Boveri, Margret, *Minaret and Pipeline: Yesterday and Today in the Near East*, trans. from the German by Louisa Marie Sieveking. London: Oxford University Press, 1939 (New York: OUP, 1940).

Churchill, Winston S., *The Second World War*, Vol. VI, *Triumph and Tragedy*. London: Cassell, 1953 (Boston: Houghton Mifflin, 1953).

A Collection of Engagements and Sanads relating to India and Neighbouring Countries, XI, ed. by C. U. Aitcheson, Delhi: Government of India, 1933.

Continuity and Change in the World Oil Industry. Beirut: Middle East Research and Publishing Center, 1970.

Crowther, James Gerald, *About Petroleum*. London and New York: Oxford University Press, 1938.

Deterding, Sir Henri, *An International Oilman* (as told to Stanley Naylor). New York: Harper, 1934.

Dickson, Violet, *Forty Years in Kuwait*. London: Allen & Unwin, 1970.

Eddy, William A., *F.D.R. Meets Ibn Saud*. New York: American Friends of the Middle East, Inc., 1954.

Forbes, Robert James, and O'Beirne, D. R., *The Technical Development of Royal Dutch–Shell 1890–1940*. Leiden, Netherlands: E. J. Brill, 1957.

Forbin, Victor, *Le Pétrole dans le monde*. Paris: Payet, 1940.

Frankel, Paul H., *The Essentials of Petroleum: A Key to Oil Economics*. London: Chapman & Hall, 1946.

———, *Mattei: Oil and Power Politics*. London: Faber & Faber, 1966 (New York: Praeger, 1966).

———, *Oil: The Facts of Life*. London: Weidenfeld & Nicolson, 1962.

Frischauer, Willi, *Onassis: A Biography of the Golden Greek Shipowner*. New York: Meredith, 1968.

Gerretson, F. C., *The History of the Royal Dutch*. 4 vols. Leiden, Netherlands: E. J. Brill, 1953.

Getty, J. Paul, *My Life and Fortunes*. New York: Meredith, 1963.

Gibb, George Sweet, and Knowlton, Evelyn H., *The History of Standard Oil Company (New Jersey)*, Vol. II, *The Resurgent Years, 1911–1927*. New York: Harper, 1956.

Gulbenkian, Nubar, *Pantaraxia: The Autobiography of Nubar Gulbenkian*. London: Hutchinson, 1965 (*Portrait in Oil: The Autobiography of Nubar Gulbenkian;* New York: Simon & Schuster, 1965).

Hardinge, Sir Arthur, *A Diplomatist in the East*. London: Jonathan Cape, 1928.

Hartshorn, Jack Ernest, *Oil Companies and Governments: An Account of the International Oil Industry in Its Political Environment*. London: Faber & Faber, 1962 (*Politics and World Oil Economics*, etc.; New York: Praeger, 1962).

Henriques, Robert, *Marcus Samuel*. London: Barrie & Rockliffe, 1960 (*Bearsted: A Biography of Marcus Samuel, 1st Viscount Bearsted and Founder of Shell Transport and Trading Company;* New York: Viking, 1960).

Henry, J. D., *Baku: An Eventful History*. London: Constable, 1906.

———, *Oil Fuel and Empire*. London: Bradbury, Agnew, 1908.

Hewins, Ralph, *The Golden Dream*. London: W. H. Allen, 1963.

———, *Mr Five Per Cent: The Biography of Calouste Gulbenkian*. London: Hutchinson, 1957 (New York: Rinehart, 1958).

Hopwood, Derek, *The Arabian Peninsula.* London: Allen & Unwin, 1972.

Howarth, David A., *The Desert King: Ibn Saud and His Arabia.* London: Collins, 1964 (New York: McGraw-Hill, 1964).

Kirk, George E., *Contemporary Arab Politics.* New York: Praeger, 1961.

Larson, Henrietta M.; Knowlton, Evelyn H.; and Popple, C. S., *The History of Standard Oil Company (New Jersey)*, Vol. III, *New Horizons, 1927–1950.* New York: Harper, 1971.

L'Espagnol de la Tramerye, Pierre, *The World Struggle for Oil.* New York: Knopf, 1924.

Longhurst, Henry, *Adventure in Oil: The Story of British Petroleum.* London: Sidgwick & Jackson, 1959.

Longrigg, Stephen Hemsley, *Oil in the Middle East: Its Discovery and Development.* 3rd ed. London: Oxford University Press, 1968 (New York: OUP, 1967).

Lutfi, Ashraf, *OPEC Oil.* Beirut: Middle East Research and Publishing Center, 1968.

Mann, Clarence, *Abu Dhabi.* Beirut: Khayat, 1964.

Marcosson, Isaac F., *The Black Golconda.* New York: Harper, 1924.

Mikdashi, Zuhayr, *A Financial Analysis of Middle East Oil Concessions, 1901–1965.* New York: Praeger, 1966.

Morris, James, *Sultan in Oman: Venture into the Middle East.* London: Faber & Faber, 1957 (New York: Pantheon, 1957).

Mosley, Leonard, *Curzon: The End of an Epoch.* London: Longmans, 1960 (*The Glorious Fault: The Life of Lord Curzon;* New York: Harcourt, 1960).

Mughraby, Muamad A., *Permanent Sovereignty over Oil Resources.* Beirut: Middle East Research and Publishing Center, 1966.

Our Industry: Petroleum, by various authors. London: British Petroleum, 1970.

Pachachi, Nadim, *The Role of OPEC in the Emergence of New Patterns in Government-Company Relationships.* London: Royal Institute of International Affairs, 1972.

Philby, Harry St. John, *Arabian Jubilee.* London: Robert Hale, 1962.

———, *Arabian Oil Ventures.* Washington: Middle East Institute, 1964.

———, *Forty Years in the Wilderness.* London: Robert Hale, 1959.

Phillips, Wendell, *Oman: A History.* London: Longmans, 1967 (New York: Reynal, 1968).

Phillips, Wendell, *Unknown Oman.* London: Longmans, 1966 (New York: McKay, 1966).

Reports of the High Commissioner on Iraq Administration (*October 1920–1924; 1925–1931*). London: His Majesty's Stationery Office.

Rihani, Ameen, *Ibn Saoud of Arabia: His People and His Land.* London: Constable, 1928.

Shamir, Shamma, *The Oil of Kuwait: Present and Future.* Beirut: Middle East Research and Publishing Center, 1959.

Stark, Freya, *The Arab Island: The Middle East 1939–1943.* New York: Knopf, 1945 (*East Is West;* London: John Murray, 1945).

Stegner, Wallace, *Discovery!* Beirut: Middle East Export Press, 1971.

Stocking, George W., *Middle East Oil.* Nashville, Tenn.: Vanderbilt University Press, 1970.

Tarbell, Ida, *The History of the Standard Oil Company.* New York: Macmillan, 1904.

Trevelyan, Sir Humphrey, *The Middle East in Revolution.* London: Macmillan, 1970 (Boston: Gambit, 1970).

Tugendhat, Christopher, *Oil: The Biggest Business.* London: Eyre & Spottiswoode, 1968 (New York: Putnam, 1968).

United States, Congress, Senate, *Hearings. Part 41 Petroleum Arrangements with Saudi Arabia,* 80th Cong., 1st sess. Washington: 1948.

———, *American Petroleum Interests in Foreign Countries. Hearings, Petroleum Resources United States,* 79th Cong., 1st sess. Washington: 1946.

———, *The Emergency Oil Lift Program and Related Oil Problems. Hearings,* 85th Cong. Washington: 1957.

United States, Congress, *Proceedings of the Senate Small Business Committee.* Washington: 1946.

United States, Federal Trade Commission, *The International Petroleum Cartel, Staff Report submitted to the Subcommittee on Monopoly of the Select Committee on Small Business,* 82nd Cong., 2nd sess. Washington: 1952.

Van der Meulen, D., *The Wells of Ibn Saud.* London: John Murray, 1957 (New York: Praeger, 1957).

Ward, T. E., *Negotiations for Oil Concessions in Bahrain, Al Hasa, the Neutral Zone, Qatar and Kuwait* (privately printed, 1969).

Williamson, J. Q., *In a Persian Oilfield.* London: Ernest Benn, 1930.

Wilson, Sir Arnold, *A Bibliography of Persia.* London: Clarendon Press, 1930.

———, *Loyalties, Mesopotamia 1914–1917: A Personal and Historical Record.* London and New York: Oxford University Press, 1930.

Wilson, Sir Arnold, *Mesopotamia 1917–1920: A Clash of Loyalties, A Personal and Historical Record.* London and New York: Oxford University Press, 1931.

———, *The Persian Gulf.* London: Allen & Unwin, 1928.

———, *S.W. Persia: A Political Officer's Diary.* London and New York: Oxford University Press, 1941.

Yamani, Ahmed Zaki, *Economics of the Petroleum Industry.* Beirut: Middle East Research and Publishing Center, 1970.

Zeine, Zeine N., *The Struggle for Arab Independence.* Beirut: Khayat, 1970.

Zischka, Antoine, *La Guerre secrète pour le pétrole.* Paris: 1933.

Index

Index

Index

Index

About the Author

LEONARD MOSLEY was born in Manchester, England, fifty-five years ago, and has been a foreign correspondent most of his working life. Before World War II he was chief correspondent for the Kemsley (now Thomson) newspaper chain, first in the United States and then in Germany, where he met most of the Nazi leaders, including Adolf Hitler. When the war broke out he covered the Battle of Britain and all the campaigns in Africa, the Middle East and India, and on D-Day he dropped into Normandy by parachute with the invading Allied forces. Since then he has traveled widely in Africa, the Middle East and the Far East, and though he now lives in the south of France, he still logs more than fifty thousand miles a year in the course of research for his books.

IRAQ
(MESOPOTAMIA)
Basra
Shatt al Arab
Khorramshahr
Abadan
KUWAIT
Al Kuwait
NEUTRAL ZONE
KUWAIT OIL CO.
Burgan
Bushire
ARABIAN OIL CO.
AMINOIL
NEUTRAL ZONE
GETTY OIL CO.
SAUDI ARABIA
ARAMCO
AL HASA PROVINCE
Jubail
PERSIAN GULF
Ras Tanura
CONTINENTAL OIL CO. OF BAHRAIN
Damman
Dhahran
Oqair
Manama
BAPCO
BAHRAIN
Abqaiq
Al Khobar
QATAR
Al Hufuf
QATAR PETROLEUM CO.
Doha
Ghawar Oil Field
AL HASA OASIS
ADMA
Riyadh
RUB AL KHALI (EMPTY QUARTER)
JEAN PAUL TREMBLAY